AF358831

Memorial Issue Dedicated to Dr. Howard D. Flack: The Man behind the Flack Parameter

Memorial Issue Dedicated to Dr. Howard D. Flack: The Man behind the Flack Parameter

Editors

Catherine E. Housecroft
Katharina M. Fromm

MDPI • Basel • Beijing • Wuhan • Barcelona • Belgrade • Manchester • Tokyo • Cluj • Tianjin

Editors
Catherine E. Housecroft
University of Basel
Switzerland

Katharina M. Fromm
University of Fribourg
Switzerland

Editorial Office
MDPI
St. Alban-Anlage 66
4052 Basel, Switzerland

This is a reprint of articles from the Special Issue published online in the open access journal *Chemistry* (ISSN 2624-8549) (available at: https://www.mdpi.com/journal/chemistry/special_issues/for_Flack).

For citation purposes, cite each article independently as indicated on the article page online and as indicated below:

LastName, A.A.; LastName, B.B.; LastName, C.C. Article Title. *Journal Name* **Year**, *Volume Number*, Page Range.

ISBN 978-3-0365-1914-2 (Hbk)
ISBN 978-3-0365-1915-9 (PDF)

Contents

About the Editors

Catherine E. Housecroft is a Professor of Chemistry in the Department of Chemistry at the University of Basel, Switzerland. Her research group, run jointly with Professor Edwin Constable, is engaged in aspects of materials, coordination, and structural and supramolecular chemistries. Much of the emphasis is upon the design and development of materials which will contribute to mankind achieving the 17 Sustainable Development Goals designated in 2015 by the United Nations General Assembly. Compounds of interest range from coordination compounds which can absorb or emit light to coordination networks which can act as hosts for small molecules.

Katharina M. Fromm was raised and educated in Germany, France, and the United States. After receiving a Ph.D. in metal-organic chemistry from the University of Karlsruhe in Germany in 1994, she joined the group of Prof. Joachim Strahle in Tübingen (solid-state chemistry) and Nobel Prize winner Prof. Jean-Marie Lehn (supramolecular chemistry) for her postdoctoral studies. In 1998, she moved to the University of Geneva for her habilitation, which she received in 2002. After a short intermission at the University of Karlsruhe with an Emmy Noether Stipend II, she was awarded a Swiss National Science Foundation Professorship at the University of Basel, allowing her to expand her research at the University of Fribourg, taking over the chair from Prof. Alexander von Zelewsky. From 2010 to end of 2019, she served as a Research Councilor for the Division "Programs" of the Swiss National Science Foundation, of which she became president in mid-2015 before being elected Vice-President of the Research Council in 2016. After her time with SNSF, she was nominated Vice-Rector for Research and Innovation in January 2020. Her main research interests deal with s-block elements and silver bioinorganic chemistry, although her activities go beyond that and include mechano-responsive polymers, nanocapsules and batteries. In 2013, she was named Fellow of the American Chemical Society (first in Europe), and became a member of the European Academy of Sciences in 2018, the same year she was also announced winner of the Prix Jaubert of the University of Geneva. In 2019, she was elected member of the Swiss Academy of Technical Sciences.

chemistry

MDPI

Editorial

Memorial Issue Dedicated to Dr. Howard D. Flack: The Man behind the Flack Parameter

Catherine E. Housecroft [1],* and Katharina M. Fromm [2]

[1] Department of Chemistry, University of Basel, Mattenstrasse 24a, BPR 1096, 4058 Basel, Switzerland
[2] Department of Chemistry, University of Fribourg, Chemin du Musée 9, 1700 Fribourg, Switzerland;
     katharina.fromm@unifr.ch
* Correspondence: catherine.housecroft@unibas.ch

Citation: Housecroft, C.E.; Fromm, K.M. Memorial Issue Dedicated to Dr. Howard D. Flack: The Man behind the Flack Parameter. *Chemistry* **2021**, *3*, 818–820. https://doi.org/10.3390/chemistry3030058

Received: 23 July 2021
Accepted: 23 July 2021
Published: 27 July 2021

Publisher's Note: MDPI stays neutral with regard to jurisdictional claims in published maps and institutional affiliations.

This Special Issue of *Chemistry* is dedicated to Dr. Howard D. Flack (1943–2017), a renowned crystallographer who transformed the way in which, by using single crystal X-ray diffraction, we are able to determine the absolute structure of a crystalline material, and thereby determine the absolute configuration of molecular species within the material. All of us who routinely use cifs are familiar with the 'Flack Parameter', and the younger generations take it for granted. Not all realize that this is a relatively recent implementation [1].

Howard Flack studied at the University of Cambridge, UK, and moved to a position of Maître-assistant in the Laboratoire de Cristallographie at the University of Geneva, Switzerland, in 1972. He remained in Geneva working as a crystallographer for his entire career, and contributed his knowledge, enthusiasm and encouragement to the establishment of crystallographic facilities at other Swiss universities. His invaluable computational and scholarly contributions to single-crystal X-ray diffraction have left an extraordinary mark on the field, as can be appreciated by the bibliography of the Howard Flack's publications contributed to this Special Issue by Constable [2].

As noted above, Howard Flack is probably most associated with the Flack parameter, which is routinely reported for all chiral molecules for which-single crystal X-ray diffraction data are reported in the literature. Watkin and Cooper provide an excellent account of the importance and applications of the Flack parameter, and conclude their paper with some advice on its use [3]. Complementing this, Constable and Housecroft have focused on the development of crystallographic methods within metal coordination chemistry, and demonstrated the role of the Flack parameter in providing a routine method for determining the absolute configuration of coordination compounds [4]. The contributions made by Howard Flack in providing tools for crystallographers to take account of crystal twinning in structure elucidation are represented in the paper by Øien-Ødegaard and Lillerud, which describes the crystal structures of three zirconium-containing metal–organic frameworks determined from twinned crystals [5].

At the heart of a crystal lattice are intermolecular interactions. Tools to analyze structures in terms of these weak interactions are widely available, starting from the molecular structure, although predicting the crystalline assembly is far from straight forward. Aakeröy and coworkers approached this challenge using energy- and informatics-based prediction models founded on molecular electrostatic potentials, hydrogen-bond energies, propensity and coordination. They have applied these criteria to the crystal structures of twelve pyrazole-based molecules in order to explore the ability to correctly predict supramolecular synthons [6]. Remaining with the theme of crystal engineering, Černý et al. present a review of the crystal chemistry of inorganic hydroborates. This detailed review deals with salts containing hydroborate ions as the only type of anion. The structural patterns that emerge from this investigation should be invaluable to those interested in the design of hydroborate-based materials [7].

An octahedral metal center coordinated by three bidentate ligands, e.g., [M(2,2′-bipyridine)$_3$]$^{n+}$, is the archetypal coordination motif that chemistry undergraduates learn can possess a Δ- or Λ-configuration. Lappin and coworkers have redetermined the single-crystal structure of Λ-[Co(en)$_3$]Δ-[Co(edta)]$_2$Cl·10H$_2$O at low temperature, and they also report the structure of racemic [Co(sep)][Co(edta)]Cl$_2$·2H$_2$O. The former provides valuable insight into chelate ring conformation and hydrogen bonded interactions in the lattice. The data serve to provide a better understanding of stereoselectivity in ion-pairing and electron transfer reactions [8]. Piguet and coworkers present a detailed investigation of pseudo-octahedral spin-crossover [Fe(L)$_3$]$^{2+}$ complexes, in which L is an unsymmetrical and sterically demanding α,α′-diimine ligand [9]. The theme of metal coordination chemistry continues with a contribution from McKenzie and coworkers, describing the chemisorption of gaseous NO by the enantiomorphic phases of the cobalt coordination compounds containing a chiral salen ligand [10]. Six-, seven- and eight-coordinate lanthanoid (Ln) metal complexes of types *cis*-[Ln(dppmO$_2$)$_2$Cl$_2$]Cl, [Ln(dppmO$_2$)$_3$Cl]Cl$_2$ and [Ln(dppmO$_2$)$_4$]Cl$_3$ are reported by Reid and coworkers; the dependence of the coordination number and geometry on the Ln^{3+} radius, and the role of the chloride ions are discussed [11].

Small molecules and their crystal structures are the focus of several contributions to this Special Issue. Bauer and Götz present the crystal structure of chloropentaphenyldisiloxane and highlight intermolecular interactions in monofunctionalized disiloxanes [12]. A new synthetic approach to arylsulfonates has been exemplified by Ngassa and coworkers by the preparation of a series of substituted nitrophenyl-4′-phenylbenzenesulfonates; structural characterizations provide insight into the intermolecular interactions in these species [13]. This structural discussion is complemented by the contribution from Seidel et al., which reports the structural characterization of (*R*)-and *rac*-2-bromo-3-methylbutanoic acid with particular focus on hydrogen-bonded motifs and conformational preferences observed in the solid-state [14]. In a related study, Lehmann and coworkers describe the determination of the absolute configuration of the flavoring agent (+)-γ-decalactone; crystal growth was achieved in situ by cryo-crystallization methods [15]. Rifamycins are an important group of antibacterial agents, and in their contribution to this Special Issue, Frampton et al. report the structural characterization and absolute stereochemistry of a semi-synthetic rifamycin which was prepared using an Alder-Ene addition [16].

A fascinating illustration of the application of Erdmann's anion, *trans*-[Co(NH$_3$)$_2$(NO$_2$)$_4$]$^-$, in the isolation and crystallization of impounded street drugs is provided by Lalancette and coworkers; using salts of cocaine, methamphetamine and methylone, single-crystal structural data were used to calculate powder X-ray diffraction patterns, giving a means for rapid screening of confiscated materials [17].

We are indebted to all the contributors to this Special Issue which provides a timely opportunity to pay tribute to Howard Flack—the man behind the Flack parameter and more.

Conflicts of Interest: The authors declare no conflict of interest.

References

1. Flack, H.D. On enantiomorph-polarity estimation. *Acta Cryst.* **1983**, *A39*, 876–881. [CrossRef]
2. Constable, E.C. The Publications of Howard Flack (1943–2017). *Chemistry* **2020**, *2*, 40. [CrossRef]
3. Watkin, D.J.; Cooper, R.I. Howard Flack and the Flack Parameter. *Chemistry* **2020**, *2*, 52. [CrossRef]
4. Constable, E.C.; Housecroft, C.E. When Stereochemistry Raised Its Ugly Head in Coordination Chemistry—An Appreciation of Howard Flack. *Chemistry* **2020**, *2*, 49. [CrossRef]
5. Øien-Ødegaard, S.; Lillerud, K.P. Twinning in Zr-Based Metal-Organic Framework Crystals. *Chemistry* **2020**, *2*, 50. [CrossRef]
6. Sandhu, B.; McLean, A.; Sinha, A.S.; Desper, J.; Aakeröy, C.B. Assessment of Computational Tools for Predicting Supramolecular Synthons. *Chemistry* **2021**, *3*, 43. [CrossRef]
7. Černý, R.; Brighi, M.; Murgia, F. The Crystal Chemistry of Inorganic Hydroborates. *Chemistry* **2020**, *2*, 53. [CrossRef]
8. Osvath, P.; Oliver, A.; Lappin, A.G. Stereosopecificity in [Co(sep)][Co(edta)]Cl$_2$·2H$_2$O. *Chemistry* **2021**, *3*, 17. [CrossRef]
9. Deorukhkar, N.; Lathion, T.; Guénée, L.; Besnard, C.; Piguet, C. The Tyranny of Arm-Wrestling Methyls on Iron(II) Spin State in Pseudo-Octahedral [Fe(didentate)$_3$] Complexes. *Chemistry* **2020**, *2*, 15. [CrossRef]
10. Møller, M.D.; Liljedahl, M.C.; McKee, V.; McKenzie, C.J. Solid Phase Nitrosylation of Enantiomeric Cobalt(II) Complexes. *Chemistry* **2021**, *3*, 41. [CrossRef]

11. Bannister, R.D.; Levason, W.; Reid, G. Bis(diphenylphosphino)methane Dioxide Complexes of Lanthanide Trichlorides: Synthesis, Structures and Spectroscopy. *Chemistry* **2020**, *2*, 60. [CrossRef]
12. Bauer, J.; Götz, T. Chloropentaphenyldisiloxane—Model Study on Intermolecular Interactions in the Crystal Structure of a Monofunctionalized Disiloxane. *Chemistry* **2021**, *3*, 33. [CrossRef]
13. Stenfors, B.A.; Staples, R.J.; Biros, S.M.; Ngassa, F.N. Synthesis and Crystallographic Characterization of X-Substituted 2,4-Dinitrophenyl-4′-phenylbenzenesulfonates. *Chemistry* **2020**, *2*, 36. [CrossRef]
14. Seidel, R.W.; Nöthling, N.; Goddard, R.; Lehmann, C.W. Structural Elucidation of Enantiopure and Racemic 2-Bromo-3-Methylbutyric Acid. *Chemistry* **2020**, *2*, 44. [CrossRef]
15. Patzer, M.; Nöthling, N.; Goddard, R.; Lehmann, C.W. Absolute Configuration of In Situ Crystallized (+)-γ-Decalactone. *Chemistry* **2021**, *3*, 40. [CrossRef]
16. Frampton, C.S.; Gall, J.H.; MacNicol, D.D. Novel Ansa-Chain Conformation of a Semi-Synthetic Rifamycin Prepared Employing the Alder-Ene Reaction: Crystal Structure and Absolute Stereochemistry. *Chemistry* **2021**, *3*, 52. [CrossRef]
17. Wood, M.R.; Mikhael, S.; Bernal, I.; Lalancette, R.A. Erdmann's Anion—An Inexpensive and Useful Species for the Crystallization of Illicit Drugs after Street Confiscations. *Chemistry* **2021**, *3*, 42. [CrossRef]

 *chemistry*

Editorial

The Publications of Howard Flack (1943–2017)

Edwin Charles Constable

Department of Chemistry, University of Basel, BPR 1096, Mattenstrasse 24a, CH-4058 Basel, Switzerland; edwin.constable@unibas.ch; Tel.: +41-61-207-1001

Received: 17 June 2020; Accepted: 18 June 2020; Published: 2 July 2020

Abstract: Howard Flack was a driving force in the development of modern crystallography. Today "the Flack parameter" has entered into the common parlance of crystallography but his influence was far wider. This article provides an overview of his scientific output and a full bibliography.

Keywords: crystallography; Howard Flack

1. Introduction

This short article provides a bibliography of the publications of Howard Flack. This is not the place to provide an appreciation of the contributions of this outstanding crystallographer, but rather to provide a convenient point of reference for his life's work. Excellent overviews of the contribution of Howard Flack are to be found in this volume and elsewhere [1,2]. The short accompanying text groups the output by theme. Published conference abstracts are included for completeness. The bibliography is as complete as I have been able to compile, but as always, the author is responsible for any omissions (and apologizes for any such omissions in advance).

2. The Scientific Output

2.1. Properties and Structure of Solid State Inorganic Materials

A number of publications or conference presentations appeared concerning the crystal, mechanical and electronic properties of solid state materials such as MoS_2 [3], $NbSe_2$ [3], HfS_2 [3], $SmAu_6$ [4], Ga_xSe_{1-x} [5], TiC [6,7], TiN [6,7], VN [8,9], Al_2O_3 [10], $Mo_4Ru_2Se_8$ [11], $Mo_{1.5}Re_{4.5}Se_8$ [11], MNi_3Al_9 (M = Y, Gd, Dy, Er) [12], $TiGePt$ [13], Nb_3Si [14] and Nb_3As [14]. Another reported structure of an inorganic material is the "simple" salt, $MgSO_3 \cdot 6H_2O$ [15]. One publication from 2003 entitled "Anti-wurtzite reoriented" [16] is noteworthy as it provides a link to the earliest studies of zinc sulfide in which the Friedel pairs were shown to have different intensities [17,18].

Two very early studies were concerned with the properties of crystalline polytetrafluoroethene [19,20].

2.2. Structural Studies on Discrete Species

Although Flack is not widely thought of as a service crystallographer, a number of small molecule structures determined by him have been reported, including $[Cl_2(SEt_2)_2Ir(\mu-Cl)_2IrCl_2(SEt_2)_2]$ [21], $[Cl_3(SEt_2)Ir(\mu-Cl)(\mu-SEt_2)IrCl_2(SEt_2)_2]$ [22], $[Ir(NCS)(NH_3)_5]Cl_2$ [23], $[Ir(NCS)(NH_3)_5](ClO_4)_2$ [24], $MeOC(Ph) = C(Ph)OH$ [25], (+)-8β-acetoxy-12-(4-bromobenzoyloxy)-13,14,15,16-tetranorlabdane [26] and other chiral organic [27] and inorganic [28] species.

2.3. Crystallographic Publications

2.3.1. Disorder

A number of early publications describe types of disorder in anthrone crystals [29–31] and present a general description of X-ray diffraction by such disordered crystals [32]. A body of publications commencing in the 1970s is concerned with new methods for making absorption corrections in diffraction measurements [30,33–37]. The crystallographer's obsession with the "ideal" spherical crystal also attracted his attention at this time [36,38–42].

2.3.2. Improving Data Quality

There is a significant body of work concerned with general aspects of improving the quality of collected crystallographic data [43–48]. Flack also made a number of additional contributions to the use of statistical methods in crystallography [49–53] including a review of the standard work by Shmueli and Weiss [54].

2.3.3. Absolute Configuration, Absolute-Structure Refinement and the Flack Parameter

Probably the most important paper published by Howard Flack was the 1983 work "On enantiomorph-polarity estimation" in which he derived an expression for the determination of the absolute structure of crystals and, therefore, the absolute configuration of any chiral molecules contained therein [55]. Flack included a parameter which he called "x", but which the community now knows as 'The Flack Parameter'. He subsequently published numerous articles on the theory and practice of solving absolute structures [56–82]. Related work dealt with the origins of chirality [83] and general reflections on the use and misuse of symmetry and chiral descriptions in chemistry [84–88].

One additional interesting publication is a review of planar-chiral five-membered metallacycles [89].

2.3.4. Technical Contributions and Software

By its nature, crystallography is a technical and sometimes mathematically demanding and intense subject and Flack made a wide range of contributions to fundamental aspects of crystallography [90–98]. A 2017 book chapter presented a concise state-of-the-art view of chemical crystallography [99].

Flack was also involved in the new knowledge culture, and contributions entitled "Crystallographic publishing in the electronic era" [100], "Internet resources for crystallography" [101] and "World Wide Web for crystallography" [102] attest to this, as does a 1996 workshop "Surfing the Crystallographic Net (CRYSNET) workshops" [103]. Another conference contribution entitled "Is your crystal representative of the bulk?" addressed one of the fundamental issues for a synthetic chemist relying on crystallography [104].

He was a co-author of many software programs including DIFRAC [105] and HUG [106], X-RAY 76 [107] X-RAY 76 (the first System offering all the software needed for an X-ray analysis) and its successor, XTAL [108,109].

Finally, I should note that one of his last publications testifies to the breadth of his engagement with, and commitment to, crystallography in its broadest sense [110]. The web-site http://crystal.flack.ch/ maintains an up-to-date publication list and links to other sources commemorating his work.

Funding: This research received no external funding.

Conflicts of Interest: The author declare no conflict of interest.

References

1. Watkin, D.; Schwarzenbach, D. Howard D. Flack (1943–2017). *J. Appl. Crystallogr.* **2017**, *50*, 666–667. [CrossRef]
2. Watkin, D.J. Special issue in memory of Howard D. Flack, who died suddenly in February 2017. *Tetrahedron Asymmetry* **2017**, *28*, 1189–1191. [CrossRef]
3. Flack, H.D. Compressibilities of some layer dichalcogenides. *J. Appl. Crystallogr.* **1972**, *5*, 137–138. [CrossRef]
4. Flack, H.D.; Moreau, J.M.; Parthé, E. Refinement of SmAu$_6$. *Acta Crystallogr. Sect. B Struct. Sci.* **1974**, *30*, 820–821. [CrossRef]
5. Flack, H.D.; Terhell, J.C.J.M.; Parthe, E. Stacking disorder in powdered GaS$_x$Se$_{1-x}$ and related compounds. *Acta Crystallogr. Sect. A Cryst. Phys. Diffr. Theor. Gen. Crystallogr.* **1978**, *34*, S327.

6. Dunand, A.; Bernardinelli, G.; Flack, H.D. Densité électronique dans TiC et TiN. *Helv. Phys. Acta* **1982**, *55*, 132–133. [CrossRef]

7. Dunand, A.; Flack, H.D.; Yvon, K. Bonding study of titanium monocarbide and titanium mononitride. I. High-precision X-ray-diffraction determination of the valence-electron density distribution, Debye-Waller temperature factors, and atomic static displacements in $TiC_{0.94}$ and $TiN_{0.99}$. *Phys. Rev. B Condens. Matter Mater. Phys.* **1985**, *31*, 2299–2315. [CrossRef]

8. Kubel, F.; Flack, H.D.; Yvon, K. Charge-density study of vanadium nitride (VN). *Chem. Scr.* **1986**, *26*, 483.

9. Kubel, F.; Flack, H.D.; Yvon, K. Electron densities in vanadium mononitride. I. High-precision x-ray-diffraction determination of the valence-electron density distribution and atomic displacement parameters. *Phys. Rev. B Condens. Matter Mater. Phys.* **1987**, *36*, 1415–1419. [CrossRef]

10. Lewis, J.; Schwarzenbach, D.; Flack, H.D. Electric field gradients and charge density in corundum, α-aluminum oxide. *Acta Crystallogr. Sect. A Cryst. Phys. Diffr. Theor. Gen. Crystallogr.* **1982**, *38*, 733–739. [CrossRef]

11. Hoenle, W.; Flack, H.D.; Yvon, K. Single crystal X-ray study of Mo_6Se_8-type selenides containing partially substituted (Mo,M)6 clusters (M = Ru, Re). *J. Solid State Chem.* **1983**, *49*, 157–165. [CrossRef]

12. Gladyshevskii, R.E.; Cenzual, K.; Flack, H.D.; Parthe, E. Structure of RNi_3Al_9 (R= Y, Gd, Dy, Er) with either ordered or partly disordered arrangement of aluminum-atom triangles and rare-earth-metal atoms. *Acta Crystallogr. Sect. B Struct. Sci.* **1993**, *49*, 468–474. [CrossRef]

13. Ackerbauer, S.-V.; Borrmann, H.; Bürgi, H.B.; Flack, H.D.; Grin, Y.; Linden, A.; Palatinus, L.; Schweizer, W.B.; Warshamanage, R.; Wörle, M. Tigept–a study of Friedel differences. *Acta Crystallogr. Sect. B Struct. Sci. Cryst. Eng. Mater.* **2013**, *69*, 457–464. [CrossRef] [PubMed]

14. Waterstrat, R.M.; Yvon, K.; Flack, H.D.; Parthe, E. Refinement of Nb_3Si and Nb_3As. *Acta Crystallogr. Sect. B Struct. Sci.* **1975**, *31*, 2765–2769. [CrossRef]

15. Flack, H. Etude de la structure cristalline du sulfite de magnésium hexahydraté, $MgSO_3.6H_2O$. *Acta Crystallogr. Sect. B Struct. Sci.* **1973**, *29*, 656–658. [CrossRef]

16. Hostettler, M.; Flack, H.D. Anti-wurtzite reoriented. *Acta Crystallogr. Sect. B Struct. Sci.* **2003**, *59*, 537–538. [CrossRef]

17. Nishikawa, S.; Matukawa, K. Hemihedry of zincblende and x-ray reflexion. *Proc. Imp. Acad. (Tokyo)* **1928**, *4*, 96–97. [CrossRef]

18. Coster, D.; Knol, K.S.; Prins, J.A. Unterschiede in der Intensität der Röntgenstrahlen-Reflexion an den beiden 111-Flächen der Zinkblende. *Z. Phys.* **1930**, *63*, 345–369. [CrossRef]

19. Flack, H.D. High-pressure phase of polytetrafluoroethylene. *J. Polym. Sci. Polym. Phys. Ed.* **1972**, *10*, 1799–1809. [CrossRef]

20. Flack, H.D. Sliding in polytetrafluoroethylene crystals. *J. Polym. Sci. Polym. Phys. Ed.* **1974**, *12*, 81–87. [CrossRef]

21. Williams, A.F.; Flack, H.D.; Vincent, M.G. Structure of di-μ-chloro-tetrachlorotetrakis(diethyl sulfide) diiridium(III). *Acta Crystallogr. Sect. B Struct. Sci.* **1980**, *36*, 1204–1206. [CrossRef]

22. Williams, A.F.; Flack, H.D.; Vincent, M.G. Structure of μ-chloro-pentachloro-μ-(diethyl sulfide)-tris(diethyl sulfide)diiridium(III). *Acta Crystallogr. Sect. B Struct. Sci.* **1980**, *36*, 1206–1208. [CrossRef]

23. Flack, H. Les structures cristallines et moléculaires des complexes thiocyanato et isothiocyanato d'iridium ou de rhodium(III). II. Isothiocyanatopenta-ammineiridium(III) dichloride. *Acta Crystallogr. Sect. B Struct. Sci.* **1973**, *29*, 2610–2611. [CrossRef]

24. Flack, H.D.; Parthé, E. The crystal and molecular structure of thiocyanato and isothiocyanato Ir and Rh complexes. I. Thiocyanatopenta-ammineiridium(III) diperchlorate, $[Ir(SCN)(NH_3)_5](ClO_4)_2$. *Acta Crystallogr. Sect. B Struct. Sci.* **1973**, *29*, 1099–1102. [CrossRef]

25. McGarrity, J.F.; Cretton, A.; Pinkerton, A.A.; Schwarzenbach, D.; Flack, H.D. (Z)-2-methoxy-1,2-diphenylvinyl alcohol, a remarkably stable enol. *Angew. Chem. Int. Ed. Engl.* **1983**, *22*, 405–405. [CrossRef]

26. Bernardinelli, G.; Dunand, A.; Flack, H.D.; Yvon, K.; Giersch, W.; Ohloff, G. Stereochemistry and absolute structure of (+)-8β-acetoxy-12-(4-bromobenzoyloxy)-13,14,15,16-tetranorlabdane, $C_{25}H_{35}BrO_4$. *Acta Crystallogr. Sect. C Cryst. Struct. Commun.* **1984**, *40*, 1911–1914. [CrossRef]

27. Mueller, P.; Bernardinelli, G.; Allenbach, Y.F.; Ferri, M.; Flack, H.D. Selectivity enhancement in the Rh(II)-catalyzed cyclopropanation of styrene with (silanyloxyvinyl)diazoacetates. *Org. Lett.* **2004**, *6*, 1725–1728. [CrossRef]

28. Naim, A.; Bouhadja, Y.; Cortijo, M.; Duverger-Nedellec, E.; Flack, H.D.; Freysz, E.; Guionneau, P.; Iazzolino, A.; Ould Hamouda, A.; Rosa, P.; et al. Design and study of structural linear and nonlinear optical properties of chiral [Fe(phen)$_3$]$^{2+}$ complexes. *Inorg. Chem.* **2018**, *57*, 14501–14512. [CrossRef] [PubMed]

29. Flack, H.D. Interpretation of anomalous streaks in crystals of anthrone. *Acta Crystallogr. Sect. A Cryst. Phys. Diffr. Theor. Gen. Crystallogr.* **1970**, *26*, 678–680. [CrossRef]

30. Flack, H.D. Refinement and thermal expansion coefficients of the structure of anthrone (20, −90 °C) and comparison with anthraquinone. *Philos. Trans. R. Soc. A* **1970**, *266*, 561–574. [CrossRef]

31. Flack, H.D. Short-range order in crystals of anthrone and in mixed crystals of anthrone-anthraquinone. *Philos. Trans. R. Soc. A* **1970**, *266*, 583–591. [CrossRef]

32. Flack, H.D. Calculation of dimensions of ordered regions in triclinic and monoclinic pseudosymmetic crystals from the intensity of diffuse scattering. *Philos. Trans. R. Soc. A* **1970**, *266*, 575–582. [CrossRef]

33. Flack, H.D. Automatic experimental absorption correction. *Acta Crystallogr. Sect. A Cryst. Phys. Diffr. Theor. Gen. Crystallogr.* **1975**, *31*, S219.

34. Flack, H.D. CAMEL JOCKEY, an absorption correction program. *J. Appl. Crystallogr.* **1975**, *8*, 520–521. [CrossRef]

35. Flack, H.D. An experimental absorption–extinction correction technique. *Acta Crystallogr. Sect. A Cryst. Phys. Diffr. Theor. Gen. Crystallogr.* **1977**, *33*, 890–898. [CrossRef]

36. Flack, H.D.; Vincent, M.G. Absorption-weighted mean path lengths for spheres. *Acta Crystallogr. Sect. A Cryst. Phys. Diffr. Theor. Gen. Crystallogr.* **1978**, *34*, 489–491. [CrossRef]

37. Flack, H.D.; Vincent, M.G.; Alcock, N.W. Absorption and extinction corrections: Standard tests. *Acta Crystallogr. Sect. A Cryst. Phys. Diffr. Theor. Gen. Crystallogr.* **1980**, *36*, 682–686. [CrossRef]

38. Vincent, M.G.; Flack, H.D. *A priori* estimates of errors in intensities for imperfectly spherical crystals. *Acta Crystallogr. Sect. A Cryst. Phys. Diffr. Theor. Gen. Crystallogr.* **1979**, *35*, 78–82. [CrossRef]

39. Flack, H.D.; Vincent, M.G. Intensity errors due to beam inhomogeneity and imperfectly spherical crystals. *Acta Crystallogr. Sect. A Cryst. Phys. Diffr. Theor. Gen. Crystallogr.* **1979**, *35*, 795–802. [CrossRef]

40. Vincent, M.G.; Flack, H.D. On the polarization factor for crystal-monochromated x-radiation. Ii. A method for determining the polarization ratio for crystal monochromators. *Acta Crystallogr. Sect. A Cryst. Phys. Diffr. Theor. Gen. Crystallogr.* **1980**, *36*, 614–620. [CrossRef]

41. Vincent, M.G.; Flack, H.D. On the polarization factor for crystal-monochromated X-radiation. I. Assessment of errors. *Acta Crystallogr. Sect. A Cryst. Phys. Diffr. Theor. Gen. Crystallogr.* **1980**, *36*, 610–614. [CrossRef]

42. Flack, H.D.; Vincent, M.G. On the polarization factor for crystal-monochromated X-radiation. III. A weighting scheme for products. *Acta Crystallogr. Sect. A Cryst. Phys. Diffr. Theor. Gen. Crystallogr.* **1980**, *36*, 620–624. [CrossRef]

43. Jacquet, J.; Very, J.M.; Flack, H.D. The 2θ determination of diffraction peaks from 'poor' powder samples. Application to biological apatite. *J. Appl. Crystallogr.* **1980**, *13*, 380–384. [CrossRef]

44. Flack, H.D. Correcting intensity data for systematic effects. In *Methods and Applications in Crystallographic Computing*; Hall, S.R., Ashida, T., Eds.; Oxford University Press: Oxford, UK, 1984; pp. 41–55.

45. Flack, H.D. On absorption and background refinement. *Z. Kristallogr.* **1988**, *182*, 93–94.

46. Schwarzenbach, D.; Flack, H.D. On the definition and practical use of crystal-based azimuthal angles. *J. Appl. Crystallogr.* **1989**, *22*, 601–605. [CrossRef]

47. Schwarzenbach, D.; Flack, H.D. On the refinement on profile, background and net intensities. *Acta Crystallogr. Sect. A Found. Crystallogr.* **1991**, *47*, 134–137. [CrossRef]

48. Blanc, E.; Schwarzenbach, D.; Flack, H.D. The evaluation of transmission factors and their first derivatives with respect to crystal shape parameters. *J. Appl. Crystallogr.* **1991**, *24*, 1035–1041. [CrossRef]

49. Flack, H.D.; Vincent, M.G.; Vincent, J.A. Testing for serial correlation in intensity data. *Acta Crystallogr. Sect. A Cryst. Phys. Diffr. Theor. Gen. Crystallogr.* **1980**, *36*, 495–496. [CrossRef]

50. Hill, R.J.; Flack, H.D. The use of the Durbin-Watson d statistic in Rietveld analysis. *J. Appl. Crystallogr.* **1987**, *20*, 356–361. [CrossRef]

51. Schwarzenbach, D.; Abrahams, S.C.; Flack, H.D.; Gonschorek, W.; Hahn, T.; Huml, K.; Marsh, R.E.; Prince, E.; Robertson, B.E.; Rollett, J.S.; et al. Statistical descriptors in crystallography: Report of the IUCR subcommittee on statistical descriptors. *Acta Crystallogr. Sect. A Found. Crystallogr.* **1989**, *45*, 63–75. [CrossRef]

52. Schwarzenbach, D.; Abrahams, S.C.; Flack, H.D.; Prince, E.; Wilson, A.J.C. Statistical descriptors in crystallography. II. Report of a working group on expression of uncertainty in measurement. *Acta Crystallogr. Sect. A Found. Crystallogr.* **1995**, *A51*, 565–569. [CrossRef]

53. Schwarzenbach, D.; Abrahams, S.C.; Flack, H.D.; Gonschorek, W.; Hahn, T.; Marsh, R.E.; Robertson, B.E.; Rollett, J.S.; Wilson, A.J.C. Statistical descriptors in crystallography. Report of the Intematioiial Union of Crystallography Subcommittee on statistical descriptors. *Z. Kristallogr.* **1988**, *182*, 241.
54. Flack, H.D. Introduction to crystallographic statistics by U. Shmueli and G. H. Weiss. *Acta Crystallogr. Sect. A Found. Crystallogr.* **1997**, *53*, 251–252. [CrossRef]
55. Flack, H.D. On enantiomorph-polarity estimation. *Acta Crystallogr. Sect. A Found. Crystallogr.* **1983**, *39*, 876–881. [CrossRef]
56. Bernardinelli, G.; Flack, H.D. Easy refinement of absolute configuration and polarity, and easy avoidance of biased positional parameters. *Acta Crystallogr. Sect. A Found. Crystallogr.* **1984**, *40*, C427. [CrossRef]
57. Bernardinelli, G.; Flack, H.D. Least-squares absolute-structure refinement. Practical experience and ancillary calculations. *Acta Crystallogr. Sect. A Found. Crystallogr.* **1985**, *41*, 500–511. [CrossRef]
58. Bernardinelli, G.; Flack, H.D. Least-squares absolute-structure refinement. A case study of the effect of absorption correction, data region, stability constant and neglect of light atoms. *Acta Crystallogr. Sect. A Found. Crystallogr.* **1987**, *43*, 75–78. [CrossRef]
59. Flack, H.D.; Bernardinelli, G. Absolute structure and absolute configuration. *Acta Crystallogr. Sect. A Found. Crystallogr.* **1999**, *55*, 908–915. [CrossRef]
60. Flack, H.D.; Bernardinelli, G. Absolute structure and absolute configuration. *Acta Crystallogr. Sect. A Found. Crystallogr.* **1999**, *55*, 105.
61. Shmueli, U.; Flack, H.D. The mean-square Friedel intensity difference in $p1$ with a centrosymmetric substructure. *Acta Crystallogr. Sect. A Found. Crystallogr.* **2007**, *63*, S165. [CrossRef]
62. Flack, H.D.; Shmueli, U. The mean-square Friedel intensity difference in $p1$ with a centrosymmetric substructure. *Acta Crystallogr. Sect. A Found. Crystallogr.* **2007**, *63*, 257–265. [CrossRef]
63. Flack, H.D.; Bernardinelli, G. Applications and properties of the Bijvoet intensity ratio. *Acta Crystallogr. Sect. A Found. Crystallogr.* **2008**, *64*, 484–493. [CrossRef] [PubMed]
64. Shmueli, U.; Schiltz, M.; Flack, H.D. Intensity statistics of Friedel opposites. *Acta Crystallogr. Sect. A Found. Crystallogr.* **2008**, *64*, 476–483. [CrossRef] [PubMed]
65. Flack, H.D.; Bernardinelli, G. The use of X-ray crystallography to determine absolute configuration. *Chirality* **2008**, *20*, 681–690. [CrossRef] [PubMed]
66. Flack, H.D. The use of X-ray crystallography to determine absolute configuration. II. *Acta Chim. Slov.* **2008**, *55*, 689–691.
67. Shmueli, U.; Flack, H.D. Concise intensity statistics of Friedel opposites and classification of the reflections. *Acta Crystallogr. Sect. A Found. Crystallogr.* **2009**, *65*, 322–325. [CrossRef]
68. Shmueli, U.; Flack, H.D. Intensity statistics of Friedel opposites and classification of reflections. *Acta Crystallogr. Sect. A Found. Crystallogr.* **2009**, *65*, S109. [CrossRef]
69. Shmueli, U.; Flack, H.D. Probability density functions of the average and difference intensities of Friedel opposites. *Acta Crystallogr. Sect. A Found. Crystallogr.* **2010**, *66*, 669–675. [CrossRef] [PubMed]
70. Shmueli, U.; Flack, H.D. Response to Olczak's comment on probability density functions of the average and difference intensities of Friedel opposites. *Acta Crystallogr. Sect. A Found. Crystallogr.* **2011**, *67*, 318. [CrossRef]
71. Flack, H.D.; Sadki, M.; Thompson, A.L.; Watkin, D.J. Practical applications of averages and differences of Friedel opposites. *Acta Crystallogr. Sect. A Found. Crystallogr.* **2011**, *67*, 21–34. [CrossRef] [PubMed]
72. Flack, H.D. Physical and spectrometric analysis: Absolute configuration determination by x-ray crystallography. In *Comprehensive Chirality*; Carreira, E.M., Yamamoto, H., Eds.; Elsevier: Amsterdam, The Netherlands, 2012; pp. 648–656. ISBN 9780080951683.
73. Flack, H.D. Virtual issue on absolute structure. *Acta Crystallogr. Sect. C Cryst. Struct. Commun.* **2012**, *68*, e13–e14. [CrossRef] [PubMed]
74. Flack, H.D. Absolute-structure reports. *Acta Crystallogr. Sect. C Cryst. Struct. Commun.* **2013**, *69*, 803–807. [CrossRef] [PubMed]
75. Parsons, S.; Flack, H.D.; Wagner, T. Use of intensity quotients and differences in absolute structure refinement. *Acta Crystallogr. Sect. B Struct. Sci. Cryst. Eng. Mater.* **2013**, *69*, 249–259. [CrossRef] [PubMed]
76. Flack, H.D. Absolute-structure determination: Past, present and future. *Chimia* **2014**, *68*, 26–30. [CrossRef]
77. Flack, H.D. Absolute structure. *Acta Crystallogr. Sect. A Found. Adv.* **2016**, *71*, s176–s176. [CrossRef]
78. Cooper, R.I.; Watkin, D.J.; Flack, H.D. Absolute structure determination using CRYSTALS. *Acta Crystallogr. Sect. C Struct. Chem.* **2016**, *72*, 261–267. [CrossRef]

79. Flack, H.D.; Bernardinelli, G.; Clemente, D.A.; Linden, A.; Spek, A.L. Centrosymmetric and pseudo-centrosymmetric structures refined as non-centrosymmetric. *Acta Crystallogr. Sect. B Struct. Sci.* **2006**, *62*, 695–701. [CrossRef]

80. Flack, H.D.; Bernardinelli, G. Centrosymmetric crystal structures described as non-centrosymmetric: An analysis of reports in Inorganica Chimica Acta. *Inorg. Chim. Acta* **2006**, *359*, 383–387. [CrossRef]

81. Parsons, S.; Pattison, P.; Flack, H.D. Analysing Friedel averages and differences. *Acta Crystallogr. Sect. A Found. Crystallogr.* **2012**, *68*, 736–749. [CrossRef]

82. Flack, H.D.; Bernardinelli, G. Reporting and evaluating absolute-structure and absolute-configuration determinations. *J. Appl. Crystallogr.* **2000**, *33*, 1143–1148. [CrossRef]

83. Flack, H.D. Louis pasteur's discovery of molecular chirality and spontaneous resolution in 1848, together with a complete review of his crystallographic and chemical work. *Acta Crystallogr. Sect. A Found. Crystallogr.* **2009**, *65*, 371–389. [CrossRef] [PubMed]

84. Flack, H.D. Chiral and achiral crystal structures. *Helv. Chim. Acta* **2003**, *86*, 905–921. [CrossRef]

85. Flack, H.D.; Bernardinelli, G. The mirror of Galadriel: Looking at chiral and achiral crystal structures. *Cryst. Eng.* **2003**, *6*, 213–223. [CrossRef]

86. Flack, H.D. Teaching about chirality in crystals and molecules. *Acta Crystallogr. Sect. A Found. Crystallogr.* **2004**, *60*, s106–s106. [CrossRef]

87. Flack, H.D. Symmetry, spectroscopy, and crystallography: The structural nexus. By Robert gGaser. *Acta Crystallogr. Sect. A Found. Adv.* **2017**, *73*, 207–207. [CrossRef]

88. Flack, H.D. Perspective and concepts: Chirality in nineteenth century science. In *Comprehensive Chirality*; Carreira, E.M., Yamamoto, H., Eds.; Elsevier: Amsterdam, The Netherlands, 2012; pp. 1–10. ISBN 9780080951683.

89. Djukic, J.P.; Hijazi, A.; Flack, H.D.; Bernardinelli, G. Non-racemic (scalemic) planar-chiral five-membered metallacycles: Routes, means, and pitfalls in their synthesis and characterization. *Chem. Soc. Rev.* **2008**, *37*, 406–425. [CrossRef]

90. Flack, H.D. A segment description for the unique set of reflections in non-centrosymmetric crystal classes. *J. Appl. Crystallogr.* **1984**, *17*, 361–362. [CrossRef]

91. Flack, H.D. The derivation of twin laws for (pseudo-)merohedry by co-set decomposition. *Acta Crystallogr. Sect. A Found. Crystallogr.* **1987**, *43*, 564–568. [CrossRef]

92. Flack, H.D.; Schwarzenbach, D. On the use of least-squares restraints for origin fixing in polar space groups. *Acta Crystallogr. Sect. A Found. Crystallogr.* **1988**, *44*, 499–506. [CrossRef]

93. Flack, H.D.; Wondratschek, H.; Hahn, T.; Abrahams, S.C. Symmetry elements in space groups and point groups. Addenda to two IUCR reports on the nomenclature of symmetry. *Acta Crystallogr. Sect. A Found. Crystallogr.* **2000**, *56*, 96–98. [CrossRef]

94. Flack, H.D.; Wörle, M. Merohedral twin interpretation spreadsheet, including command lines forshelxl. *J. Appl. Crystallogr.* **2013**, *46*, 248–251. [CrossRef]

95. Flack, H.D. How to determine the space group of a twinned crystal or one with metric specialization. *Acta Crystallogr. Sect. A Found. Adv.* **2015**, *71*, s505. [CrossRef]

96. Flack, H.D. Methods of space-group determination–a supplement dealing with twinned crystals and metric specialization. *Acta Crystallogr. Sect. C Struct. Chem.* **2015**, *71*, 916–920. [CrossRef] [PubMed]

97. Flack, H.D. Patterson functions. *Z. Kristallogr.* **2015**, *230*, 743–748. [CrossRef]

98. Flack, H.D. The revival of the Bravais lattice. *Acta Crystallogr. Sect. A Found. Adv.* **2015**, *71*, 141–142. [CrossRef]

99. Schwarzenbach, D.; Flack, H.D. Structure refinement (solid state diffraction). In *Encyclopedia of Spectroscopy and Spectrometry*, 3rd ed.; Lindon, J.C., Tranter, G.E., Koppenaal, D.W., Eds.; Elsevier: Amsterdam, The Netherlands, 2017; Volume 4, pp. 319–324. ISBN 978-0-12-803224-4.

100. McMahon, B.; Strickland, P.R.; Flack, H.D.; Epelboin, Y.; Cranswick, L.M.D. Crystallographic publishing in the electronic era. *Acta Crystallogr. Sect. A Found. Crystallogr.* **1999**, *55*, 585. [CrossRef]

101. Epelboin, Y.; Flack, H.D. Internet resources for crystallography. *Acta Crystallogr. Sect. A Found. Crystallogr.* **1996**, *52*, C77–C77. [CrossRef]

102. Flack, H.D. World Wide Web for crystallography. *J. Res. Natl. Inst. Stand. Technol.* **1996**, *101*, 375–380. [CrossRef] [PubMed]

103. Flack, H.D. Surfing the crystallographic net (crysnet) workshops. *Acta Crystallogr. Sect. A Found. Crystallogr.* **1996**, *52*, C578–C578. [CrossRef]

104. Flack, H.D.; Bernardinelli, G. Is your crystal representative of the bulk? *Acta Crystallogr. Sect. A Found. Crystallogr.* **2005**, *61*, c316–c316. [CrossRef]

105. Flack, H.D.; Blanc, E.; Schwarzenbach, D. DIFRAC, single-crystal diffractometer output-conversion software. *J. Appl. Crystallogr.* **1992**, *25*, 455–459. [CrossRef]

106. Cooper, R.I.; Watkin, D.J.; Flack, H.D. HUG and SQUEEZE: Using crystals to incorporate resonant scattering in the SQUEEZE structure-factor contributions to determine absolute structure. *Acta Crystallogr. Sect. C Struct. Chem.* **2017**, *73*, 845–853. [CrossRef] [PubMed]

107. Stewart, J.M.; Machin, P.A.; Dickinson, C.W.; Ammon, H.L.; Heck, H.S.; Flack, H. *The XRAY system-version of 1976*; Tech. Rep. Tr-446.; Computer Science Center, University of Maryland: College Park, MD, USA, 1976.

108. International Union of Crystallography. Available online: https://www.iucr.org/resources/cif/software/xtal (accessed on 17 June 2020).

109. Hall, S.R.; du Boulay, D.J.; Olthof-Hazekamp, R. *Xtal3.7 System*; University of Western Australia: Perth, WA, Australia, 2000.

110. Shmueli, U.; Flack, H.D.; Spence, J.C.H. Methods of space-group determination. In *International Tables for Crystallography Volume A, Edition 6: Space-Group Symmetry*; Aroya, M.I., Ed.; Kluwer Academic Publishers: Dordrecht, The Netherlands, 2016; Chapter 1.6.; pp. 107–131.

chemistry

Review

Howard Flack and the Flack Parameter

David John Watkin * and Richard Ian Cooper

Chemical Crystallography Laboratory, Department of Chemistry, University of Oxford, Mansfield Road, Oxford OX1 3TA, UK; richard.cooper@chem.ox.ac.uk
* Correspondence: david.watkin@chem.ox.ac.uk

Received: 25 August 2020; Accepted: 21 September 2020; Published: 23 September 2020

Abstract: The Flack Parameter is now almost universally reported for all chiral materials characterized by X-ray crystallography. Its elegant simplicity was an inspired development by Howard Flack, and although the original algorithm for its computation has been strengthened by other workers, it remains an essential outcome for any crystallographic structure determination. As with any one-parameter metric, it needs to be interpreted in the context of its standard uncertainty.

Keywords: Howard Flack; Flack parameter; structure analysis; X-ray crystallography

---

1. Introduction

The formation of Howard Flack (1943–2017) was as a chemist, but he was also an able mathematician who turned this skill to crystallographic problems (Figure 1). Most of his contributions to crystallography will pass unnoticed by chemists relying on X-ray analysis to robustly characterise new materials and are probably not well known even to many professional structure analysts. He did not grind out an endless stream of papers, but instead produced a small number of carefully written, carefully thought out works in the domains of data acquisition, space group theory and crystallographic twinning.

Much of Flack's output has gone into the infrastructure of crystallography and need not concern structural chemists, but two of his insights have an immediate impact on ordinary structure analyses—correcting the observed data for absorption effects, and evaluating the absolute structure of chiral materials. An excellent technical review of the physics and mathematics behind absolute structure determination can be found in [1].

Figure 1. Howard Flack, courtesy of his widow Evelyne.

2. Impact on Service Crystallography

2.1. Background

Modern atomic-resolution small molecule X-ray structure analyses produce an enormous number of experimental observations even from relatively simple chemical substances. This wealth of data has enabled two procedures first described by Flack to become universally accepted as part of routine structure analyses.

Unlike specular reflection of light by a mirror, interaction between a crystal and X-rays is an interference phenomenon, which means that 'reflections' (actually diffracted beams) only occur when the crystal is orientated to satisfy Bragg's Law. Every diffracted beam (identified by the Miller Indices *h,k,l*) has a unique relationship with the unit cell of the crystal and an intensity dependent upon the internal make-up of the sample. Diffraction occurs at the Bragg Angles (θ) given by:

$$\left(\frac{\sin \theta}{\lambda}\right)^2 = \text{h}G^*\text{h}^t \tag{1}$$

where G^* is the reciprocal metric tensor (characteristic of the crystal) and h is the row vector of indices. *h,k,l* are positive and negative integers permutated up until the maximum $\sin \theta = 1$. In practice there are experimental limitations below this maximum, but even so, X-ray crystal structure analysis is *data rich*, that is to say, there are very many more observations available than there are parameters describing the structure. For much of the 1960s, 70s and 80s most X-ray data were measured on instruments which observed reflections one at a time (serial diffractometers). Depending upon the space group of the sample, some (and perhaps many) of the permutated indices refer to *equivalent reflections*. Equivalent reflections are ones for which the theoretical intensity should be identical even though the indices are different. Thus, in the space group *Pc* the reflection *h,k,l* will be equivalent to *h,–k,l*. Because of the time taken to measure data, it was common practice to only measure one of a group of equivalent reflections, thus, in the example above, halving the data collection time. When more than one equivalent reflections are measured, after correction for experimental geometry and short comings, they are averaged (merged) by crystallographic analysis software to provide a set of *unique* observed reflections. R_{merge} and R_{int} are measures of the self-consistency of groups of equivalent reflections. Current area detector diffractometers measure hundreds of reflections quasi-simultaneously, so that many equivalent reflections can be observed.

In a structure determination, the analyst postulates an atomic model (usually via a *structure solution* program), generally with 9 parameters per independent atom. A structure factor can be calculated from this model, and the parameters optimized to match I_c with I_o (see Appendix A) by the method of nonlinear least squares. In a modern analysis, the observation: parameter ratio is often in the range 10 to 20, equivalent to fitting a straight line ($y = mx + c$) to between 20 to 40 observations.

Just as the image of an object remains visible when a mirror is rotated about its normal, so diffraction continues to occur as the crystal is rotated about the normal to the diffracting plane. This means that on a suitably engineered instrument, (e.g., a four circle diffractometer), the rotational orientation (ψ) of the crystal can be varied for a given Miller index. Before the widespread use of area detectors, about 25 years ago, only a single measurement was generally made for each unique Miller index. Modern instruments are normally programmed to make between 5 and 20 observations of equivalent reflection at different ψ values. These independent measurement of the same quantity are termed redundancy, or multiplicity of observation, MoO. Measurements on a typical organometallic crystal might take a less than an hour on a laboratory instrument, or just a few minutes at a synchrotron source, yielding a data set of tens of thousands of observations.

2.2. Correction for Absorption

One problem facing all X-ray structure analysts is 'correcting' their observed data for experimental effects. Some of these, such as the Lorentz and polarisation corrections, have been well characterised analytically since the 1920's. More problematic has been computing corrections for absorption effects. X-rays are attenuated as they pass through a medium according to Beer's Law:

$$I_o = I_i e^{-\mu t}, \tag{2}$$

where I_o is the observed X-ray intensity, I_i is the incident intensity, μ is the absorption coefficient (computed from the elemental composition [2]) and t is the path length through the sample. Unless the crystal can be ground to a sphere (which is still sometimes done for special purposes), the exponential means that exact mathematical calculations depend upon very accurate measurements of the crystal shape and size—still not easily done even with modern digital microscopes.

For an X-ray beam travelling along the long axis of a prismatic crystal, the intensity of the emergent beam will reduce and then increase again as the crystals is rotated through an angle ψ about a vector perpendicular to the long axis. In the 1960s, methods based on experimental observations were developed for making empirical corrections for absorption by tracking the intensity of a chosen reflection as ψ was systematically varied [3,4]. These methods created a kind of calibration curve for the sample. Flack recognised that these curves could be better represented by a smooth mathematical function [5]. Because of the periodic nature of the curves, one natural function of choice was the Fourier series. Flack's careful experimental verification of the method, which included making thousands of observations on a serial diffractometer, demonstrated the strength of the concept. Today, area detector diffractometers inevitably measure many reflections at several different values of ψ. While these differing ψ values almost never correspond to systematic curve tracking, their great abundance enables them to be used is a similar manner. Blessing replaced the Fourier series by spherical harmonics [6], and this has become the basis of all experimental data corrections up to modern times. It turns out that this empirical method of correcting for absorption can also be used to correct for a range of other experimental problems, and since modern area-detector diffractometers provide a large number of equivalent observations (high MoO), the corrections are generally applied without any special action from the analyst. The robust correction of the observations is an important contributor to a robust determination of the absolute structure of a crystal.

2.3. Determination of Absolute Structure

However, for most chemists working in chiral chemistry, it is Flack's parameter that will most evidently impact them. It had been known since the 1930's that under specific conditions the diffraction of X-rays from a non-centrosymmetric crystal displays small effects which reflect the symmetry of the crystal, and hence the symmetry of the materials the crystal was built from [7]. Of special interest to the chemist working with chiral materials is the fact that for crystals of these materials, the diffracted intensities of a pair of reflections h,k,l and $-h,-k,-l$ (which are identical for crystals in centrosymmetric space groups) are subtly different [8]. The difference between the members of these Friedel pairs are generally very small and so easily masked by experimental issues. The magnitude of these Friedel differences depends upon the resonant scattering of the elements in the material and the wavelength of the X-rays. Early attempts to determine the absolute structure of a material depended upon the analyst identifying enantiomer sensitive pairs and then carefully remeasuring them. Hamilton's 'R-factor Ratio Test' tried to use all the observable reflections [9], but was difficult to apply reliably. A breakthrough was Roger's proposal that the absolute structure of a material could be estimated, together with a reliability index, by the determination of one additional parameter (which he called η) in the model [10]. Unfortunately, the derivative of this parameter with respect to the calculated structure factor is discontinuous over the physical range of the parameter, meaning that its interpretation was unsound except for truly enantio-pure materials. Flack became interested in this mathematical problem, and within two years had reformulated it in terms of a new parameter (which he called 'x', see Section 4 for derivation) which had a continuous derivative, and thus could be used in the analysis of crystals which, while containing only chirality-preserving symmetry operations, contained a mixture of domains of two enantiomers—in crystallography called '*twinning by inversion*' [11]. When x is zero, the material is enantiopure and the atomic coordinates determined from the data correspond to the actual structure. If x is unity, then the atomic model is the inverse of the actual structure—the model is 'inverted'.

3. Twinning by Inversion

It has been estimated that when solutions contain exactly equal amounts of both enantiomers of a compound form crystals, approximately 90% will form racemic crystals, containing exactly equal numbers of each enantiomer [12]. These may be related by a centre of inversion—the centrosymmetric crystals drawn from the set labelled CA in Figure 2, or by chirality-inverting non-centrosymmetric operations in non-centrosymmetric space groups (crystal classes labelled NA in Figure 2) for example space group *Pc*.

Occasionally, as Pasteur had the good fortune to observe, the two enantiomers preferentially separate out into enantiopure crystals, a phenomenon called spontaneous resolution. The enantiopure constituent molecules restrict the choice of space group to one of the 65 Sohncke groups which contain exclusively chirality-preserving symmetry operations, i.e., rotations and translations (from the crystal classes labelled NC in Figure 2).

In rare cases both enantiomers may crystallize together in a Sohncke space group as independent molecules which are not related by a symmetry operation of the crystal. Approximately 50 examples of this type were identified in Flack's review of chiral and achiral crystal structures in 2003 [13]. Materials exhibiting this phenomenon have subsequently been termed kryptoracemates, as their racemic nature is hidden by the space group symmetry. A recent survey identified 724 examples in the crystallographic literature [14].

In addition to these cases, it is also possible to form a solid solution of enantiomers. The crystal grows without chiral selectivity so that either enantiomer can occupy a molecular site within a crystal structure resulting in a partially disordered solid solution of enantiomers [15].

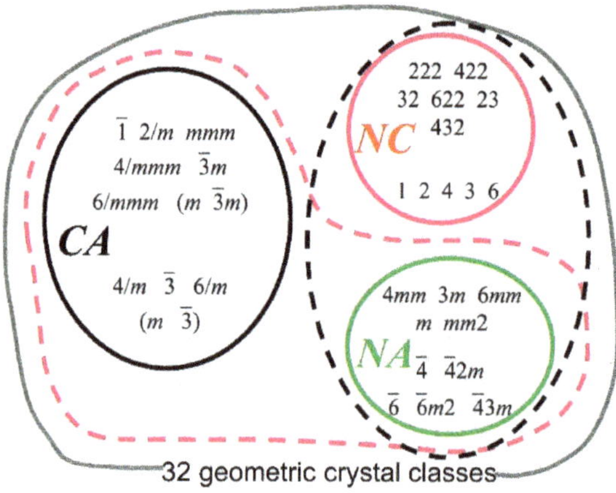

Figure 2. The 32 geometric crystal classes. CA means Centrosymmetric & Achiral, NA means Non-centrosymmetric & Achiral, NC means Non-centrosymmetric & chiral. Note that the symbols in the achiral classes contain either an m, $\bar{4}$ or $\bar{6}$ operator, allowing pairs of enantiomers to be related by a chirality-inverting symmetry operation which is not a centre of symmetry. Figure by courtesy of Evelyne Flack [16].

If the solution contains one and only one enantiomer, it will crystallise in a Sohncke space group. In this case, the determination of Flack's parameter will enable the stereochemistry of the material to be estimated. Note well that the stereochemistry is only estimated. Refinement, by least squares, of all the parameters needed to characterise the material (the positions of the atoms in the unit cell, their motion about their mean positions, and Flack's parameter) provides both values for these parameters and estimates of their standard uncertainties (written u or s.u.), previously called standard deviations. This is really important. For example, the standard uncertainties in atomic positions enable one to compute the standard uncertainty in the distance between them, which in its turn enables a rational decision to be made as to whether two bond lengths are significantly different. Flack's parameter is also determined with a standard uncertainty, enabling the analyst to assign a level of confidence to the proposed absolute structure.

It is not clear when 'Flack's x parameter' became simply 'The Flack parameter'. For the 20 years following its formulation, the determination of the Flack parameter required extreme attention to experimental details. For first row elements the magnitude of the resonant scattering is small for copper radiation ($\lambda = 1.54$ Å) and even smaller for molybdenum radiation ($\lambda = 0.71$ Å), and so it was generally held that the absolute structure of light atom materials required copper radiation. It was not uncommon at that time to form a derivative by inserting an atom with larger resonant scattering (such as chlorine) into the molecule. One snag with this approach was that heavier atoms tend to have higher absorption coefficients—though the absorption corrections described above tend to minimize the impact on the data quality. The millennium saw the widespread use of area detectors for the measurement of diffraction data at temperatures close to the boiling point of liquid nitrogen. Working at this temperature reduces the motion of the atoms in the molecules, and greatly improves the quality of the experimental observations. These instruments measure the intensities of very many Bragg reflections quasi-simultaneously rather than measuring the reflections one-by-one. As well as providing the many observations needed for the absorption corrections, these instruments also measured many of the Bijvoet (commonly used as a synonym for Friedel) pairs of reflections needed for the determination of Flack's parameter (see Appendix B). The determination of absolute structure by the determination of the Flack Parameter became common place.

If the enantiopurity of the sample is unquestionably known by other experimental methods, then an estimate of the Flack parameter by any diffraction calculation is likely to be sound. Detailed evaluation of modern data from light atoms structures of known chirality has shown that even analyses using molybdenum radiation are unlikely to point to an incorrect absolute structure [17].

Problems arise if the enantiopurity is uncertain. Unless instructed otherwise, most X-ray analysts search through the sample for the 'best' looking crystal, i.e., one with bright faces, roughly isometric shape, no inclusions, etc. The outcome of the analysis is the absolute structure of that particular crystal. Nothing can be said about the nature of the other crystals in the batch, nor of the solution from which the crystals were grown. Other, non-crystallographic, methods must be used to verify that the absolute structure of the sample crystal is indeed the same as that of the whole batch. Although the crystal used in the X-ray analysis may measure less than $0.1 \times 0.1 \times 0.1$ mm, it is now possible to verify by NMR, chromatography or optical methods that the single crystallographic specimen really represents the bulk sample [18]. Without a references sample or sophisticated simulations such methods cannot directly determine the absolute configuration of enantiomers, however these methods are extremely valuable for determining the enantiomeric excess of samples which can be used as a valuable piece of prior information in a crystallographic absolute structure determination.

Thompson et al. tabulated all the possible outcomes from samples of unknown purity [19]. The bottom line is that for a sample which is not enantiopure, these cases may occur singly or together:

(1)　Spontaneous resolution, with separate crystals forming of each enantiomer.
(2)　Pairs of enantiomers may crystallize together form racemic crystals, with any excess concentration of one enantiomer forming enantiopure chiral crystals.
(3)　Formation of inversion twinned crystals, which contain contiguous domains of sufficient size to diffract coherently, some of which are related by inversion.
(4)　The material is achiral but forms chiral crystals.

The crystallographer can do little about cases 1 and 2 except pay attention to the morphology of the crystals in the sample. In case 1 crystals may grow, as they did for Pasteur, having small facets which are mirror images in the two enantiomeric crystals. In case 2 it is very likely that the achiral crystals will look quite different to the enantiopure chiral crystals. It is now widely accepted that case 3 is uncommon for organic materials, but may be an issue for inorganic or organo-metallic materials where chirality of the sample may arise from the arrangement of the building blocks in the crystal [20]. Where twinning by inversion does exist, it may be clearly identifiable under an optical microscope, or it may be on a sub-microscopic scale with small zones of each enantiomer distributed throughout the crystal. Case 4 is illustrated by silica. The building block of quartz, SiO_4, is achiral but these motifs can be assembled in crystals to form left or right handed spirals. The hand of the spiral sometimes reveals itself as small facets on the surface of well-formed crystals but is extremely rare [21].

4. Interpretation of the Flack Parameter

The equation defining the Flack Parameter [11] is simply:

$$I_o \cong I_c = (1-x)I_s^+ + xI_s^-,\tag{3}$$

where x is a kind of partition or mixing coefficient with extreme values of zero or unity. I_o is the observed structure amplitude, I_s is the amplitude computed for a single enantiomer, I_c is the amplitude computed for the twin. The crystallographer develops an atomic model representing the structure (i.e., one which can be displayed in a graphics program) and from this can compute structure amplitudes (I_s^+ and I_s^-) for the structure and its inverse.

x can be included as a parameter in the structure refinement, or Equation (3) can be solved for x by least squares from all the reflections once $I_c^\pm$ are known. The first method is that used by Flack in his original work [11]. Sheldrick developed a post-refinement method for estimating the parameter as a one-off computation once the main refinement was completed which he called the "Hole in One" method [22]:

$$I_o^+ - I_s^+ \cong x\left(I_s^- - I_s^+\right)\tag{4}$$

Later Hooft et al. [23] re-wrote the expression to use the differences of Friedel pairs and developed a Bayesian statistical post-refinement method:

$$D_o \cong (1-2x)D_s,\tag{5}$$

where $D_0 = I_0^+ - I_0^-$ and similarly for D_s. Parsons included the averages of the observed and of the computed Friedel pairs as a way of reducing the influence of experimental errors—the method now widely known as Parsons' Quotients in Equation (6) [24]:

$$Q_o \cong (1-2x)Q_s,\tag{6}$$

where $Q_o = \left(I_0^+ - I_0^-\right)/\frac{1}{2}\left(I_0^+ + I_0^-\right)$ and similarly for Q_s.

For a while these post-refinement methods were treated with suspicion because they lacked the estimates of covariance between the atomic parameters and Flack's parameter which classical full matrix refinement provided. Leverage analysis of the full matrix for refinement of the Flack parameter together with the atomic parameters identifies those reflections most influential in determining x [25]. Ironically, it turns out that relatively few reflections carry strong information about the Flack parameter—very careful remeasurement of these reflections should reduce the standard uncertainty of x. Eventually, an in-depth analysis by Parsons et al. showed that except in marginal cases the covariance was small, and that post refinement methods were generally safe to use [26].

Most crystallographic programs enable you to compute (with greater or lesser ease) the Flack parameter by both direct refinement in the main analysis (Flack's original method), and by one or more of the post refinement methods. Generally, the methods give substantially the same result, except the standard uncertainty of the direct refinement can be about twice as large as post refinement methods. Where there is a significant difference either in the value of x itself or in the standard uncertainty, this can sometimes be traced back to problems with the analysis [27]. These problems are masked in the main refinement because the observations are given weights which include both a contribution from the observed variance of the reflection and a modifier chosen so that the weighted residual, $w\left(F_o^2 - F_c^2\right)^2$, is fairly uniform across all the data. This modifier is justified in that it accounts for unidentified short comings in both the data and the model, and yields a linear normal probability plot. Post refinement methods generally give a linear normal probability plot using the unmodified variances.

5. Conclusions

Practical Advice Concerning the Flack Parameter

It is crucial to make it clear to the X-ray analyst from the outset that the absolute structure needs to be determined. Although modern instruments routinely yield exceptionally high-quality data, data collection parameters can be further optimized for the determination of absolute structures. Since the X-ray analysis will usually be made on just one single crystal selected from the material submitted, it is important to provide an estimate the sample's enantiopurity. This will warn the analyst about the need to look out for crystals with an anomalous morphology. Where possible, recover the actual crystal used in the X-ray analysis and verify that it represents the bulk sample. It is no longer fashionable to make heavy atom derivatives since it is preferable to avoid unnecessary additional synthetic processes. However, it does not seem to be widely appreciated that the heavy atom does not have to be part of the material under investigation. Samples containing solvent of crystallization where the solvent contains a heavy atom (e.g., CH_2Cl_2) will generate larger Bijvoet differences.

The determination of a parameter from many thousands of observations by the method of least squares means that as well as getting an estimate of the parameter value, one also gets an estimate of its reliability—the standard uncertainty, e.g., 0.05 ± 0.01, written by crystallographers as 0.05(1).

The standard uncertainty is key to interpreting the value of x. If the parameter is determined with a large s.u. then the value of x is unreliable and caution should be used when drawing conclusions about the enantiopurity of the sample. Flack and Bernardinelli argued that for a material known to be enantiopure, a standard uncertainty of less than 0.08 is sufficient for the assignment of absolute structure to be well determined [28]. A chirally sensitive HPLC or NMR experiment can establish the ratio of enantiomers in a sample, but not the absolute configuration of the dominant enantiomer. If the enantiopurity is unknown at the time of the crystallographic analysis, the s.u. needs to be below 0.04 to indicate a confident outcome. Even here, there is room for uncertainty. If the refinement gives $x = 0.10(4)$, this can be interpreted as x lying within 3 s.u. of zero, so the absolute configuration is confirmed. However, it can also be interpreted as meaning that the sample contains up to 10% of the opposite enantiomer in the form of a racemic twin. X-ray crystallography can provide a robust indication of the absolute structure of the bulk of the single crystal used in the experiment, not the absolute structure of the bulk of a batch crystalline sample, and it can only provide an estimate of the enantiopurity (with standard uncertainty).

Author Contributions: Conceptualization and production, R.I.C. and D.J.W. All authors have read and agreed to the published version of the manuscript.

Funding: This research received no external funding.

Conflicts of Interest: The authors declare no conflict of interest.

Appendix A. Observed and Calculated Structure Amplitudes

If an atomic model exists for a crystalline material, a calculated structure factor can be computed from:

$$F = A + iB$$

and hence:

$$F^2 = A^2 + B^2$$

Note that the structure factor is a complex quantity, having both a magnitude and a phase. The squared term is the calculated structure intensity which is proportional to the observed structure intensity. The magnitude can easily be derived from the intensity, but in general the phases are unknown.

In the older literature, the observed intensity usually referred to the physically measured magnitude squared of a given reflection and was represented by the symbol I_o. This was converted to an observed squared structure factor by the application of various experiment-dependent correction factors, represented by the symbol F_o^2, and the corresponding term computed from the atomic model was represented by F_c^2.

Increasingly (and confusingly), following the usage in the macromolecular community, I_o is used to represent $F_o{}^2$, and similarly for I_c. We have followed this convention in this manuscript, with the additional extension that I_s represents the squared structure amplitude computed for an enantiopure material, and I_c allows for twinning. See Equation (3).

Appendix B. Friedel and Bijvoet Pairs

A definition gaining traction is that Friedel pairs are strictly related by inversion (e.g., h,k,l and $-h,-k,-l,$), where as a Bijvoet pair is any symmetry equivalent related by a symmetry operator of the second kind (e.g., h,k,l and $h,-k,l$) [29].

References

1. Parsons, S. Determination of absolute configuration using X-ray diffraction. *Tetrahedron Asymmetry* **2017**, *28*, 1304–1313. [CrossRef]
2. Creagh, D.C.; Hubbel, J.H. X-ray absorption (or attenuation) coefficients. In *International Tables for Crystallography*; Prince, E., Ed.; Springer: Heidelberg, Germany, 2006; Volume C, Section 4.2.4; pp. 220–229.
3. North, A.C.T.; Phillips, D.C.; Mathews, F.S. A semi-empirical method of absorption correction. *Acta Cryst.* **1968**, *24*, 351–359. [CrossRef]
4. Kopfmann, G.; Huber, R. A method of absorption correction by X-ray intensity measurements. *Acta Cryst.* **1968**, *24*, 348–351. [CrossRef]
5. Flack, H.D. Automatic absorption correction using intensity measurements from azimuthal scans. *Acta Cryst.* **1974**, *30*, 569–573. [CrossRef]
6. Blessing, R.H. An empirical correction for absorption anisotropy. *Acta Cryst.* **1995**, *51*, 33–38. [CrossRef] [PubMed]
7. Coster, D.; Knol, K.S.; Prins, J.A. Unterschiede in der intensität der röntgenstrahlen-reflexion an den beiden 111-flächen der zinkblende. *Z. Physik.* **1930**, *63*, 345–369. [CrossRef]
8. Bijvoet, J.M.; Peerdeman, A.F.; Van Bommel, A.J. Determination of the absolute configuration of optically active compounds by means of X-rays. *Nature* **1951**, *168*, 271–272. [CrossRef]
9. Hamilton, W.C. Significance tests on the crystallographic R factor. *Acta Cryst.* **1965**, *18*, 502–510. [CrossRef]
10. Rogers, D. On the application of hamilton's ratio test to the assignment of absolute configuration and an alternative test. *Acta Cryst.* **1981**, *A37*, 734–741. [CrossRef]
11. Flack, H.D. On enantiomorph-polarity estimation. *Acta Cryst.* **1983**, *39*, 876–881. [CrossRef]
12. Collet, A.; Ziminski, L.; Garcia, C.; Vigne-Maeder, F. Discrimination in crystalline enantiomer systems: Facts, interpretations, and speculations. In *Supramolecular Stereochemistry*; Siegel, J.S., Ed.; NATO ASI Series (Series C: Mathematical and Physical Sciences); Springer: Dordrecht, The Netherlands, 1995; Volume 473.
13. Flack, H. Chiral and achiral crystal structures. *Helvetica Chim. Acta* **2003**, *86*, 905–921. [CrossRef]
14. Clevers, S.; Coquerel, G. Kryptoracemic compound hunting and frequency in the cambridge structural Database. *CrystEngComm* **2020**. [CrossRef]
15. Brandel, C.; Petit, S.; Cartigny, Y.; Coquerel, G. Structural aspects of solid solutions of enantiomers. *Curr. Pharm. Des.* **2016**, *22*, 4929–4941. [CrossRef]
16. Flack, H.D. *The Mirror of Galadriel: Looking at Chiral and Achiral Crystal Structures*; The Chemistry Research Laboratory: Oxford, UK, 2005.
17. Thompson, A.L.; Watkin, D.J. CRYSTALS enhancements: Absolute structure determination. *J. Appl. Cryst.* **2011**, *44*, 1017–1022. [CrossRef]

18. Tranter, G.; Le Pevelen, D.D. Chiroptical spectroscopy and the validation of crystal structure stereochemical assignments. *Tetrahedron Asymmetry* **2017**, *28*, 1192–1198. [CrossRef]

19. Thompson, A.L.; Watkin, D.J. X-ray crystallography and chirality: Understanding the limitations. *Tetrahedron Asymmetry* **2009**, *20*, 712–717. [CrossRef]

20. Flack, H.D. Absolute-structure reports. *Acta Cryst.* **2013**, *C69*, 803–807. [CrossRef] [PubMed]

21. The Quartz Page. Available online: http://www.quartzpage.de/crs_twins.html (accessed on 25 May 2020).

22. Sheldrick, G.M. *SHELX-76 Program for Crystal Structure Determination*; University of Cambridge: Cambridge, UK, 1976.

23. Hooft, R.W.W.; Straver, L.H.; Spek, A.L. Determination of absolute structure using Bayesian statistics on Bijvoet differences. *J. Appl. Crystallogr.* **2008**, *41*, 96–103. [CrossRef]

24. Parsons, S.; Flack, H. Precise absolute-structure determination in light-atom crystals. *Acta Cryst.* **2004**, *60*, s61. [CrossRef]

25. Parsons, S.; Wagner, T.; Presly, O.; Wood, P.A.; Cooper, R.I. Applications of leverage analysis in structure refinement. *J. Appl. Crystallogr.* **2012**, *45*, 417–429. [CrossRef]

26. Parsons, S.; Flack, H.D.; Wagner, T. Use of intensity quotients and differences inabsolute structure refinement. *Acta Cryst.* **2013**, *B69*, 249–259.

27. Watkin, D.J.; Cooper, R.I. Why direct and post-refinement determinations of absolute structure may give different results. *Acta Cryst.* **2016**, *B72*, 661–683. [CrossRef] [PubMed]

28. Flack, H.; Bernardinelli, G. Reporting and evaluating absolute-structure and absolute-configuration determinations. *J. Appl. Cryst.* **2000**, *33*, 1143–1148. [CrossRef]

29. Vijayan, M.; Ramaseshan, S. Isomorphous replacement and anomolous scattering. In *International Tables for Crystallography Volume B: Reciprocal Space*; Shmueli, U., Ed.; International Union of Crystallography: Chester, UK, 2010; pp. 282–296.

 *chemistry*

Review

When Stereochemistry Raised Its Ugly Head in Coordination Chemistry—An Appreciation of Howard Flack

Edwin C. Constable * and Catherine E. Housecroft

Department of Chemistry, University of Basel, BPR 1096, Mattenstrasse 24a, CH-4058 Basel, Switzerland; catherine.housecroft@unibas.ch
* Correspondence: edwin.constable@unibas.ch; Tel.: +41-61-207-1001

Academic Editor: Spyros P. Perlepes
Received: 24 August 2020; Accepted: 8 September 2020; Published: 12 September 2020

Abstract: Chiral compounds have played an important role in the development of coordination chemistry. Unlike organic chemistry, where mechanistic rules allowed the establishment of absolute configurations for numerous compounds once a single absolute determination had been made, coordination compounds are more complex. This article discusses the development of crystallographic methods and the interplay with coordination chemistry. Most importantly, the development of the Flack parameter is identified as providing a routine method for determining the absolute configuration of coordination compounds.

Keywords: crystallography; Flack parameter; chirality; coordination chemistry

---

1. Introduction

When the authors first studied chemistry, the accepted mantra was that crystallography could not be used routinely to determine the absolute configuration of a compound. The basis for this assertion was subsequently shown to be incorrect, but routine determination of the absolute configuration of a compound by crystallography remained in the realms of the relatively exotic. In 1966, a comprehensive listing identified 54 organic structures, the absolute configurations of which had been determined by crystallographic methods [1]. Over the next few years, more determinations were reported, with an additional 40, 39 and 133 absolute structures identified in 1968 [2], 1969 [3] and 1970 [4], respectively. For organic compounds, classical correlation methods relying on real or virtual chemical transformation were still used to relate the configuration to one of the known absolute configurations. Indirect methods of determining the absolute configuration relied upon esoteric spectroscopic methods or empirical correlations such as the octant rule [5]. In 1983, the situation was changed by a paradigm-shifting publication by Howard Flack [6].

This article is not a detailed account of the crystallographic background, but rather takes the opportunity to survey the impact of crystallographic methods on the investigation of coordination compounds and to appreciate the broader contributions of Howard Flack. Nevertheless, a short introduction to the crystallographic complexities is included. Optical activity and subsequently chirality have had a profound influence on the development of coordination chemistry, from the time of Werner onwards.

2. Chirality

2.1. Through a Glass, Darkly

From the earliest documented history, mirrors have excited and fascinated Mankind. In the New Testament of the Christian Bible, the phrase "For now we see through a glass, darkly" refers to a poorly discerned image in a mirror. Mirrors, images in mirrors, and mirror images influenced Western art and literature from the early Renaissance to modern times [7]. The 19th Century C.E. imagination was energized and inspired by mirror images and their relationship to "reality". One of the best known literary works on this theme is the 1871 work *Through the Looking-Glass, and What Alice Found There* by Lewis Carroll (Figure 1) [8].

Figure 1. Mankind has long been fascinated by objects and their mirror images. In this illustration by John Tenniel from *Through the Looking-Glass, and What Alice Found There* we see the subtley different reality in the enantiomeric world. (Public domain image. Source https://en.wikipedia.org/wiki/Through_the_Looking-Glass#/media/File:Aliceroom3.jpg).

By the end of the 19th Century C.E., chemists were familiar with the fact that some compounds exhibited the phenomenon of optical activity. Jacobus Henricus van't Hoff [9,10] and Joseph Achille LeBel [11] had independently explained the phenomenon of optical activity as being a consequence of four groups attached to a carbon atom being oriented in space in the form of a tetrahedron. In his paper, "A suggestion looking to the extension into space of the structural formulas at present used in chemistry and a note upon the relation between the optical activity and the chemical constitution of organic compounds", van't Hoff used the word *asymmetric* to describe a carbon atom with four different groups attached. Both van't Hoff and LeBel showed that when four different groups were bonded in a tetrahedral arrangement about a carbon atom, the object and its mirror image were non-superposable. Although it is critical to distinguish between the observable phenomenon (optical activity) and its origin (dissymmetry), it was slowly becoming clear by the end of the 19th Century, that optical activity should be regarded as arising from a molecular dissymmetry rather than an asymmetry associated with a single atomic centre [12–18].

Although most people "knew" what a mirror image was and understood, on some level at least, the consequent left–right inversion, it fell to William Thomson (1824–1907), better known under his later title of Lord Kelvin to make a scientific definition that brought clarity to the concept of mirror images. Kelvin is well known for his works in physical chemistry and thermodynamics, indeed he is commemorated in the S.I. unit of temperature, the kelvin. Nevertheless, for us his critical contribution, which was not widely recognized or adopted at the time, is the introduction of the terms chiral and chirality; "I call any geometrical figure, or group of points, *chiral*, and say it has chirality, if its image in a plane mirror, ideally realized, cannot be brought to coincide with itself. Two equal and similar right hands are homochirally similar. Equal and similar right and left hands are heterochirally similar or 'allochirally' similar (but heterochirally is better). These are also called 'enantiomorphs,' after a usage introduced, I believe, by German writers. Any chiral object and its image in a plane mirror are heterochirally similar" [19]. The word 'chiral' is derived from the Greek word for hand, $\chi\varepsilon\iota\rho$. Kelvin's dismissal of enantiomorphs is a little disingenuous. Not only was this description being used by the United Kingdom chemical community in the 1890s [20,21], but also had a long tradition in the German literature [22–24] being introduced by Naumann in 1856 [25] and treated at length in Schoenflies' 1891 text on crystallography [26]. The IUPAC has subsequently made recommendations on the basic terminology of stereochemistry [27] and Gal has published extensively on the early etymology of stereochemical terms [28–32].

As mentioned above, Kelvin's proposal was widely ignored by the community and according to a Scifinder search, the first subsequent mention of the word chiral in the chemical literature occurred in the early 1920s [33,34] and with no further mention up to 1950, when Raman used the term in the context of quartz crystals [35]. The rehabilitation of the word seems to stem from letters to *Nature* in the late 1950s [36,37]. It is interesting to note that in the first two publications introducing the Cahn–Ingold–Prelog system, the words chiral and chirality are not used [38,39] and only in the third paper do they appear extensively, with the comment "This useful word [chirality] was brought to our attention by Professor K. Mislow, who referred us to Webster's Dictionary (2nd Edition), where *chiral* is defined as *Of, or pertaining to the hand, specifically turning the plane of polarisation of light to either hand*"[40].

2.2. Some Definitions

In introducing the concept of chirality, Kelvin made the link between molecular dissymmetry and the experimental observable of optical activity. It cannot be stressed strongly enough that chirality is a property of an object as a whole rather than something associated with a portion of that object, such as an "asymmetric atom". The term chirality equates exactly with the term dissymmetry (*dissymmétrie*) originally used by Pasteur [28,29,41–43]. Note that dissymmetric and asymmetric are not equivalent—asymmetric means without symmetry whereas dissymmetric is more specific and means "lacking improper symmetry elements". In simpler language, dissymmetric means chiral. The IUPAC definition of asymmetric is very clear [44]:

> *"Lacking all symmetry elements (other than the trivial one of a one-fold axis of symmetry), i.e., belonging to the symmetry point group* C_1. *The term has been used loosely (and incorrectly) to describe the absence of a rotation–reflection axis (alternating axis) in a molecule, i.e., as meaning chiral, and this usage persists in the traditional terms asymmetric carbon atom, asymmetric synthesis, asymmetric induction, etc".*

If an object is chiral, then the object and its mirror image are not identical and cannot be superposed. The IUPAC uses the word superpose rather than the more familiar English word superimpose.

The concepts of chirality and symmetry are closely related. This article is not intended as a primer in group theory, and the requirements for chirality are presented without clarification of the symmetry operations that are involved. The reader is referred to standard texts on group theory for further education and elucidation [45].

For a molecule to be chiral, it must have no symmetry elements of the second kind, such as a mirror plane, a centre of inversion or a rotation–reflection axis. The presence of any one of these symmetry elements precludes chirality. It is not sufficient to only look for a mirror plane—even if no mirror plane is present, the presence of an inversion centre will ensure that an object is not chiral. This reinforces the primary criterion for chirality—that of non-superposability of the object on its mirror image. If an object is superposable on its mirror image, it is described as being *achiral*. Such objects possess a mirror plane, a centre of inversion, or a rotation–reflection axis. And remember, nothing can surpass a good quality model or diagram to determine whether a molecule is chiral. The IUPAC definition of chiral is concise and precise [44]:

> *"The geometric property of a rigid object (or spatial arrangement of points or atoms) of being non-superposable on its mirror image; such an object has no symmetry elements of the second kind (a mirror plane, $\sigma = S_1$, a centre of inversion, $i = S_2$, a rotation-reflection axis, S_{2n}). If the object is superposable on its mirror image the object is described as being achiral."*

3. The Importance of Chirality in Coordination Chemistry

3.1. It all Began with Werner

The observation of different optical forms of chiral coordination compounds played a critical role in the development and acceptance of Werner's coordination theory and the establishment of the octahedral geometry of six-coordinate metal complexes. Werner reported the resolution of salts of *cis*-$[CoCl(en)_2(NH_3)]^{2+}$ [46], *cis*-$[CoBr(en)_2(NH_3)]^{2+}$ [46], *cis*-$[Co(en)_2(NO_2)_2]^+$ [47], *cis*-$[CoCl(en)_2(NO_2)]^+$ [48], *cis*-$[CoCl_2(en)_2]^+$ [49], $[Co(en)_3]^{3+}$ [50], $[Rh(en)_3]^{3+}$ [51] and the preparation by ligand exchange reactions of resolved precursors of optically active salts or by seeding of *cis*-$[Co(en)_2(H_2O)(NH_3)]^{3+}$ [46], *cis*-$[Co(CO_3)(en)_2]^+$ [52], *cis*-$[Co(C_2O_4)(en)_2]^+$ [53,54], *cis*-$[CoBrCl(en)_2]^+$ [55], $[(en)_2Co(NH_2)(NO_2)Co(en)_2]^{4+}$ [56], $[(en)_2Co(NH_2)(O_2)Co(en)_2]^{4+}$ [57] and, finally the "all-inorganic" compound $[Co\{(OH)_2Co(NH_3)_4\}_3]^{6+}$ [58]. For determining the optical rotation, Werner utilized an F. Schmidt and Haensch polarimeter [59], and originally used the *d* and *l* descriptors to describe the compounds with positive and negative rotation of 656.3 nm wavelength linearly polarized light, respectively. He subsequently reported optical rotatory dispersion (ORD) spectra which are simply plots of the variation in optical rotation of linearly polarized light with wavelength [60]. The development of the successful resolution methods is described in the doctoral thesis of Victor King [61]. A number of excellent surveys of this aspect of Werner's work have been published [62–64].

3.2. Non-Crystallographic Approaches to Determining the Absolute Configuration of Metal Complexes

Although Werner had successfully obtained the large selection of optically pure compounds described above, they were only defined by a relative configuration. Subsequent workers adopted the practice of using the $(+)$D or $(-)$D notation to identify the clockwise or anticlockwise rotation at the sodium D-line (589 nm). The situation regarding absolute configuration was worse than that with organic compounds, as only few transformations between optically active species were known and there was no understanding of rules for retention or inversion at octahedral centres. This meant that it was not possible to use mechanistic principles to correlate the relative configuration in a series of complexes. An example of the complexity of the situation is seen in a classical publication from John Bailar Jr. in which the the reaction of $(-)$D-*cis*-$[CoCl_2(en)_2]Cl$ with liquid ammonia at 196 K gave $(-)$D-*cis*-$[Co(en)_2(NH_3)_2]Cl$ $([\alpha]_D^{298} -32°)$, whereas reaction with a saturated solution of ammonia in methanol at 298 K gave $(+)$D-*cis*-$[Co(en)_2(NH_3)_2]Cl$ $([\alpha]_D^{298} +31°)$ [65].

In the 1930s, Matthieu attempted with only poor success to correlate the relative (or indeed absolute) configuration of complexes [66–79] with the sign of the Cotton effects that they exhibited [80–82]. If a compound under investigation has an absorption band in the region of the optical rotatory dispersion spectrum, an anomalous dispersion effect is seen. This is called the Cotton effect and refers to the change in sign of the optical rotatory dispersion close to an absorption band. Close to the region where light is absorbed, the magnitude of the optical rotation varies rapidly with wavelength, passes through zero at the absorption maximum and continues varies with wavelength but with an opposite sign.

The first successful attempts at determining the absolute configuration of octahedral transition metal complexes were made by Werner Kuhn in the 1930s [83,84]. Kuhn used a coupled oscillator model to calculate the sign of the Cotton effect. Kuhn assigned the absolute configuration of Δ to the $(-)$D-$[Co(C_2O_4)_3]^{3-}$ anion [83] and the $(-)$D-$[Co(en)_3]^{3+}$ cation (Figure 2) [84].

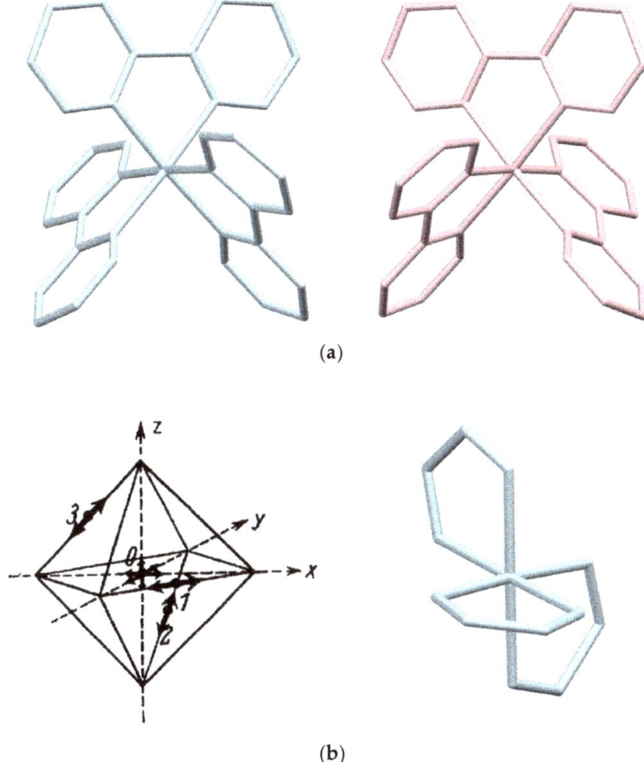

Figure 2. (**a**) IUPAC recommends the notation Δ and Λ for denoting the absolute configuration of octahedral metal complexes. The Δ (pale blue) and Λ (pink) enantiomers of an $[M(bpy)_3]^{n+}$ complex (bpy = 2,2′-bipyridine) are presented here. (**b**) The absolute configuration proposed by Kuhn and Bein for $(-)_D$-$[Co(C_2O_4)_3]^{3-}$ and a representation using the same colour coding as in (a) to show the Δ configuration (only the carbon and the coordinated oxygen atoms of the oxalate ligands are shown) [83]. Figure 2b ©1934 Kuhn, W.; Bein, K. *Z. Anorg. Allg. Chem.* **1934**, *216*, 321–348 reproduced with kind permission of John Wiley and Sons.

4. A Brief History of Crystallography

4.1. X-rays—The Early Days

The availability of routine X-ray crystallographic methods for the determination of solid state structures has had a profound effect on the way in which coordination (and other) chemists characterize their compounds. It is necessary to add a small caveat to curb the enthusiasm of the chemist—typically a single crystal structure provides information about the arrangement of atoms, molecules and ions *within the crystal studied* but, in the absence of other methods, gives no information about the bulk material.

X-rays were discovered in 1895 by Wilhelm Conrad Roentgen when he identified a new type of radiation from a Crookes tube [85,86]. Laue demonstrated that the wavelengths of X-rays were commensurate with the spacings between atoms, ions and molecules in crystals and that, therefore, an optical diffraction was to be expected [87–89]. This critical work by Laue was paradigm shifting for chemistry—suddenly, atoms became objects which could be detected and quantified in terms of size and spatial position. By 1913, the Braggs had built their first single-crystal X-ray spectrometer with a gold leaf electroscope as the detector and the crystal could be rotated in the X-ray beam.

A number of excellent reviews on the early history of X-ray diffraction have been published and the reader is referred to these to learn more about this fascinating story [90–96].

4.2. From X-rays to Chemical Crystallography

Chemical crystallography began when William Henry Bragg and his son William Lawrence Bragg showed that the diffraction data could be solved by ad hoc trial-and-error methods to yield models of the spatial arrangement of atoms and ions in crystalline solids such as NaCl, KCl, KBr, KI, ZnS and diamond [97–105].

4.3. Initial Approaches to the Phase Problem

X-ray detectors measure the intensity of radiation but not the phase of that radiation. Any phase change due to scattering of radiation is almost the same for all atoms. The underlying physics of X-ray diffraction was developed by Darwin and Bragg from 1914 onwards [106–108]. The diffraction data correspond to the amplitude of the Fourier transform of the unit cell electron density in the unit cell. Friedel pairs of diffraction spots are Bragg reflections which are reflected through the origin and related by Friedel's law, which states that they have equal amplitudes and opposite phases [109].

The electron density can be obtained by Fourier synthesis if the phase is known. The introduction of Fourier analyses [110], and subsequent elaboration by Patterson with his eponymous function, started the transition of crystallography to become a more routine technique and also provided the methods that could be used for phase evaluation of the diffraction data [111,112]. The Patterson method recognizes that although the phase information is needed to locate the peaks in electron density within a unit cell, the magnitudes of the structure factors contain information about the spacings. The Patterson map shows peaks at all positions corresponding to an interatomic vector rather than the position of the atoms. Patterson methods can only be used for relatively small molecules (<50 atoms) and are not appropriate for the ab initio determination of absolute configuration.

The next development was the study of series of compounds which only differed by the replacement of one atom in the structure. The method relied upon the replaced atom being on a special position and having a different scattering factor. Within these constraints, the effect is the modification of the structure factors by values dependent on the phase of the reflection [113]. One of the earliest applications was to the alums, and diffraction data from the compounds $AB(SO_4)_2 \cdot 12H_2O$ (A = NH_4, K, Rb, Cs, Tl; B = Cr, Al) were used to determine the structures using this isomorphous replacement method [114]. Probably the most spectacular early success used the related heavy atom method in an isomorphous series for solving and refining the centrosymmetric phthalocyanine complexes [M(pc)] (M = Ni, Pt) (Figure 3) and consequently, the correct phasing of H_2Pc [115–117]. One of the early applications to a chiral molecule was reported in the structure of cholesteryl iodide. In this compound, the iodine is not on a special position and the step-wise work-flow is worth rehearsing to demonstrate the skill and tenacity of the early crystallographers: (1) measure the intensities of all reflections, (2) locate the heavy atoms from the Patterson map, (3) calculate the phases from the heavy atom contributions, (4) use chemical knowledge to choose correct distances and angles, and (5) recalculate phases on all the atoms and repeat the Fourier summations.

Figure 3. The projection of the nickel phthalocyanine molecule presented by Robertson in his pioneering work on the crystallography of phthalocyanines [116]. ©1937 Robertson, J.M.; Woodward, I. *J. Chem. Soc.* **1937**, 219–230 reproduced with kind permission of The Royal Society of Chemistry.

4.4. The Bijvoet Method

Most crystallographers accepted Friedel's law that the Bragg pairs had equal amplitudes and opposite phases and considered that X-ray methods could not be used directly for the determination of absolute configuration without chemical modification or isomorphous replacement. However, a number of physicists and crystallographers remained open-minded. If the X-ray radiation has a wavelength close to the absorption edge of an atom in the compound, a small phase change occurs in the scattered X-rays from these atoms, phenomenologically similar to the Cotton effect. As a consequence, the diffraction pattern is no longer centrosymmetric but has pairs of spots with unequal intensities. This effect was first reported in studies of zinc blende (ZnS) using W L_β (1.2447–1.3017 Å) [118] or Au $L_{\alpha 1}$ (0.8638 Å) [119] radiation, which are close to the Zn K edge (1.2837 Å). However, these results were forgotten by the community for over a decade.

At the end of the 1940s, the Dutch crystallographer Johannes Bijvoet developed a general isomorphous replacement method for non-centrosymmetric structures and successfully solved the structure of strychnine using the sulfate and selenate salts [120–125]. However, the most significant contribution of Bijvoet was in rediscovering that X-ray analysis could determine absolute configurations using the anomalous scattering of X-rays of a wavelength close to an X-ray absorption of an atom in the compound. This effect was observed in sodium rubidium tartrate (Figure 4) using Zr K_α radiation, which is close to the K-edge of rubidium, and in the early 1950's, Bijvoet published a series of landmark papers, commencing in 1951 with "Determination of absolute configuration of optically active compounds by means of X-rays" [95,126–128].

A further simplification was introduced by Mathieson who, making use of the known absolute configurations that were available from the Bijvoet method, proposed the use of diastereoisomers in which one of the stereochemical centres was absolutely defined. This avoided the need to use X-ray radiation of a wavelength close to an absorption edge in the compound and utilized the heavy atom method with auxiliaries of known absolute configuration such as (*R*) or (*S*)-chloroiodoacetate [129].

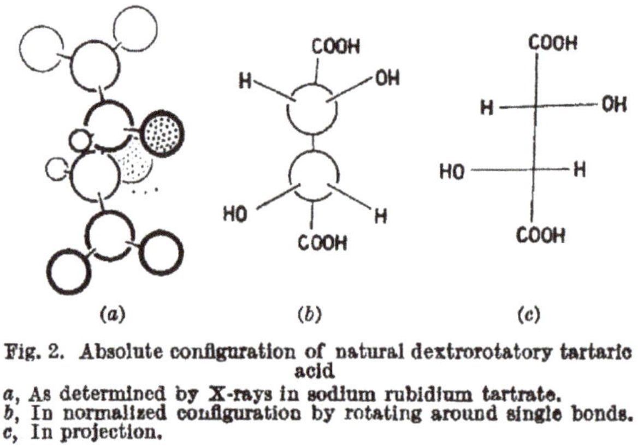

Figure 4. The absolute configuration of tartaric acid as depicted by Bijvoet after the determination of the structure of sodium rubidium tartrate. [116,127]. ©1951 Bijvoet, J.M.; Peerdeman, A.F.; van Bommel, A.J. *Nature* **1951**, *168*, 271–272, reproduced with kind permission of Springer Nature. The absolute configuration confirmed the random choice selected by Fischer.

In parallel, direct methods were being developed to permit the extraction of the phase information from the crystallographic structure factors and crystallography entered its modern phase [130,131]. Direct methods estimate and then test and select the initial phases and were initially introduced for centrosymmetric space groups and subsequently extended to non-centrosymmetric space groups [132,133].

5. Chirality and Crystallography

Chirality, crystallography and symmetry are related in intimate and subtle ways that lead to fascination and frustration. The nuances and manifestations of these relationships delighted and occupied Howard Flack through much of his career. In this section, we rehearse a few of the consequences of these relationships for coordination compounds.

5.1. Absolutism

Chemists tend to talk of *absolute configuration* meaning the spatial arrangement of the atoms of in a chiral molecular entity and denoted by a stereochemical descriptor such as R/S, P/M, D/L or Δ/Λ. In contrast, crystallographers use the term *absolute structure*, introduced in 1984 by Peter Jones [134], to describe spatial arrangement of atoms in a non-centrosymmetric crystal [135]. The term absolute structure refers specifically to the crystalline state, whereas the broader term absolute configuration can be applied to any phase or to solutions.

5.2. Chiral Space Groups and Chiral Molecules

There are 230 three-dimensional space groups. Of these 22 are chiral space groups comprising 11 enantiomorphic pairs ($P4_1$-$P4_3$: $P4_122$-$P4_322$: $P4_12_12$-$P4_32_12$: $P3_1$-$P3_2$: $P3_121$-$P3_221$: $P3_112$-$P3_212$: $P6_1$-$P6_5$: $P6_122$-$P6_522$: $P6_2$-$P6_4$: $P6_222$-$P6_422$: $P4_132$-$P4_332$). It would be tempting to expect that a chiral molecule would crystallize in a chiral space group, but that would be too simple, and would not take the infinite perversity of nature into account. In Section 2.2, we stated that for an object to be chiral, it must possess no symmetry elements of the second kind. Of the 230 space groups, there are 65 (including the 22 chiral groups) which only possess operations of the first kind (rotations, rotation–translations, and translations) and these 65 are known collectively as the Sohncke groups [136]. Enantiopure chiral molecules must crystallize in one of these 65 Sohncke space groups.

Returning to the perversity of nature, achiral molecules can also occur in any of the 230 space groups—the packing of achiral objects may be in a chiral or an achiral manner. A systematic survey of achiral molecules in non-centrosymmetric space groups was reported in 2005 [137]. Similarly, a crystal composed of equal numbers of the two enantiomers of a molecule (a racemate) may occur in any of the 230 space groups, although examples in the Sohncke set are very rare.

5.3. The Flack Parameter

By the early 1980s, the ratios of crystallographic R or R_w values for structures refined with alternative absolute configurations were being used to assign absolute configurations and absolute structures, although the statistical methods were debated [138,139]. In 1983, Flack published a paper entitled "On Enantiomorph-Polarity Estimation" which revolutionized the field of structure determination of chiral molecules and structures [6]. Rogers had introduced a parameter η which he proposed as a good method for distinguishing between refinements of structures with opposite configurations. Flack pointed out an inherent problem with the use of η and introduced a new parameter that he called x and defined in terms of the structure factor for reflection $\mathbf{h}$ in Equation (1).

$$\left|F(\mathbf{h},x)\right|^2 \;=\; (1-x)\left|F(\mathbf{h})\right|^2 \;+\; \left|F(-\mathbf{h})\right|^2 \tag{1}$$

If the experimental data and the model used for refinement have the same chirality, the value of x is 0, and if they have opposite configurations, then x has a value of 1. A value of 0.5 indicates a racemic crystal with equal amounts of both enantiomer. This single publication revolutionized the determination of absolute structures. Flack also recognized that there was no need to formally solve the structures in both configurations and in a note added in proof to the original paper, he notes that "The refinement of … x has now been added as a permanent feature in our implementation of the X-RAY76 system … x is varied automatically in the final stages of refinement with a non-centrosymmetric structure." The direct refinement of the Flack parameter together with other structural parameters is now routine. To date, this paper has been cited 11,161 times. The parameter x rapidly, and universally, became known as the Flack Parameter. To date, the Flack parameter does not yet belong to the data indexed by the Cambridge Structural Database (CSD) [140,141].

5.4. Some Musings on Racemates, Spontaneous Resolution and Other Complexities

Although the introduction of the Flack parameter transformed the mechanics and quantification of the determination of absolute configurations, it is our opinion that a second paper by Flack is equally important. The 2003 article entitled "Chiral and Achiral Crystal Structures" provides the clearest and most comprehensive compilation of the language of chirality and crystallography that we know. He commences by identifying three origins of chirality in crystal structures: (i) the molecular components (ii) the crystal structure itself and (iii) the symmetry group of the structure. The really valuable part of the publication concerns the quantification of a descriptor (racemic, racemate) that is often used loosely or incorrectly by the chemical community [142]. These descriptors simply refer to equimolar amounts of opposite enantiomers with no restriction to phase. If the two enantiomers are not present in equal amounts, the correct description is an enantiomeric mixture.

Particularly important is the identification of a *racemic conglomerate*, obtained when the crystallization of a solution of a racemic compound results in spontaneous resolution and the generation of equal numbers (strictly equal weights) of enantiopure crystals each only containing components with only one chirality.

He then identified the term *ordered racemic crystal structure* or racemic structure to describe a crystal containing an ordered array of equal numbers of the different enantiomers. The term anomalous racemate had been used to describe ordered crystals in which the ratio of the two enantiomers was not 1:1, and Flack proposed a new description of *M:N mixed enantiomeric crystal structure* or M:N enantiomeric structure, where M:N is the ratio of the two enantiomers present. He further proposed the term *disordered mixed enantiomeric structure* to replace pseudoracemate in describing crystals in which each position can be occupied by a molecule of either configuration. This publication is recommended to anyone who wishes to think more deeply about the consequences of chirality in the solid state.

6. Chiral Coordination Compounds

Although chirality has been so important in the development of coordination chemistry, it is surprising how few monographs or comprehensive reviews exist. The standard works on stereochemistry concentrate on organic compounds [143]. For coordination chemistry, the bibles are von Zelewsky's 1996 work "Stereochemistry of Coordination Compounds" [144] and Hawkins earlier work "Absolute Configuration of Metal Complexes" [145]. A more organometallic-oriented, but extremely useful and relevant, presentation is found in the 2008 book "Chirality in Transition Metal Chemistry" by Amouri and Gruselle [146]. Two earlier reviews complement these to provide an overview of the period at which crystallography was starting to provide information about absolute stereochemistry and a volume of the ACS Symposium Series from 1980 that was dedicated to the "Stereochemistry of Optically Active Transition Metal Complexes" provides an excellent overview of the state of the art at that time [147–149]. Volume 12 of Topics in Stereochemistry was entitled "Topics in Inorganic and Organometallic Stereochemistry"and for the real afficionados, the chapter "Conformational Analysis and Steric Effects in Metal Chelates" provides a masterly overview of conformational effects within chelate rings and is both a *tour de force* and a challenge [150].

6.1. The First Absolute Determination

The first determinations of the absolute configuration of metal complexes using the Bijvoet method was reported in 1955 by Yoshihiko Saito who studied Λ- and Δ-[Co(en)$_3$]Cl$_3$·0.5NaCl [151]. In a 1974 review, Saito surveyed the literature up to 1972 and reported that, in the intervening 23 years, the absolute configuration of salts of only an additional 53 metal cations had been determined through structural characterization [148].

The list of compounds characterized provides a snap-shot of the contemporary coordination chemistry. The first class included those which required the "pure" Bijvoet anomalous dispersion approach for compounds such as [ML$_3$]$^{n+}$ (L = chelating bidentate ligand) or *cis*-[ML$_2$X$_2$]$^{n+}$ (L = chelating bidentate ligand, X = monodentate ligand). The structural elucidation of the diamines into related polyamines with polymethylene spacers resulted in new classes of chiral complexes in which the chirality arises from the dissymmetric arrangement of the ligand donor atoms about the metal centre (see the general references at the beginning of this section). These latter compounds can be seen as the direct progenitors of macrocyclic chemistry and the grandparents of supramolecular chemistry. Interesting as these compounds are, we return to our main theme when we consider the remaining types of compounds characterized.

We start with a small digression into the consequences of multiple stereogenic centres in a compound. The classical method of resolving a chiral coordination compound, for example a cation C$^+$, is to form salts with a chiral anion A$^-$. Four possible compounds can be formed, of which [(Δ-C)(Δ-A)] and [(Λ-C)(Λ-A)] form a pair of enantiomers as do [(Δ-C)(Λ-A)] and [(Λ-C)(Δ-A)] (Figure 5). All other relationships between the combinations are as diastereoisomers. Enantiomers have identical physical properties (solubility, melting point, NMR spectra) and only differ in their interactions with other chiral agents (for example polarized light). On the other hand, diastereoisomers differ in physical properties as the spatial interactions between, for example, (Δ-C) with (Λ-A) and (Δ-A) will be different (think about putting your left foot into a left shoe and into a right shoe—the thermodynamics of the pairing are different). A typical resolution method might involve treating a racemic mixture of cations (Λ-C) and (Δ-C) with (Λ-A) and hoping that the solubility of the diastereoisomeric salts [(Δ-C)(Λ-A)] and [(Δ-C)(Δ-A)] might be sufficiently large that one selectively precipitates or crystallizes. Salts of this type provided the next class that were studied extensively in this first period of determining the absolute configuration of coordination compounds. Very typically, chiral organic anions, or anions containing coordinated chiral organic ligands, were used for the precipitation. After Bijvoet established the absolute configuration of tartrate, known chemical transformations allowed correlation to a large number of other "simple" chiral organic compounds. As a consequence, the absolute configuration of the organic component of the anion was generally known when the structure of the diastereoisomeric salt [(C)(A)] was determined and the configuration at the cation followed from the correct assignment of the (known) configuration to the anion.

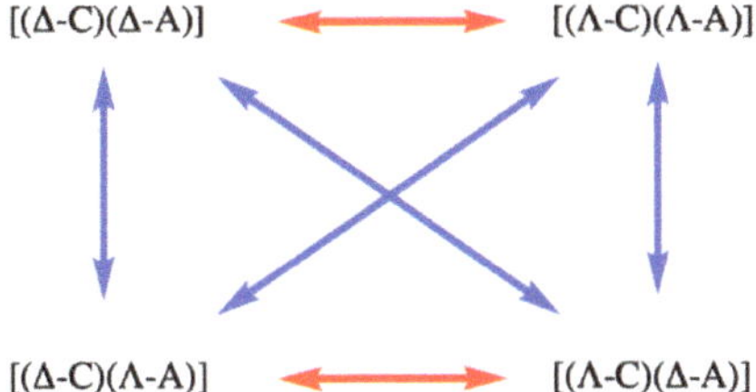

Figure 5. The four possible combinations of a chiral cation and a chiral anion: [(Δ-C)(Δ-A)] and [(Λ-C)(Λ-A)] form a pair of enantiomers (denoted by a red double-headed arrow) as do [(Δ-C)(Λ-A)] and [(Λ-C)(Δ-A)]; all other relationships are as diastereoisomers (denoted by a blue double-headed arrow).

The final class of compounds studied belonged to the next level of structural development in which the ligands themselves are chiral. Consider propane-1,2-diamine ($H_2NCH_2C^*H(Me)NH_2$, pn) in which C2 (indicated with an asterisk) is a stereogenic centre; as a result the ligand pn is typically encountered as one of the two pure enantiomers *R*-pn or *S*-pn, as the racemate containing equal amounts of *R*-pn and *S*-pn or the enantiomeric mixture x(*R*-pn):1–x(*S*-pn). We now consider the formation of the octahedral complex $[M(pn)_3]^{n+}$.

Starting with enantiomerically pure *R*-pn or *S*-pn, four compounds could be obtained: Δ-[M(*S*-pn)$_3$]$^{n+}$ and Λ-[M(*R*-pn)$_3$]$^{n+}$ (a pair of enantiomers) and a second pair of enantiomers Λ-[M(*S*-pn)$_3$]$^{n+}$ and Δ-[M(*R*-pn)$_3$]$^{n+}$. All other relationships are diastereoisomeric. Thus, reaction with *R*-pn will give two chemically distinct diastereoisomers Λ-[M(*R*-pn)$_3$]$^{n+}$ and Δ-[M(*R*-pn)$_3$]$^{n+}$.

The more perverse readers will now ask what happens with the racemic ligand? In this case, in addition to the four homoleptic complexes already mentioned, we now have the possibility of the heteroleptic complexes Δ-[M(*S*-pn)$_2$(*R*-pn)]$^{n+}$ and Λ-[M(*R*-pn)$_2$(*S*-pn)]$^{n+}$ (a pair of enantiomers) as well as Δ-[M(*R*-pn)$_2$(*S*-pn)]$^{n+}$ and Λ-[M(*S*-pn)$_2$(*R*-pn)]$^{n+}$ (a second pair of enantiomers). But it gets worse! The two nitrogen donor atoms of each pn ligand are not chemically equivalent as one is attached to the stereogenic C2 and the other to C1. As a consequence, there are also the complexes with a facial or meridional arrangement of the C*-NH_2 donors! Once again, this is not the place to follow the fascinating stereochemistry of "simple" systems like this.

We make one further foray into the world of stereochemical complexity revealed in these early crystallographic studies. Our discussion commenced with the definition of the Δ- and Λ-configuration using $[M(bpy)_3]^{n+}$ complexes as an example (Figure 2). Why did we not start with Werner's $[M(en)_3]^{n+}$ compounds? Whereas the chelate ring in bpy complexes is planar, that in en complexes is non-planar and chiral. The configuration is denoted by the descriptor δ or λ as defined in Figure 6. Although individual δ or λ chelate rings are enantiomeric, the situation is different in $[M(en)_3]^{n+}$ when we have the possibilities of Δ-[M(δ-en)$_3$]$^{n+}$ and Λ-[M(λ-en)$_3$]$^{n+}$ (a pair of enantiomers), Λ-[M(δ-en)$_3$]$^{n+}$ and Δ-[M(λ-en)$_3$]$^{n+}$ (a second pair of enantiomers) and Δ-[M(δ-en)$_2$(λ-en)]$^{n+}$ and Λ-[M(λ-en)$_2$(δ-en)]$^{n+}$ (a third pair of enantiomers) as well as Δ-[M(λ-en)$_2$(δ-en)]$^{n+}$ and Λ-[M(δ-en)$_2$(λ-en)]$^{n+}$ (a fourth pair of enantiomers), with each pair of enantiomers having a different thermodynamic stability. The early crystallographic studies confirmed that the favoured crystal forms were typically Λ-[M(δ-en)$_3$]$^{n+}$ and Δ-[M(λ-en)$_3$]$^{n+}$.

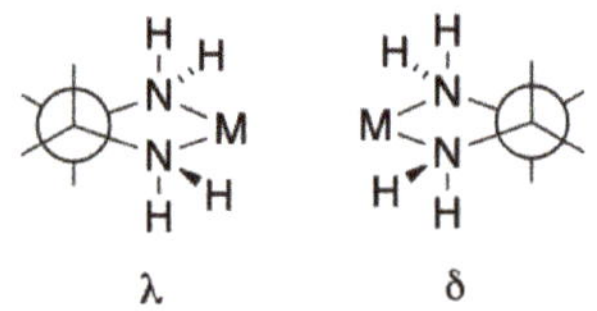

λ   δ

Figure 6. The two enantiomeric forms of a five-membered chelate ring involving an en-like ligand are denoted by the descriptors δ or λ.

6.2. The Influence of Flack

It is in the area of dissymmetric metal complexes that we see the real influence of Flack and his parameter. If a metal complex crystallizes in one of the Sohncke space groups, most crystallographic software will automatically generate the Flack parameter (hopefully close to 0 or 1) to indicate whether the choice of absolute configuration is correct. A Flack parameter value close to 0.5 is an indication that something is amiss, typically an inversion twin. What is remarkable is how often the chemist, as opposed to the crystallographer, does not comment on the absolute configuration of the complex. There are many examples of spontaneous resolution of tris(chelate)metal complexes in the CSD which are not specifically identified in their associated publications.

Today, we have a veritable wealth of chiral systems and different types of chirality that Werner and Pasteur could only have dreamed of. The Flack parameter finds widespread use in the study of novel chiral systems such as helicates, cyclic helicates and knotted systems exhibiting topological chirality and has been used to establish the absolute configuration of molecular trefoil knots assembled in self-sorting processes [152]. The use of the Flack parameter to establish the asymmetric crystallization of coordination networks and metal-organic frameworks is an interesting new development [153–155].

And what of the future? Routine inclusion of the Flack parameter in CIF files should be encouraged, if only to stimulate further discussion of chirality aspects within the inorganic and coordination chemistry communities! The merits and demerits of determining the Flack parameter during the refinement or post-refinement are currently being discussed [156]. Finally, data-mining activities would be dramatically improved by the inclusion of the Flack parameter in the standard data in Crystal Structure Databases. In the course of writing this article, we became aware of the difficulty of answering simple questions such as "How many chiral coordination compounds have been structurally characterized?" and "How many coordination compounds exhibit asymmetric crystallization?".

7. Concluding Remarks

This article is an attempt to express our thanks to Howard Flack for providing a tool that made our lives as coordination chemists more interesting and rewarding and, at the same time, to place this innovation of the Flack parameter in its historical perspective.

Author Contributions: This article was conceived and written jointly by E.C.C. and C.E.H. All authors have read and agreed to the published version of the manuscript.

Funding: This work received no external funding.

Acknowledgments: As always, we give our thanks to the various library and abstracting services which have aided us in identifying and sourcing material.

Conflicts of Interest: The authors declare no conflict of interest.

References

1. Allen, F.H.; Rogers, D. A reference list of organic structures whose absolute configurations have been determined by X-ray fluorescence. *Chem. Commun.* **1966**, 838–841. [CrossRef]
2. Allen, F.H.; Neidle, S.; Rogers, D. A reference list of organic structures whose absolute configurations have been determined by X-ray methods. Part 2. *Chem. Commun.* **1968**, 308–310. [CrossRef]

3. Allen, F.H.; Neidle, S.; Rogers, D. A reference list of organic structures whose absolute configurations have been determined by X-ray methods. Part 3. *J. Chem. Soc. D* **1969**, 452–454. [CrossRef]

4. Neidle, S.; Rogers, D.; Allen, F.H. A reference list of organic structures whose absolute configurations have been determined by X-ray methods. Part IV. *J. Chem. Soc. C* **1970**, 2340. [CrossRef]

5. Lightner, D.A. Chapter 5 Determination of Absolute Configuration by Cd. Applications of the Octant Rule and the Exciton Chirality Rule. In *Analytical Applications of Circular Dichroism*; Purdie, N., Brittain, H.G., Eds.; Elsevier: Amsterdam, The Netherlands, 1994; pp. 131–174.

6. Flack, H.D. On enantiomorph-polarity estimation. *Acta Crystallogr. Sect. A Found. Crystallogr.* **1983**, *39*, 876–881. [CrossRef]

7. Yiu, Y. The Mirror and Painting in Early Renaissance Texts. *Early Sci. Med.* **2005**, *10*, 187–210. [CrossRef]

8. Carroll, L. *Through the Looking Glass, and What Alice Found There*; Macmillan and Co.: London, UK, 1871.

9. van't Hoff, J.H. Sur les Formules de Structure dans L'Espace. *Arch. Neerl. Sci. Exactes Nat.* **1874**, *9*, 445.

10. van't Hoff, J.H. *Voorstel Tot Uitbreiding Der Tegenwoordige in De Scheikunde Gebruikte Structuurformules in De Ruimte, Benevens Een Daarmee Samenhangende Opmerking Omtrent Het Verband Tusschen Optisch Actief Vermogen En Chemische Constitutie Van Organische Verbindingen*; Greven: Utrecht, The Netherlands, 1874.

11. Le Bel, J.A. Sur les relations qui existent entre les formules atomiques des corps organiques et le pouvoir rotatoire de leurs dissolutions. *Bull. Soc. Chim. Fr.* **1874**, *22*, 337–347.

12. Riddell, F.; Robinson, M.J. J.H. van't Hoff and J.A. Le Bel — their historical context. *Tetrahedron* **1974**, *30*, 2001–2007. [CrossRef]

13. Ramberg, P.J. *Chemical Structure, Spatial Arrangement*; Routledge: London, UK, 1988.

14. Drayer, D.E. The Early History of Stereochemistry: From the Discovery of Molecular Asymmetry and the First Resolution of a Racemate By Pasteur to the Asymmetrical Chiral Carbon of Van't Hoff and Le Bel. In *Drug Stereochemistry: Analytical Methods and Pharmacology*, 3rd ed.; Jóźwiak, K., Lough, W.J., Wainer, I.W., Eds.; Informa Healthcare: London, UK, 2012; pp. 1–16.

15. Drayer, D.E. The Early History of Stereochemistry: From the Discovery of Molecular Asymmetry and the First Resolution of a Racemate by Pasteur to the Asymmetrical Chiral Carbon of Van't Hoff and Le Bel. *Clin. Res. Regulat. Aff.* **2001**, *18*, 181–203. [CrossRef]

16. Meijer, E.W.; van't Hoff, J.H. Hundred Years of Impact on Stereochemistry in The Netherlands. *Angew. Chem. Int. Ed.* **2001**, *40*, 3783–3789. [CrossRef]

17. Cintas, P. On the Origin of Tetrahedral Carbon: A Case for Philosophy of Chemistry? *Found. Chem.* **2002**, *4*, 149–161. [CrossRef]

18. Ochiai, H. Philosophical Foundations of Stereochemistry. *HYLE* **2015**, *21*, 1–18.

19. Kelvin, W.T. *The Molecular Tactics of a Crystal*; Clarendon Press: Oxford, UK, 1894.

20. Purdie, T. Resolution of lactic acid into its optically active components. *J. Chem. Soc. Trans.* **1893**, *63*, 1143–1157. [CrossRef]

21. Kipping, F.S.; Pope, W.J. Studies of the terpenes and allied compounds. The sulphonic derivatives of camphor. Part I. *J. Chem. Soc. Trans.* **1893**, *63*, 548–604. [CrossRef]

22. Einhorn, A. Ueber die Beziehungen des Cocaïns zum Atropin. *Ber. Dtsch. Chem. Ges.* **1890**, *23*, 1338–1344. [CrossRef]

23. Zelinsky, N. Ueber die Stereoisomerie der Dimethyldioxyglutarsäuren. *Ber. Dtsch. Chem. Ges.* **1891**, *24*, 4006–4017. [CrossRef]

24. Gal, J. Carl Friedrich Naumann and the introduction of enantio terminology: A review and analysis on the 150th anniversary. *Chirality* **2007**, *19*, 89–98. [CrossRef]

25. Naumann, C.F. *Elemente Der Theoretischen Krystallographie*; W. Engelmann: Leipzig, Germany, 1856.

26. Schoenflies, A. *Krystallsysteme Und Krystallstructur*; Teubner: Leipzig, Germany, 1891.

27. Moss, G.P. Basic terminology of stereochemistry (IUPAC Recommendations 1996). *Pure Appl. Chem.* **1996**, *68*, 2193–2222. [CrossRef]

28. Gal, J. Louis Pasteur, Chemical Linguist: Founding the Language of Stereochemistry. *Helv. Chim. Acta* **2019**, *102*, e1900098. [CrossRef]

29. Gal, J. Louis Pasteur, language, and molecular chirality. I. Background and Dissymmetry. *Chirality* **2011**, *23*, 1–16. [CrossRef] [PubMed]

30. Gal, J. Stereochemical vocabulary for structures that are chiral but not asymmetric: History, analysis, and proposal for a rational terminology. *Chirality* **2011**, *23*, 647–659. [CrossRef] [PubMed]

31. Gal, J. Molecular Chirality in Chemistry and Biology: Historical Milestones. *Helv. Chim. Acta* **2013**, *96*, 1617–1657. [CrossRef]

32. Gal, J. Molecular Chirality: Language, History, and Significance. In *Differentiation of Enantiomers I*; Schurig, V., Ed.; Springer: Cham, Switzerland, 2013; pp. 1–20.

33. Larmor, J. On Electro-crystalline properties as conditioned by atomic lattices. *Proc. R. Soc. Lond. Ser. A* **1921**, *99*, 1–10. [CrossRef]

34. Larmor, J. The structural significance of optical rotatory quality. *Rep. Br. Ass. Advmt. Sci.* **1922**, 351–352.

35. Raman, C.V. Crystals of quartz with iridescent faces. *Proc. Indian Acad. Sci. Sect. A* **1950**, *31A*, 275–279. [CrossRef]

36. Whyte, L.L. Chirality. *Nature* **1957**, *180*, 513. [CrossRef]

37. Whyte, L.L. Chirality. *Nature* **1958**, *182*, 198. [CrossRef]

38. Cahn, R.S.; Ingold, C.K. Specification of configuration about quadricovalent asymmetric atoms. *J. Chem. Soc.* **1951**, 612. [CrossRef]

39. Cahn, R.S.; Ingold, C.K.; Prelog, V. The specification of asymmetric configuration in organic chemistry. *Experientia* **1956**, *12*, 81–94. [CrossRef]

40. Cahn, R.S.; Ingold, C.; Prelog, V. Specification of Molecular Chirality. *Angew. Chem. Int. Ed. Engl.* **1966**, *5*, 385–415. [CrossRef]

41. Gal, J. The discovery of biological enantioselectivity: Louis Pasteur and the fermentation of tartaric acid, 1857–a review and analysis 150 yr later. *Chirality* **2008**, *20*, 5–19. [CrossRef] [PubMed]

42. Gal, J. When did Louis Pasteur present his memoir on the discovery of molecular chirality to the Académie des sciences? Analysis of a discrepancy. *Chirality* **2008**, *20*, 1072–1084. [CrossRef] [PubMed]

43. Gal, J. In defense of Louis Pasteur: Critique of Gerald Geison's deconstruction of Pasteur's discovery of molecular chirality. *Chirality* **2019**, *31*, 261–282. [CrossRef]

44. Mcnaught, A.D.; Wilkinson, A.; IUPAC. *Compendium of Chemical Terminology*, 2nd ed.; (the "Gold Book"); Blackwell Scientific Publications: Oxford, UK, 1997; Online Version (2019-) Created By S.J. Chalk; ISBN 0-9678550-9-8. [CrossRef]

45. Cotton, F.A. *Chemical Applications of Group Theory, 3rd*; Wiley: New York, NY, USA, 1990.

46. Werner, A. Zur Kenntnis des asymmetrischen Kobaltatoms. I. *Ber. Dtsch. Chem. Ges.* **1911**, *44*, 1887–1898. [CrossRef]

47. Werner, A. Zur Kenntnis des asymmetrischen Kobaltatoms. II. *Ber. Dtsch. Chem. Ges.* **1911**, *44*, 2445–2455. [CrossRef]

48. Werner, A. Zur Kenntnis des asymmetrischen Kobaltatoms. III. *Ber. Dtsch. Chem. Ges.* **1911**, *44*, 3272–3278. [CrossRef]

49. Werner, A. Zur Kenntnis des asymmetrischen Kobaltatoms. IV. *Ber. Dtsch. Chem. Ges.* **1911**, *44*, 3279–3284. [CrossRef]

50. Werner, A. Zur Kenntnis des asymmetrischen Kobaltatoms. V. *Ber. Dtsch. Chem. Ges.* **1912**, *45*, 121–130. [CrossRef]

51. Werner, A. Über Spiegelbildisomerie bei Rhodium-Verbindungen. I. *Ber. Dtsch. Chem. Ges.* **1912**, *45*, 1228–1236. [CrossRef]

52. Werner, A.; McCutcheon, T.P. Zur Kenntnis des asymmetrischen Kobaltatoms. VI. *Ber. Dtsch. Chem. Ges.* **1912**, *45*, 3281–3287. [CrossRef]

53. Werner, A.; Shibata, Y. Zur Kenntnis des asymmetrischen Kobaltatoms. VII. *Ber. Dtsch. Chem. Ges.* **1912**, *45*, 3287–3293. [CrossRef]

54. Werner, A. Zur Kenntnis des asymmetrischen Kobaltatoms XI. Über Oxalo-diäthylendiamin-kobaltisalze und eine neue Spaltungsmethode für racemische anorganische Verbindungen. *Ber. Dtsch. Chem. Ges.* **1914**, *47*, 2171–2182. [CrossRef]

55. Werner, A.; Tschernoff, G. Zur Kenntnis des asymmetrischen Kobaltatoms. VIII. *Ber. Dtsch. Chem. Ges.* **1912**, *45*, 3294–3301. [CrossRef]

56. Werner, A. Zur Kenntnis des asymmetrischen Kobaltatoms IX. *Ber. Dtsch. Chem. Ges.* **1913**, *46*, 3674–3683. [CrossRef]

57. Werner, A. Zur Kenntnis des asymmetrischen Kobaltatoms X. *Ber. Dtsch. Chem. Ges.* **1914**, *47*, 1961–1979. [CrossRef]

58. Werner, A. Zur Kenntnis des asymmetrischen Kobaltatoms XII. Über optische Aktivität bei kohlenstofffreien Verbindungen. *Ber. Dtsch. Chem. Ges.* **1914**, *47*, 3087–3094. [CrossRef]

59. Jensen, W.B. *Polarimeters*; Oesper Museum Booklets on the History of Chemical Apparatus, University of Cincinnati: Cincinnati, OH, USA, 2014.

60. Hargreaves, M.K. Optical Rotatory Dispersion: Its Nature and Origin. *Nature* **1962**, *195*, 560–566. [CrossRef]

61. King, V.L. Über Spaltungsmethoden Und Ihre Anwendung Auf Komplexe Metall-Ammoniakverbindungen. Ph.D. Dissertation, University of Zürich, Zürich, Switzerland, 1912.

62. Kauffman, G.B. The Discovery of Optically Active Coordination Compounds: A Milestone in Stereochemistry. *Isis* **1975**, *66*, 38–62. [CrossRef]

63. Kauffman, G.B. A Stereochemical Achievement of the First Order: Alfred Werner's Resolution of Cobalt Complexes, 85 Years Later. *Bull. Hist. Chem.* **1997**, *20*, 50–59.

64. Ernst, K.H.; Berke, H. Optical activity and Alfred Werner's coordination chemistry. *Chirality* **2011**, *23*, 187–189. [CrossRef] [PubMed]

65. Bailar, J.C.; Haslam, J.H.; Jones, E.M. The Stereochemistry of Complex Inorganic Molecules. III. The Reaction of Ammonia with levo-Dichlorodiethylenediaminocobaltic Chloride. *J. Am. Chem. Soc.* **1936**, *58*, 2226–2227. [CrossRef]

66. Mathieu, J.-P. Activité optique et solubilité de quelques cobaltammines. *C. R. Hebd. Seances Acad. Sci.* **1934**, *199*, 278–280.

67. Mathieu, J.-P. Configuration de quelques complexes hexacoordinés optiquement actifs. *C. R. Hebd. Seances Acad. Sci.* **1934**, *198*, 1598–1600.

68. Mathieu, J.-P. Absorption, activité optique et configuration de complexes minéraux. *C. R. Hebd. Seances Acad. Sci.* **1935**, *201*, 1183–1184.

69. Mathieu, J.-P. Werner complexes; optical activity and the configuration of ions of the type MA3. *J. Chim. Phys.* **1936**, *33*, 78–96. [CrossRef]

70. Mathieu, J.-P. Recherches expérimentales sur le dichroisme circulaire et sur quelques applications physico-chimiques de ce phénomène. *Ann. Phys.* **1935**, *11*, 371–460. [CrossRef]

71. Mathieu, J.-P. Recherches sur les complexes de Werner activité optique et configuration des ions du type MeA3. *J. Chim. Phys.* **1936**, *3*, 476–498. [CrossRef]

72. Mathieu, J.-P. The Werner complexes. Optical activity and configuration of the ions containing the groups M en2 and M ox2. *Bull. Soc. Chim. Fr. Mem.* **1936**, *3*, 476–498.

73. Mathieu, J.-P. The Werner complexes. Absorption of the hexacoördinates of cobalt and chromium in aqueous solution. *Bull. Soc. Chim. Fr. Mem.* **1936**, *3*, 463–475.

74. Mathieu, J.-P. Experiments on the complexes of Werner. Substitution in the optically active complex chlorides. *Bull. Soc. Chim. Fr. Mem.* **1937**, *4*, 687–700.

75. Mathieu, J.-P. Recent views on the stereochemistry of complex inorganic compounds. *Bull. Soc. Chim. Fr. Mem.* **1938**, *5*, 725–805.

76. Mathieu, J.-P. The Werner complexes-absorption and optical activity of cobalt compounds with a double nucleus. *Bull. Soc. Chim. Fr. Mem.* **1938**, *5*, 105–113.

77. Mathieu, J.-P. Researches on Werner complexes. Cobaltammines containing optically active amino acids. *Bull. Soc. Chim. Fr. Mem.* **1939**, *6*, 873–882.

78. Mathieu, J.-P. Researches on Werner complexes. Optical activity and configuration of platinum. IV. Triethylenediamine ion. *Bull. Soc. Chim. Fr. Mem.* **1939**, *6*, 1258–1259.

79. Mathieu, J.-P. Activité Optique Naturelle. In *Handbuch der Physik/Encyclopedia of Physics*; Spektroskopie II/Spectroscopy II; Madelung, O., Ed.; Springer: Berlin/Heidelberg, Germany, 1957; pp. 333–432.

80. Cotton, A. Recherches sur l'absorption et la Dispersion de la lumière par les Milieux Doués du Pouvoir Rotatoire. *Ann. Chim. Phys. Sér. 7* **1896**, *8*, 347–432. [CrossRef]

81. Cotton, A. Absorption inégale des rayons circulaires droit et gauche dans certains corps actifs. *C. R. Hebd. Seances Acad. Sci.* **1895**, *120*, 989–991.

82. Cotton, A. Dispersion rotatoire anomale des corps absorbants. *C. R. Hebd. Seances Acad. Sci.* **1895**, *120*, 1044–1046.

83. Kuhn, W.; Bein, K. Konfiguration Und Optische Drehung Bei Anorganischen Komplexverbindungen, Z. *Anorg. Allg. Chem.* **1934**, *216*, 321–348. [CrossRef]

84. Kuhn, W. Das Problem der absoluten Konfiguration optisch aktiver Stoffe. *Sci. Nat.* **1938**, *26*, 289–296. [CrossRef]

85. Röntgen, W.C. Ueber eine neue Art von Strahlen. *Ann. Phys.* **1898**, *300*, 12–17. [CrossRef]

86. Röntgen, W.C. Ueber eine neue Art von Strahlen (Vorläufige Mittheilung). *Sonderabbdruck Sitz. Würzburger Physik.Medic. Ges.* **1895**, *1896*, 2–20. [CrossRef]

87. Friedrich, W.K.; Laue, M. Interferenz -Erscheinungen bei Röntgenstrahlen (mit 5 Tafeln). *Sitz. Bayer. Akad. Wiss.* **1912**, 303–322.

88. Laue, M. Eine quantitative Prüfung der Theorie für die Interferenz-Erscheinungen bei Röntgenstrahlen. *Sitz. Bayer. Akad. Wiss.* **1912**, 363–373.

89. Eckert, M. Max von Laue and the discovery of X-ray diffraction in 1912. *Ann. Phys.* **2012**, *524*, A83–A85. [CrossRef]

90. Hendrickson, W.A. Evolution of diffraction methods for solving crystal structures. *Acta Crystallogr. Sect. A Found. Crystallogr.* **2013**, *69*, 51–59. [CrossRef]

91. Ewald, P.P. *Fifty Years of X-Ray Diffraction*; Oosthoek's Uitgevers Mij. N.V.: Utrecht, The Netherlands, 1962; ISBN 978-1-4615-9961-6.

92. Bragg, W.L. *The Development of X-Ray Analysis*; G. Bell: London, UK, 1975; ISBN 0486673162.

93. Bacon, G.E. *X-Ray and Neutron Diffraction*, 1st ed.; Ter Haar, D., Ed.; Pergamon: Oxford, UK, 1966; ISBN 0080119999.

94. Hildebrandt, G. The Discovery of the Diffraction of X-rays in Crystals—A Historical Review. *Cryst. Res. Technol.* **1993**, *28*, 747–766. [CrossRef]

95. Bijvoet, J.M.; Bernal, J.D.; Patterson, A.L. Forty years of X-ray diffraction. *Nature* **1952**, *169*, 949–950. [CrossRef]

96. Bijvoet, J.M.; Burgers, W.G.; Hägg, G. (Eds.) *Early Papers on Diffraction of X-Rays By Crystals*; Oosthoek's Uitgevers Mij. N.V.: Utrecht, The Netherlands, 1969; Volume 1, ISBN 978-1-4615-6880-3.

97. Bragg, H.B.; Bragg, W.L. The structure of the diamond. *Proc. R. Soc. Lond. Ser. A* **1913**, *89*, 277–291. [CrossRef]

98. Bragg, W.L.; Bragg, W.L. The Structure of the Diamond. *Nature* **1913**, *91*, 557. [CrossRef]

99. Bragg, W.L. The Specular Reflection of X-rays. *Nature* **1912**, *90*, 410. [CrossRef]

100. Bragg, W.L.; James, R.W.; Bosanquet, C.H. The intensity of reflexion of X-rays by rock-salt.—Part II. *Philos. Mag. (1798–1977)* **1921**, *42*, 1–17. [CrossRef]

101. Bragg, W.L.; James, R.W.; Bosanquet, C.H. The intensity of reflexion of X-rays by rock-salt. *Philos. Mag. (1798–1977)* **1921**, *41*, 309–337. [CrossRef]

102. Bragg, W.L.; James, R.W.; Bosanquet, C.H. The distribution of electrons around the nucleus in the sodium and chlorine atoms. *Philos. Mag. (1798–1977)* **1922**, *44*, 433–449. [CrossRef]

103. Bragg, W.L. The Diffraction of Short Electromagnetic Waves by a Crystal. *Proc. Camb. Philos. Soc.* **1913**, *17*, 43–57.

104. Bragg, W.L. The structure of some crystals as indicated by their diffraction of X-rays. *Proc. R. Soc. Lond. Ser. A* **1913**, *89*, 248–277. [CrossRef]

105. Bragg, W.L. The analysis of crystals by the X-ray spectrometer. *Proc. R. Soc. Lond. Ser. A* **1914**, *89*, 468–489. [CrossRef]

106. Darwin, C.G. LXXVIII. The theory of X-ray reflexion. Part II. *Philos. Mag. (1798–1977)* **1914**, *27*, 675–690. [CrossRef]

107. Darwin, C.G. XXXIV. The theory of X-ray reflexion. *Philos. Mag. (1798–1977)* **1914**, *27*, 315–333. [CrossRef]

108. Bragg, W.L.; Darwin, C.G.; James, R.W. LXXXI. The intensity of reflexion of X-rays by crystals. *Philos. Mag. (1798–1977)* **1926**, *1*, 897–922. [CrossRef]

109. Friedel, G. Sur les symétries cristallines que peut révéler la diffraction des rayons X. *C.R. Acad. Sci. Paris* **1913**, *157*, 1533–1536.

110. Bragg, W.L. The determination of parameters in crystal structures by means of Fourier series. *Proc. R. Soc. Lond. Ser. A* **1929**, *123*, 537–559. [CrossRef]

111. Patterson, A.L. A Fourier Series Method for the Determination of the Components of Interatomic Distances in Crystals. *Phys. Rev.* **1934**, *46*, 372–376. [CrossRef]

112. Patterson, A.L. A Direct Method for the Determination of the Components of Interatomic Distances in Crystals. *Z. Kristallogr. Kristallgeom. Kristallphys. Kristallchem.* **1935**, *90*, 517–542. [CrossRef]

113. Robertson, J.M. X-ray analysis and application of Fourier series methods to molecular structures. *Rep. Prog. Phys.* **1937**, *4*, 332–367. [CrossRef]

114. Cork, J.M. The crystal structure of some of the alums. *Philos. Mag. (1798–1977)* **1927**, *4*, 688–698. [CrossRef]

115. Robertson, J.M. An X-ray study of the phthalocyanines. Part II. Quantitative structure determination of the metal-free compound. *J. Chem. Soc.* **1936**, 1195–1209. [CrossRef]

116. Robertson, J.M.; Woodward, I. An X-ray study of the phthalocyanines. Part III. Quantitative structure determination of nickel phthalocyanine. *J. Chem. Soc.* **1937**, 219–230. [CrossRef]

117. Robertson, J.M.; Woodward, I. An X-ray study of the phthalocyanines. Part IV. Direct quantitative analysis of the platinum compound. *J. Chem. Soc.* **1940**, 36–48. [CrossRef]

118. Nishikawa, S.; Matukawa, K. Hemihedry of Zincblende and X-Ray Reflexion. *Proc. Imp. Acad. Jpn.* **1928**, *4*, 96–97. [CrossRef]

119. Coster, D.; Knol, K.S.; Prins, J.A. Unterschiede in der Intensität der Röntgenstrahlen-reflexion an den beiden 111-Flächen der Zinkblende. *Z. Phys.* **1930**, *63*, 345–369. [CrossRef]

120. Bokhoven, C.; Schoone, J.C.; Bijvoet, J.M. The Fourier synthesis of the crystal structure of strychnine sulphate pentahydrate. *Acta Crystallogr.* **1951**, *4*, 275–280. [CrossRef]

121. Bijvoet, J.M. Phase determination in direct Fourier synthesis of crystal structure. *Proc. Sect. Sci. K. Ned. Akad. Wet.* **1949**, *52*, 313–314.

122. Bokhoven, C.; Schoone, J.C.; Bijvoet, J.M. On the crystal structure of strychnine sulphate and selenate. III. [001] projection. *Proc. Sect. Sci. K. Ned. Akad. Wet.* **1949**, *52*, 120–121.

123. Bokhoven, C.; Schoone, J.C.; Bijvoet, J.M. On the crystal structure of strychnine sulphate and selenate. I. Cell dimensions and spacegroup. *Proc. Sect. Sci. K. Ned. Akad. Wet.* **1947**, *50*, 825.

124. Bokhoven, C.; Schoone, J.C.; Bijvoet, J.M. On the crystal structure of strychnine sulphate and selenate. II. [010] projection and structure formula. *Proc. Sect. Sci. K. Ned. Akad. Wet.* **1948**, *51*, 990.

125. Groenewege, M.P.; Peerdeman, A.F. Johannes Martin Bijvoet. 23 January 1892–4 March 1980. *Biogr. Mem. Fellows R. Soc.* **1983**, *29*, 26–41. [CrossRef]

126. Bijvoet, J.M.; van Bommel, A.J.; Peerdeman, A.F. Determination of absolute configuration of optically active compounds by means of X-rays. *Proc. Sect. Sci. K. Ned. Akad. Wet.* **1951**, *54*, 16–19. [CrossRef]

127. Bijvoet, J.M.; Peerdeman, A.F.; van Bommel, A.J. Determination of the Absolute Configuration of Optically Active Compounds by Means of X-Rays. *Nature* **1951**, *168*, 271–272. [CrossRef]

128. Bijvoet, J.M.; Peerdeman, A.F.; van Bommel, A.J. Structure of Optically Active Compounds in the Solid State. *Nature* **1954**, *173*, 888–891. [CrossRef]

129. Mathieson, A.M. The determination of absolute configuration by the use of an internal reference asymmetric centre. *Acta Crystallogr.* **1956**, *9*, 317. [CrossRef]

130. Karle, J.; Hauptman, H. The phases and magnitudes of the structure factors. *Acta Crystallogr.* **1950**, *3*, 181–187. [CrossRef]

131. Karle, J.; Hauptman, H. A theory of phase determination for the four types of non-centrosymmetric space groups 1P222, 2P22, 3P12, 3P22. *Acta Crystallogr.* **1956**, *9*, 635–651. [CrossRef]

132. Woolfson, M.M. Direct methods in crystallography. *Rep. Prog. Phys.* **1971**, *34*, 369–434. [CrossRef]

133. Ladd, M.F.C.; Palmer, R.A. (Eds.) *Theory and Practice of Direct Methods in Crystallography*; Springer: New York, NY, USA, 1980.

134. Jones, P.G. The determination of absolute structure. I. Some experiences with the Rogers η refinement. *Acta Crystallogr. Sect. A Found. Crystallogr.* **1984**, *40*, 660–662. [CrossRef]

135. Flack, H.D. Absolute-structure determination: Past, present and future. *Chimia* **2014**, *68*, 26–30. [CrossRef]

136. Sohncke, L. *Entwickelung Einer Theorie Der Krystallstruktur*; B.G. Teubner: Leipzig, Germany, 1879.

137. Pidcock, E. Achiral molecules in non-centrosymmetric space groups. *Chem. Commun.* **2005**, 3457–3459. [CrossRef] [PubMed]

138. Hamilton, W.C. Significance tests on the crystallographic R factor. *Acta Crystallogr.* **1965**, *18*, 502–510. [CrossRef]

139. Rogers, D. On the application of Hamilton's ratio test to the assignment of absolute configuration and an alternative test. *Acta Crystallogr. Sect. A Cryst. Phys. Diffr. Theor. Gen. Crystallogr.* **1981**, *37*, 734–741. [CrossRef]

140. CCDC the Cambridge Structural Database (CSD). Available online: https://www.ccdc.cam.ac.uk (accessed on 20 July 2020).

141. Groom, C.R.; Bruno, I.J.; Lightfoot, M.P.; Ward, S.C. The Cambridge Structural Database. *Acta Crystallogr. Sect. B Struct. Sci. Cryst. Eng. Mater.* **2016**, *72*, 171–179. [CrossRef]

142. Flack, H. Chiral and Achiral Crystal Structures. *Helv. Chim. Acta* **2003**, *86*, 905–921. [CrossRef]

143. Eliel, E.L.; Wilen, S.H.; Mander, L.N. *Stereochemistry of Organic Compounds*; Wiley-Interscience: New York, NY, USA, 1994.

144. von Zelewsky, A. *Stereochemistry of Coordination Compounds*; John Wiley & Sons Inc.: Chichester, UK, 1996.

145. Hawkins, C.J. *Absolute Configuration of Metal Complexes*; Wiley-Interscience: New York, NY, USA, 1971.

146. Amouri, H.; Gruselle, M. *Chirality in Transition Metal Chemistry*; John Wiley & Sons Inc.: Chichester, UK, 2008.

147. Gillard, R.D.; Mitchell, P.R. The Absolute Configuration of Transition Metal Complexes. In *Structure and Bonding*; Springer: Berlin/Heidelberg, Germany, 1970; Volume 7, pp. 46–86.

148. Saito, Y. Absolute configurations of metal complexes determined by X-ray analysis. *Coord. Chem. Rev.* **1974**, *13*, 305–337. [CrossRef]

149. Saito, Y. Absolute Configuration of Transition Metal Complexes. In *ACS Symposium Series: Stereochemistry of Optically Active Transition Metal Compounds*; American Chemical Society: Washington, DC, USA, 1980; pp. 13–42.

150. Buckingham, D.A.; Sargeson, A.M. Conformational Analysis and Steric Effects in Metal Chelates. In *Topics in Stereochemistry*; Allinger, N.L., Eliel, E.L., Eds.; Wiley: Chichester, UK, 1971; Volume 6, pp. 219–277.

151. Saito, Y.; Nakatsu, K.; Shiro, M.; Kuroya, H. Determination of the absolute configuration of optically active complex ion, $[Coen_3]^{3+}$, by means of X-rays. *Acta Crystallogr.* **1955**, *8*, 729–730. [CrossRef]

152. Zhong, J.; Zhang, L.; August, D.P.; Whitehead, G.F.S.; Leigh, D.A. Self-Sorting Assembly of Molecular Trefoil Knots of Single Handedness. *J. Am. Chem. Soc.* **2019**, *141*, 14249–14256. [CrossRef]

153. Zhang, J.; Chen, S.; Nieto, R.A.; Wu, T.; Feng, P.; Bu, X. A tale of three carboxylates: Cooperative asymmetric crystallization of a three-dimensional microporous framework from achiral precursors. *Angew. Chem. Int. Ed. Engl.* **2010**, *49*, 1267–1270. [CrossRef]

154. Zheng, W.; Wei, Y.; Xiao, X.; Wu, K. Spontaneous asymmetric crystallization of a quartz-type framework from achiral precursors. *Dalton Trans.* **2012**, *41*, 3138–3140. [CrossRef]

155. Yuan, S.; Deng, Y.-K.; Xuan, W.-M.; Wang, X.-P.; Wang, S.-N.; Dou, J.-M.; Sun, D. Spontaneous chiral resolution of a 3D (3,12)-connected MOF with an unprecedented ttt topology consisting of cubic $[Cd_4(\mu_3\text{-}OH)_4]$ clusters and propeller-like ligands. *Cryst. Eng. Comm.* **2014**, *16*, 3829–3833. [CrossRef]

156. Watkin, D.J.; Cooper, R.I. Why direct and post-refinement determinations of absolute structure may give different results. *Acta Crystallogr. Sect. B Struct. Sci. Cryst. Eng. Mater.* **2016**, *72*, 661–683. [CrossRef]

 *chemistry*

Article

Twinning in Zr-Based Metal-Organic Framework Crystals

Sigurd Øien-Ødegaard * and Karl Petter Lillerud

Centre for Materials Science and Nanotechnology, University of Oslo, P.O. box 1126 Blindern, 0318 Oslo, Norway
* Correspondence: sigurdoi@kjemi.uio.no

Received: 19 August 2020; Accepted: 14 September 2020; Published: 16 September 2020

Abstract: Ab initio structure determination of new metal-organic framework (MOF) compounds is generally done by single crystal X-ray diffraction, but this technique can yield incorrect crystal structures if crystal twinning is overlooked. Herein, the crystal structures of three Zirconium-based MOFs, that are especially prone to twinning, have been determined from twinned crystals. These twin laws (and others) could potentially occur in many MOFs or related network structures, and the methods and tools described herein to detect and treat twinning could be useful to resolve the structures of affected crystals. Our results highlight the prevalence (and sometimes inevitability) of twinning in certain Zr-MOFs. Of special importance are the works of Howard Flack which, in addition to fundamental advances in crystallography, provide accessible tools for inexperienced crystallographers to take twinning into account in structure elucidation.

Keywords: MOFs; crystallography; twinning

1. Introduction

Metal-organic frameworks (MOFs) are porous solids consisting of inorganic nodes linked by organic multidentate ligands (e.g., through strongly coordinating groups such as carboxylates or Lewis bases) to form extended networks [1]. The combination of diverse coordination chemistry and the structural richness of organic ligands enable a vast number of possible MOF structures and network topologies [2]. Due to their remarkable ability to harbor a large range of chemically functional groups in pores that are accessible to guest species, MOFs are studied mainly (albeit not exclusively) for their properties as adsorbents [3,4] and catalysts [5]. In particular, MOFs based on zirconium oxide clusters are frequently reported due to their relatively high stability and structural diversity [6]. These MOFs are normally obtained as single crystals by adding growth modulators (monocarboxylic acids) to the synthesis liquor [7].

The principal method of MOF structure determination is X-ray diffraction methods, and single crystal X-ray diffraction (SC-XRD) in particular [8]. Despite being the most powerful technique to solve crystal structures, a successful SC-XRD experiment normally depends on a relatively large single crystal of high quality.

The phenomenon in which a crystal consists of two or more separate domains, and these domains are related by a symmetry operation that is not present in the space group of the crystal, is known as twinning [9]. Twinned crystals represent a commonly encountered problem in crystallography and will often prevent a successful structure determination. The observed reflections of all the present domains can be interpreted as originating from the same crystal instead of separate entities, thus obscuring the true symmetry (and thus the structure) of the crystal. Crystalline MOFs and related network structures are prone to interpenetration (intergrown separate networks that may occur if the void fraction of the structure is large) and twinning, because twin domain interfaces are enabled by the flexibility of the MOF's building units (i.e., geometrical flexibility of (1) the linkers' points of connectivity due to free rotation around C-C bonds and (2) the possible existence of metal oxoclusters with a different arrangement of coordination sites in twin boundaries).

In our ongoing crystallographic characterization of Zr-based MOFs, several cases of crystal twinning or related problems were discovered, which can be categorized into three groups:

(a) A crystal consisting of two or more domains related by seemingly arbitrary rotation matrices (although not technically twinning), is frequently encountered when investigating MOF crystals. It occurs when two randomly oriented crystals in close proximity grow into each other and forms an interface, and is frequently observed in static syntheses where crystal growth mainly occurs on the bottom of the synthesis vessel. In these cases, automatic indexing fails to provide a meaningful unit cell, but the relationship can usually be determined by manual inspection and sorting of the reflections in reciprocal space.

(b) In certain cases, automatic indexing found a hexagonal supercell due to partial overlap between the reflections from the twin domains. The twin law was found to be the so-called "spinel law", $2_{[111]}$ (where the two twin domains are related by a two-fold rotation about the body diagonal of a cubic unit cell), which is a case of twinning by reticular merohedry [10]. This twinning mode was observed in UiO-67, UiO-67-Me$_2$ (1, discussed herein) and Zr-stilbene dicarboxylate. In all of these cases, the crystals displayed a specific morphology, resembling intergrown octahedra with two shared (1 1 1) faces (Figure 1).

(c) In certain cases, twinning by syngonic merohedry was observed as a consequence of intrinsic features of the MOF. The examples presented herein are obtained from MOFs featuring partial lattice interpenetration (3) and a phase transitions from dynamic to static orientation disorder upon cooling of the sample (2).

Figure 1. Crystals of UiO-67-Me2, showing the presence of twinning by the spinel law.

The following sections describe how these twinning issues has been resolved for specific Zr-MOF samples, and how a correct assessment of twin laws has revealed structural features with significant implication for the MOFs' properties. The strategies described herein to detect and resolve twinning will presumably be useful to researchers encountering the issue while working with MOFs and other crystalline network solids.

2. Materials and Methods

2.1. General Materials and Synthesis Methods

All chemicals, unless otherwise stated, were obtained reagent grade from Sigma-Aldrich and used without further purification. The linkers for 1 and 2 (3,3′-dimethyl-4,4′-biphenyldicarboxylic acid (Me$_2$-H$_2$bpdc) and 2′,3′′-dimethyl-[1,1′:4′,1′′:4′′,1′′′-quaterphenyl]-4,4′′′-dicarboxylic acid (Me$_2$-H$_2$qpdc)) were prepared using previously published protocols [11]. Crystallization was performed under static conditions using conical glass flasks that were treated with NaOH$_{(aq)}$ 30% wt. overnight then thoroughly rinsed and dried before the reaction. This treatment seems to decrease the MOF's tendency to nucleate rapidly on the glass surface. During crystallization the flasks were capped with loose lids to prevent pressure build-up and accumulation of decomposition products of DMF (mainly formic acid and dimethylamine) and HCl.

Synthesis of UiO-67-Me$_2$ (1): To 13 mL of dimethylformamide (DMF), 56 μL (3.1 mmol) H$_2$O, 241 mg (1.03 mmol) ZrCl$_4$, and 3.79 g (31.0 mmol) benzoic acid was added and stirred until a clear solution was obtained, and subsequently heated to 120 °C. 280 mg (1.03 mmol) Me$_2$-H$_2$bpdc was then added, a clear solution was quickly obtained, transferred to a clean conical flask and kept at 120 °C for 72 h. The solid crystalline product was isolated and washed, once briefly with DMF at 100 °C, once with dry DMF at room temperature and three times with dry 2-propanol. The MOF crystals were kept in dry 2-propanol prior to activation and XRD measurement.

Synthesis of Zr-stilbene dicarboxylate (2): To 50 mL of DMF, 70 μL (3.9 mmol) H$_2$O, 301 mg (1.29 mmol) ZrCl$_4$ and 4.73 g (38.8 mmol) benzoic acid was added and stirred until a clear solution was obtained, and subsequently heated to 120 °C. 346 mg (1.29 mmol) trans-stilbene-4,4′-carboxylic acid was then added, a clear solution was obtained and the solution was transferred to a clean conical flask and kept at 120 °C for 72 h. Large octahedral crystals were isolated and washed. First briefly with DMF at 100 °C, then dry DMF at RT, then three times with tetrahydrofuran (THF), and lastly three times with n-hexane. The crystals were kept in dry n-hexane prior to measurement.

Synthesis of UiO-69-Me$_2$ (3): To 20 mL of DMF, 19 μL (1.03 mmol) H$_2$O, 120 mg (0.52 mmol) ZrCl$_4$ and 1.89 g (15.5 mmol) benzoic acid was added and stirred until a clear solution was obtained, and subsequently heated to 120 °C. 218 mg (0.52 mmol) Me$_2$-H$_2$qpdc was then added, a clear solution was obtained, transferred to a clean conical flask and kept at 120 °C for 72 h. The solid crystalline product was isolated and washed.

2.2. X-ray Crystallography

The crystals of 1 and 3 were dried in air at 200 °C for 2 h prior to measurement to ensure that the pores were free of physiosorbed solvent and water. The crystals were mounted on MiTeGen polymer loops using a minimum amount of Paratone oil. Data collection for 1–3 was performed on a Bruker D8 Venture diffractometer equipped with a Photon 100 CMOS detector using Mo Kα radiation (λ = 0.71073 Å) at 100 K. 3 was also measured at beamline ID11 at the ESRF synchrotron (Grenoble, France), equipped with a Frelon2 detector, using a wavelength of λ = 0.31120 Å.

Due to the poor chemical stability of 2, it was kept in dry hexane after synthesis. Before measurement, several crystals were placed on a glass slide and the hexane was allowed to evaporate. The crystals were then mounted on MiTeGen polymer loops using a minimum amount of Paratone oil. 2 was also measured at beamline I911-3 at the MAX2 synchrotron (Lund, Sweden) [12].

All frames were integrated, and the reflection intensities were scaled and evaluated using the APEX3 suite from Bruker AXS, consisting of SAINT, SADABS and XPREP [13]. The structures were solved with XT [14] and refined with XL [15], using Olex2 as graphical user interface [16]. Space group determination of twinned MOFs was done using comprehensive tables by Howard Flack [17]. A summary of the crystal data, data collection and structure refinement details can be found in Table 1, and full structural information can be found in the Supplementary Materials.

Table 1. Summary of crystallographic data and refinement indicators for reported MOFs 1–3.

Crystal Data	UiO-67-Me$_2$ (1)	Zr-Stilbene dc (2)	UiO-69-Me$_2$ (3)
Chemical formula	$C_{384}O_{128}Zr_{24} \cdot 32(O)$	$C_{342.25}H_{211.09}O_{128}Zr_{24}$	$C_{648}H_{408}O_{128}Zr_{24} \cdot$ $0.66(C_{648}H_{408}O_{128}Zr_{24}) \cdot$ $16(O)$
M$_r$	2340.28	8584.82	20,698.65
Crystal system, space group	Cubic, Fm$\bar{3}$m	Cubic, Pn$\bar{3}$	Cubic, F$\bar{4}$3m
Temperature (K)	100	100	100
a (Å)	26.8903 (12)	30.0322 (6)	38.995 (2)
V (Å^3)	19,444 (3)	27,087.0 (16)	59,298 (10)
Radiation type	Mo Kα	Synchrotron, λ = 0.760 Å	Synchrotron, λ = 0.3112 Å
μ (mm^{-1})	0.35	0.25	0.17
Crystal size (mm)	$0.06 \times 0.06 \times 0.02$	$0.2 \times 0.2 \times 0.2$	$0.14 \times 0.14 \times 0.14$
Data collection			
Diffractometer	Bruker D8 Venture, CMOS detector	MD2 microdiffractometer with MK3 mini-kappa	ESRF ID11
Absorption correction	Multi-scan	Multi-scan	Multi-scan
No. of measured, independent and observed [I > 2σ(I)] reflections	4388, 4388, 4155	199,694, 11,196, 10,761	180,671, 19,944, 17,103
R$_{int}$	0.036	0.04	0.033
$(\sin \theta/\lambda)_{max}$ (Å^{-1})	0.649	0.623	0.961
Refinement			
R[F^2 > 2σ(F^2)], wR(F^2), S	0.048, 0.174, 1.13	0.029, 0.090, 1.10	0.047, 0.154, 1.08
No. of reflections	4388	11,196	19,944
No. of parameters	59	216	226
No. of restraints	0	18	304
$\Delta\rho_{max}$, $\Delta\rho_{min}$ (e Å^{-3})	1.03, −1.05	0.91, −0.46	2.17, −2.60

3. Results and Discussion

3.1. UiO-67-Me$_2$

UiO-67-Me$_2$, consisting of the Zr$_6$ oxocluster and the Me$_2$-bpdc linker, is isostructural to UiO-67, with nearly identical lattice parameters and PXRD pattern. The MOF readily forms large single crystals during synthesis, by the addition of 30 molar equivalents of benzoic acid (in respect to Zr) as modulator. This MOF has been found to have slightly better stability to water, and higher affinity to methane than UiO-67, presumably due to the steric shielding of cluster-adjacent hydrophilic sites by the methyl groups [18].

A significant number of the crystals that were screened had visibly distorted morphology (Figure 1). The crystals were seemingly intergrown, as if one crystal was sprouting out of the facets of its parent. In all of these crystals, the initial indexing resulted in a larger unit cell than expected, a primitive hexagonal cell, closely related to the reduced cell of UiO-67. (Reduced unit cell of UiO-67: a = b = c = 19.0 Å, α = β = γ = 60.0°. Hexagonal cell: a = b = 19.0 Å, c = 47.0 Å, a = b = 90°, c = 120°).

Upon inspection of the reciprocal lattice, it was apparent that the crystals were twinned by reticular merohedry, the twin law being the so-called "spinel law", 2$_{[111]}$. This twin law translates to a two-fold rotation about the [1 1 1] axis, coinciding with a three-fold rotation-reflection axis (S$_6$ or 3) in the ideal structure. This is a common twinning mode in cubic close packed crystals, as it is associated with stacking error of the close packed layers.

The twin law is easily visualized by reciprocal lattice displays of the reflections of the hk0 layer (Figure 2). The cause for twinning or the exact structure at the interface of this MOF remains unknown, but there are other Zr-based oxoclusters that facilitate such a boundary [19].

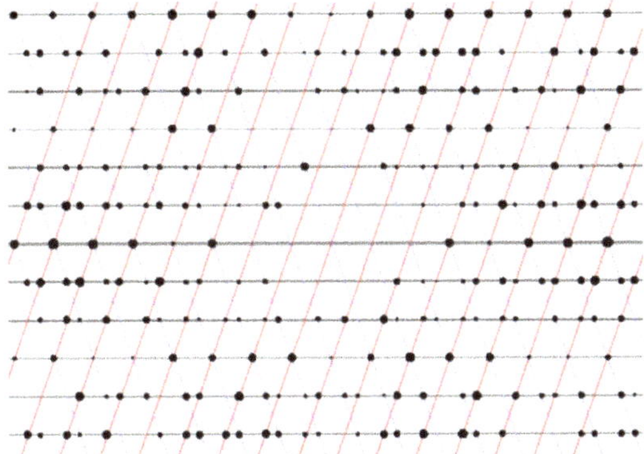

Figure 2. Schematic display of the hk0 layer of the diffraction of 1, twinned by $2_{[111]}$. The lattices of the twin domains are shown in red and blue, respectively.

3.2. Zr-stilbene Dicarboxylate

Zr-stilbene dicarboxylate (2) has previously been reported in its disordered form (space group Fm-3 m), in which the crystal symmetry is the same as for the UiO MOFs [20,21]. These MOFs consist of the same Zr_6 oxoclusters in an fcc arrangement, connected by linear ditopic linkers. However, the stilbene linker (being non-linear) is not compatible with this symmetry, implying that it is randomly disordered over two rotational conformers. The disorder could be dynamic, in which the linkers would rotate freely and continuously throughout the sample because of the low energy barrier of rotation. In this case, one would expect a threshold temperature at which the rotation is no longer feasible, and where the individual linkers adopt a static conformation. Such a static phase could exist with linkers assuming random conformers (preserving the overall $Fm\bar{3}$ m disordered symmetry) or ordered conformations resulting in a symmetry change.

Repeated measurements at room temperature and after flash cooling samples to 100 K showed the randomly oriented linker structure ($Fm\bar{3}$ m) previously reported [21]. To investigate whether a low-symmetry phase could be obtained, individual scans along [1 0 0] were acquired while slowly cooling the crystal by 5 K/min. The appearance of new diffraction peaks (forbidden in face centered cubic crystals) occurred around 160 K (Figure 3).

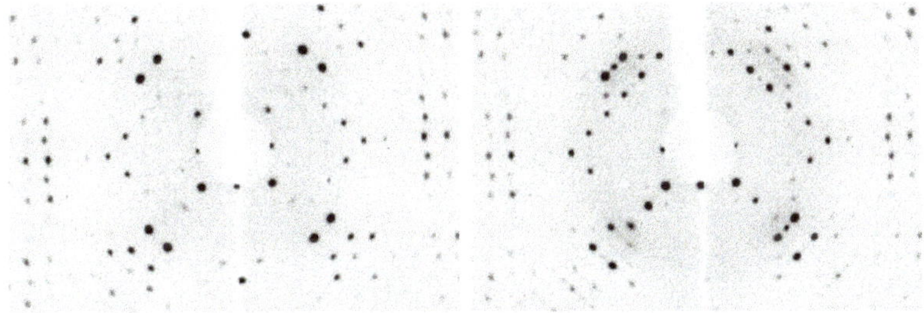

Figure 3. Oriented diffraction images of **2** along the [1 0 0] direction acquired at room temperature (**left**) and at 100 K (**right**) after slow cooling of the crystal (by 5 K/min).

Solving the structure using the data from the slowly cooled crystal revealed a structure in the primitive cubic space group Pn$\bar{3}$ (shown in Figure 4a–c). In this structure, the conformation of the linkers are unambiguous; the space group does not include mirror planes intersecting the linkers as in Fm$\bar{3}$m. When the crystal is brought back to room temperature, the face centered phase is observed again (shown in Figure 4d–f). This reversible phase transition can be used to study the molecular mechanics of these nonlinear linkers, which have also been shown to facilitate linker exchange and lattice expansion in these MOFs [22].

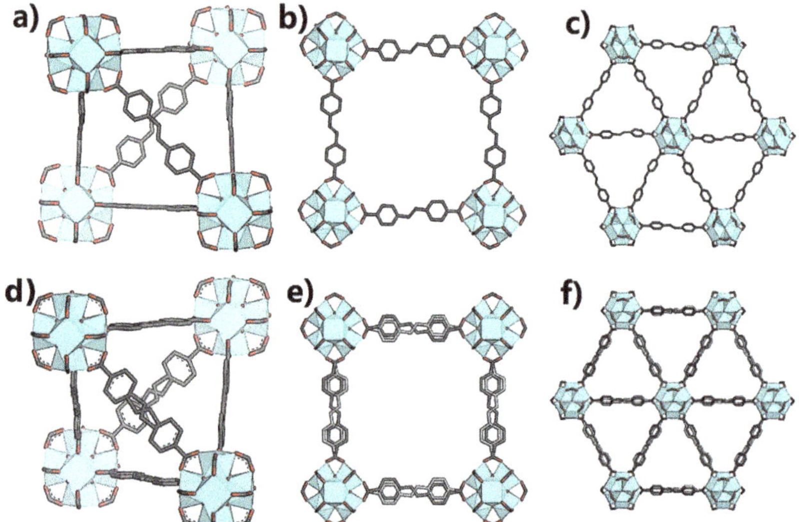

Figure 4. Partial crystal structures of **2** in Pn$\bar{3}$ (**a–c**) and Fm$\bar{3}$m (**d–f**) showing the ordered and disordered conformations of the linker.

Inspecting the initial solution of the Pn$\bar{3}$ structure, it displayed several warning signs: Very high *R*-values and presence of significant noise in the Fourier difference maps, large negative peaks in particular. Looking at the crystal structure along [1 0 0], one could envision two different orientations of the linker "kink" being present in the crystal (Figure 5). In fact, are separate domains of the crystal with opposite oriented linkers related to each other by a two-fold rotation about the [1 1 0] face diagonal. Thus, the twin law to implement in the crystal structure refinement is the matrix (0 −1 0 −1 0 0 0 0 −1). Applying this twin law to the refinement significantly improves the fit, and the twin fractions refine freely to 0.5.

3.3. Interpenetrated UiO-69-Me$_2$

UiO-69-Me$_2$ (**3**) was synthesized as single crystals using the linker 2′,3″-dimethyl-[1,1′:4′,1″:4″, 1‴-quaterphenyl]-4,4‴-dicarboxylic acid (Me$_2$-H$_2$qpdc) [11]. Using 30 equivalents of benzoic acid in respect to ZrCl$_4$ as modulator, single crystals of up to 1 mm could be obtained. The large octahedral single crystals featured a distinct opaque pattern originating from the center of each crystal and out to each face (Figure 6).

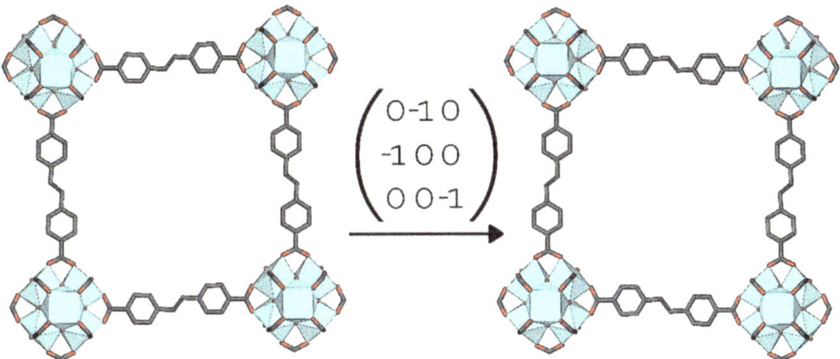

Figure 5. The twin law of **2** in the Pn$\bar{3}$ space group.

Figure 6. Single crystals of **3**, featuring an opaque pattern from the center and out towards each face.

High quality SC-XRD data was acquired using synchrotron radiation, and the subsequent structure solution revealed a structure composed of doubly interpenetrated fcu networks analogous to the PIZOF materials [23]. Based on analysis of reflection intensities, the structure appears to have the space group Fd$\bar{3}$m, as is observed for the other family of two-fold interpenetrated Zr-MOFs [24]. The structure could indeed be solved in this space group, but the refinement accuracy came out very poor ($R_1 > 13.5\%$).

When a structure appears centrosymmetric it could in reality be a twin of two equally large non-centrosymmetric domains [9,25]. Consequently, the structure was solved in the lower-symmetry space group F$\bar{4}$3m. The difference between Fd$\bar{3}$m and F$\bar{4}$3m is one symmetry element, a two-fold rotation relating the closest neighboring clusters from the two MOF lattices (intersecting 1/8 1/8 1/8). When the symmetry element is present, the two lattices must be identical for the symmetry to be true. However, this MOF feature only partial interpenetration, so the secondary lattice has a lower occupancy coefficient than the main lattice (in this case 0.66). When this coefficient is not equal, the real symmetry must be the lower F$\bar{4}$3m.

The reduction in symmetry reveals a curious detail: Neighboring clusters from the separate lattices have a clear tendency to orient opposite μ_3-O/OH functionality towards each other. The fully occupied lattice tend to orient the μ_3-O towards the partially occupied neighboring cluster's OH. This preference would violate the two-fold rotational symmetry of Fd$\bar{3}$m. There are two possible settings of the secondary lattice within the main, which requires a twin law (e.g., twinning by inversion) to resolve (Figure 7). This twinning feature contributes to the high apparent symmetry of the data, leading to the initial erroneous space group assignment.

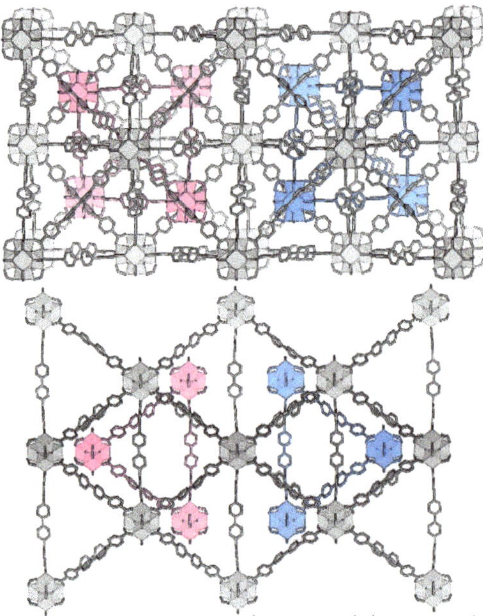

Figure 7. The twinning in UiO-69-Me$_2$ (3) can be visualized by displaying the two possible settings of the interpenetrating lattice.

4. Conclusions

The results reported herein clearly show that twinning should always be considered in the single-crystal structure determination of MOFs and other network structures, in particular in cases where unexpected features arise during structure solution and/or refinement. Details such as partial interpenetration or structural heterogeneity within the sample may easily be overlooked if twinning is not considered. It is likely that several of the many reported MOF structures that display poor refinement values or chemically unlikely features are in fact non-treated twins.

Supplementary Materials: The following are available online at http://www.mdpi.com/2624-8549/2/3/50/s1, Crystal structure data of compounds **1–3**. The crystallographic data for this paper (CCDC) can also be obtained free of charge via www.ccdc.cam.ac.uk/data_request/cif, by emailing data_request@ccdc.cam.ac.uk, or by contacting The Cambridge Crystallographic Data Centre, 12 Union Road, Cambridge CB2 1EZ, UK; Fax: +44-1223-336033.

Author Contributions: Conceptualization, S.Ø.-Ø.; methodology, S.Ø.-Ø.; validation, S.Ø.-Ø.; formal analysis, S.Ø.-Ø.; investigation, S.Ø.-Ø.; resources, S.Ø.-Ø.; data curation, S.Ø.-Ø.; writing—original draft preparation, S.Ø.-Ø.; writing—review and editing, S.Ø.-Ø.; visualization, S.Ø.-Ø.; supervision, K.P.L.; project administration, K.P.L.; funding acquisition, K.P.L. All authors have read and agreed to the published version of the manuscript.

Funding: This research was funded by the European Union's Horizon 2020 research and innovation programme under grant agreement number 685727 (ProDIA).

Acknowledgments: Knut T. Hylland is acknowledged for providing the organic linkers for the synthesis of compounds **1** and **3**.

Conflicts of Interest: The authors declare no conflict of interest.

References

1. Yuan, S.; Feng, L.; Wang, K.; Pang, J.; Bosch, M.; Lollar, C.; Sun, Y.; Qin, J.; Yang, X.; Zhang, P.; et al. Stable Metal–Organic Frameworks: Design, Synthesis, and Applications. *Adv. Mater.* **2018**, *30*, 1704303. [CrossRef]

2. Öhrström, L. Let's Talk about MOFs—Topology and Terminology of Metal-Organic Frameworks and Why We Need Them. *Crystals* **2015**, *5*, 154–162. [CrossRef]

3. Furukawa, H.; Gandara, F.; Zhang, Y.-B.; Jiang, J.; Queen, W.L.; Hudson, M.R.; Yaghi, O.M. Water Adsorption in Porous Metal-Organic Frameworks and Related Materials. *J. Am. Chem. Soc.* **2014**, *136*, 4369–4381. [CrossRef] [PubMed]

4. Li, J.; Wang, X.; Zhao, G.; Chen, C.; Chai, Z.; Alsaedi, A.; Hayat, T.; Wang, X. Metal–organic framework-based materials: Superior adsorbents for the capture of toxic and radioactive metal ions. *Chem. Soc. Rev.* **2018**, *47*, 2322–2356. [CrossRef] [PubMed]

5. Dhakshinamoorthy, A.; Li, Z.; Garcia, H. Catalysis and photocatalysis by metal organic frameworks. *Chem. Soc. Rev.* **2018**, *47*, 8134–8172. [CrossRef] [PubMed]

6. Bai, Y.; Dou, Y.; Xie, L.-H.; Rutledge, W.; Li, J.-R.; Zhou, H.-C. Zr-based metal–organic frameworks: Design, synthesis, structure, and applications. *Chem. Soc. Rev.* **2016**, *45*, 2327–2367. [CrossRef]

7. Schaate, A.; Roy, P.; Godt, A.; Lippke, J.; Waltz, F.; Wiebcke, M.; Behrens, P. Modulated synthesis of Zr-based metal-organic frameworks: From nano to single crystals. *Chem.-Eur. J.* **2011**, *17*, 6643–6651. [CrossRef]

8. Gandara, F.; Bennett, T.D. Crystallography of metal-organic frameworks. *IUCrJ* **2014**, *1*, 563–570. [CrossRef]

9. Parsons, S. Introduction to twinning. *Acta Crystallogr. Sect. D* **2003**, *59*, 1995–2003. [CrossRef]

10. Klapper, H.; Hahn, T. The application of eigensymmetries of face forms to X-ray diffraction intensities of crystals twinned by 'reticular merohedry'. *Acta Crystallogr. Sect. A* **2012**, *68*, 82–109. [CrossRef]

11. Hylland, K.T.; Øien-Ødegaard, S.; Lillerud, K.P.; Tilset, M. Efficient, Scalable Syntheses of Linker Molecules for Metal-Organic Frameworks. *Synlett* **2015**, *26*, 1480–1485. [CrossRef]

12. Ursby, T.; Unge, J.; Appio, R.; Logan, D.T.; Fredslund, F.; Svensson, C.; Larsson, K.; Labrador, A.; Thunnissen, M.M.G.M. The macromolecular crystallography beamline I911-3 at the MAX IV laboratory. *J. Synchrotron Radiat.* **2013**, *20*, 648–653. [CrossRef] [PubMed]

13. *Bruker APEX3, SAINT, SADABS, XPREP, 2019.1-1*; Bruker AXS Inc.: Madison, WI, USA, 2019.

14. Sheldrick, G. SHELXT—Integrated space-group and crystal-structure determination. *Acta Crystallogr. Sect. A* **2015**, *71*, 3–8. [CrossRef]

15. Sheldrick, G. Crystal structure refinement with SHELXL. *Acta Crystallogr. Sect. C* **2015**, *71*, 3–8. [CrossRef] [PubMed]

16. Dolomanov, O.V.; Bourhis, L.J.; Gildea, R.J.; Howard, J.A.K.; Puschmann, H. OLEX2: A complete structure solution, refinement and analysis program. *J. Appl. Crystallogr.* **2009**, *42*, 339–341. [CrossRef]

17. Flack, H. Methods of space-group determination—A supplement dealing with twinned crystals and metric specialization. *Acta Crystallogr. Sect. C* **2015**, *71*, 916–920. [CrossRef] [PubMed]

18. Øien-Ødegaard, S.; Bouchevreau, B.; Hylland, K.; Wu, L.; Blom, R.; Grande, C.; Olsbye, U.; Tilset, M.; Lillerud, K.P. UiO-67-type Metal-Organic Frameworks with Enhanced Water Stability and Methane Adsorption Capacity. *Inorg. Chem.* **2016**, *55*, 1986–1991. [CrossRef]

19. Cliffe, M.J.; Castillo-Martínez, E.; Wu, Y.; Lee, J.; Forse, A.C.; Firth, F.C.N.; Moghadam, P.Z.; Fairen-Jimenez, D.; Gaultois, M.W.; Hill, J.A.; et al. Metal–Organic Nanosheets Formed via Defect-Mediated Transformation of a Hafnium Metal–Organic Framework. *J. Am. Chem. Soc.* **2017**, *139*, 5397–5404. [CrossRef]

20. Cavka, J.H.; Jakobsen, S.; Olsbye, U.; Guillou, N.; Lamberti, C.; Bordiga, S.; Lillerud, K.P. A New Zirconium Inorganic Building Brick Forming Metal Organic Frameworks with Exceptional Stability. *J. Am. Chem. Soc.* **2008**, *130*, 13850–13851. [CrossRef]

21. Marshall, R.J.; Hobday, C.L.; Murphie, C.F.; Griffin, S.L.; Morrison, C.A.; Moggach, S.A.; Forgan, R.S. Amino acids as highly efficient modulators for single crystals of zirconium and hafnium metal-organic frameworks. *J. Mater. Chem. A* **2016**, *4*, 6955–6963. [CrossRef]

22. Feng, L.; Yuan, S.; Qin, J.-S.; Wang, Y.; Kirchon, A.; Qiu, D.; Cheng, L.; Madrahimov, S.T.; Zhou, H.-C. Lattice Expansion and Contraction in Metal-Organic Frameworks by Sequential Linker Reinstallation. *Matter* **2019**, *1*, 156–167. [CrossRef]

23. Lippke, J.; Brosent, B.; von Zons, T.; Virmani, E.; Lilienthal, S.; Preuße, T.; Hülsmann, M.; Schneider, A.M.; Wuttke, S.; Behrens, P.; et al. Expanding the Group of Porous Interpenetrated Zr-Organic Frameworks (PIZOFs) with Linkers of Different Lengths. *Inorg. Chem.* **2017**, *56*, 748–761. [CrossRef] [PubMed]
24. Schaate, A.; Roy, P.; Preuße, T.; Lohmeier, S.J.; Godt, A.; Behrens, P. Porous Interpenetrated Zirconium–Organic Frameworks (PIZOFs): A Chemically Versatile Family of Metal–Organic Frameworks. *Chem.-Eur. J.* **2011**, *17*, 9320–9325. [CrossRef] [PubMed]
25. Ferguson, A.; Liu, L.; Tapperwijn, S.J.; Perl, D.; Coudert, F.-X.; Van Cleuvenbergen, S.; Verbiest, T.; van der Veen, M.A.; Telfer, S.G. Controlled partial interpenetration in metal–organic frameworks. *Nat. Chem.* **2016**, *8*, 250–257. [CrossRef] [PubMed]

Sample Availability: Not available.

Article

Assessment of Computational Tools for Predicting Supramolecular Synthons

Bhupinder Sandhu, Ann McLean, Abhijeet S. Sinha, John Desper and Christer B. Aakeröy *

Department of Chemistry, Kansas State University, Manhattan, KS 66506, USA;
bhupinder.sandhu@bms.com (B.S.); amclean@sco.edu (A.M.); sinha@ksu.edu (A.S.S.); desperj65@gmail.com (J.D.)
* Correspondence: aakeroy@ksu.edu

Abstract: The ability to predict the most likely supramolecular synthons in a crystalline solid is a valuable starting point for subsequently predicting the full crystal structure of a molecule with multiple competing molecular recognition sites. Energy and informatics-based prediction models based on molecular electrostatic potentials (MEPs), hydrogen-bond energies (HBE), hydrogen-bond propensity (HBP), and hydrogen-bond coordination (HBC) were applied to the crystal structures of twelve pyrazole-based molecules. HBE, the most successful method, correctly predicted 100% of the experimentally observed primary intermolecular-interactions, followed by HBP (87.5%), and HBC = MEPs (62.5%). A further HBC analysis suggested a risk of synthon crossover and synthon polymorphism in molecules with multiple binding sites. These easy-to-use models (based on just 2-D chemical structure) can offer a valuable risk assessment of potential formulation challenges.

Keywords: hydrogen-bond propensity; hydrogen-bond coordination; supramolecular synthon; hydrogen-bond energies; Cambridge Structural Database; molecular electrostatic potential; pyrazoles

Citation: Sandhu, B.; McLean, A.; Sinha, A.S.; Desper, J.; Aakeröy, C.B. Assessment of Computational Tools for Predicting Supramolecular Synthons. *Chemistry* **2021**, *3*, 612–629. https://doi.org/10.3390/chemistry3020043

Academic Editors: Catherine Housecroft and Katharina M. Fromm

Received: 25 March 2021
Accepted: 25 April 2021
Published: 3 May 2021

Publisher's Note: MDPI stays neutral with regard to jurisdictional claims in published maps and institutional affiliations.

1. Introduction

A key question in crystal engineering is, given a molecular structure, can we predict its crystal structure [1]? At the core of the crystal structure of most organic molecular solids is the supramolecular synthon [2–5], a "structural unit within supermolecules which can be formed and/or assembled by known or conceivable synthetic operations involving intermolecular interactions", introduced by Desiraju in 1995 [2]. A reason for why this idea is so important to crystal engineering is the fact that detailed knowledge and control of intermolecular interactions is as vital to this field as is control of the covalent bond to molecular synthesis [6–18]. Furthermore, the synthon can serve as a valuable starting point for identifying the most likely ways in which molecules will aggregate [19,20]. This means that an important step towards predicting a crystal structure often involves finding the most likely synthons in molecules with competing molecular recognition sites, Scheme 1.

Many groups have proposed practical tools and avenues for a priori prediction of synthons in crystal structures of organic molecular solids. Some of these are based on electrostatics [21–23] and lattice energy calculations [24–26], with the focus purely on thermodynamic (enthalpic) factors [27]. For example, Hunter et al. converted maxima/minima on calculated molecular electrostatic potential surfaces into hydrogen-bond energies to determine synthon preference in multicomponent systems [28,29]. The preferred connectivity patterns of a molecule in the solid state in such cases is determined using Etter guidelines which states that the best hydrogen bond acceptor binds to the best donor [22]. The required ranking of best donor-acceptor pairs can be determined using molecular electrostatic potentials [21,23,30–42].

One way of accounting for kinetic factors that inevitably influence crystallization and assembly is to systematically analyze large swaths of structural information in the Cambridge Structural Database (CSD) [43]. Although an individual crystal structure does not provide information about kinetics of the seed formation and crystal growth, it is reasonable to assume that if a particular structural array is abundant in a database, it may reflect thermodynamic stability of a given crystal packing array, as well as a kinetic preference of its formation. Aakeröy, co-workers and others have used structural informatics tools such as hydrogen-bond propensity [44], and hydrogen-bond coordination [45] developed by the Cambridge Crystallographic Data Centre (CCDC) to understand the synthon predictions in the co-crystal formation [23,46–48]. These tools have considerable potential for validation purposes in both single component and multi-component systems. Yet, despite these efforts, there is still a need for more additional studies that systematically analyze and compare the abilities and reliability of different methods for predicting which synthon(s) is/are most likely to appear in crystal structures of high-value organic chemicals such as pharmaceuticals and agrochemicals.

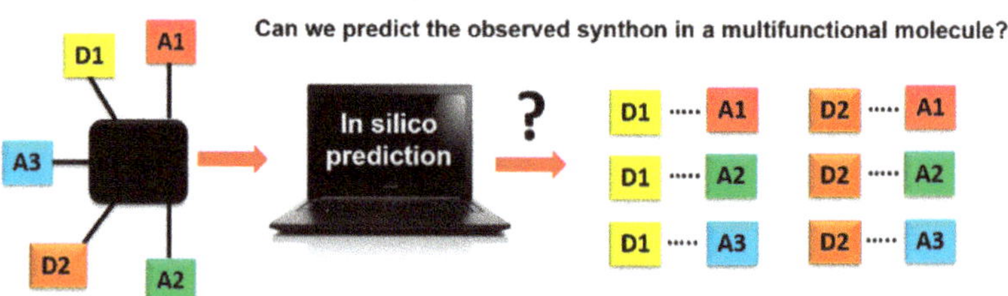

Scheme 1. Schematics of using an in-silico approach for predicting synthons in a multifunctional molecule. Hydrogen-bond donor, **D**, and hydrogen-bond acceptor, **A**.

In order to address this issue, we have employed several known methods for synthon prediction in order to map out the structural landscape of a family of relatively flexible pyrazole containing molecules (**P1–P12**) capable of forming specific and competing intermolecular interactions, Scheme 2. We selected a pyrazole backbone because many compounds comprising this chemical functionality are known to possess a wide range of biological activities such as anti-microbial, anti-fungal, anti-tubercular, anti-inflammatory, anti-convulsant, anticancer, anti-viral, and so on [49–57]. The pyrazole-amide functionality is also present in some pharmaceutical related compounds such as Entrectinib, Graniseton, and Epirizol as well as antifungal compounds such as Furametpyr, Penthiopyrad and Tolfempyrad [58]. Due to the presence of multifunctional groups, these molecules are always at risk of synthon polymorphism. Therefore, knowledge gained from a successful use of tools such as molecular electrostatic potentials, hydrogen-bond energies, hydrogen-bond propensity and hydrogen-bond coordination, for predicting synthon appearances could have significant practical applications.

The target molecules, **P1–P12** can be divided into two groups: Group 1 (**P1–P8**) includes molecules with two hydrogen-bond (HB) donors (pyrazole NH and amide NH) and two HB acceptors (pyrazole N and C=O of amide), leading to four possible/likely synthons; A-D, Scheme 3. It can be assumed that each molecule forms a combination of two synthons to satisfy all hydrogen-bond donors and acceptors leading to two possible synthon combinations; (A + D) or (B + C)).

Group 2 (**P9–P12**) comprises molecules with two HB donors (pyrazole NH and amide NH) and three HB acceptors (pyrazole N, carbonyl C=O and pyridine N) groups. Two additional synthons can therefore be postulated in this group, E and F, leaving a total of six possibilities, Scheme 3.

The overall outline for testing different predictive methods against experimental data is summarized in Scheme 4. We used four protocols for determining the most likely synthon in the crystal structures of **P1–P12**; molecular electrostatic potentials (MEPs), hydrogen-bond energies (HBE), hydrogen-bond propensities (HBP) and hydrogen-bond coordination (HBC). The prediction from each method was then compared to experimental data. The overall goal was to identify which approach is likely to deliver robust and transferable guidelines for estimating/predicting which hydrogen bonds are most likely to appear in molecular solids when there are numerous potential avenues for assembly.

Scheme 2. Target molecules (P1–P12) employed in this study.

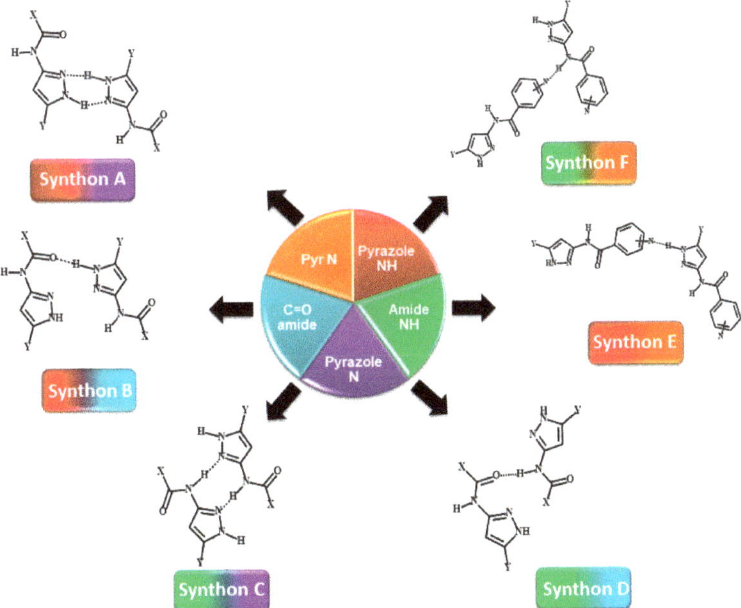

Scheme 3. Representations of six postulated synthons in a generic pyrazole-amide; A, N-H (pyrazole) . . . N (pyrazole); B, N-H (pyrazole) . . . C=O (amide); C, N-H (amide) . . . N (pyrazole); D, N-H (amide) . . . C=O (amide); E, NH (pyrazole) . . . N (aromatic) and F, NH (amide) . . . N (aromatic). (Y = H, CH$_3$; X = methyl, ethyl, or benzyl).

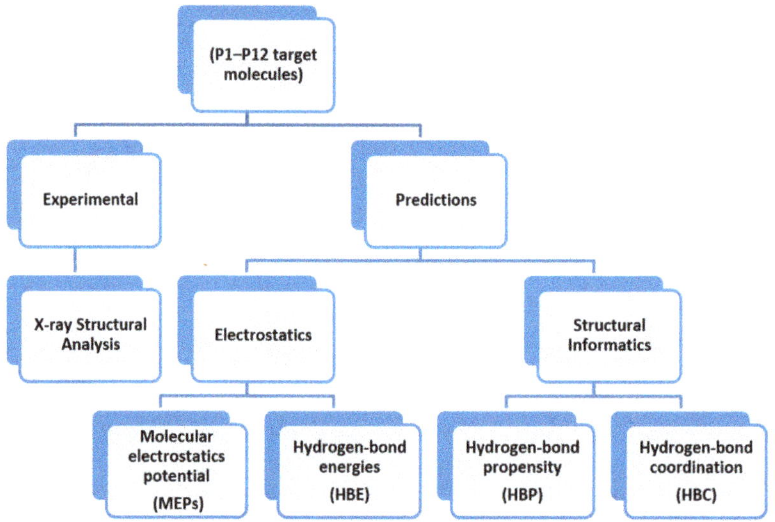

Scheme 4. The road map for synthon predictions.

The study was undertaken to address the following questions,

- Which method is more suitable for predicting the synthon outcome in the crystal structures of **P1–P12**?
- Is a combination of prediction methods better than individual methods?
- Which synthon is most optimal in group 1 (**P1–P8**) and how does adding an acceptor group affect the choice of synthon in **P9–P12**?
- Which molecules present the larger risk of synthon polymorphism, and which method is most suitable for predicting synthon polymorphism in this group of molecules?

2. Materials and Methods

2-Amino-pyrazole, 2-amino-5-methyl-pyrazole, acetic anhydride, propionic anhydride and benzoyl chloride were purchased from Sigma Aldrich (Milwaukee, WI, USA) and utilized without further purification. Synthetic procedures and characterization of all molecules are provided in the Supporting Information (SI). Melting points were measured using Fisher-Johns melting point apparatus. ^{1}H NMR data were collected on a Varian Unity plus 400 MHz spectrophotometer in DMSO.

Additionally, Cambridge Structural Database (Version 5.38 and Mercury 3.9) based structural informatics tools such as hydrogen-bond propensity (HBP), and hydrogen-bond coordination (HBC) were used [59–62] to predict the synthons in the crystal structures of **P1–P12**.

2.1. Molecular Electrostatic Potentials (MEPs)

Molecular electrostatic potential surfaces of **P1–P12** were generated with DFT B3LYP level of theory using 6-311++G** basis set in vacuum. All calculations were carried out using Spartan'08 software (Wavefunction, Inc., Irvine, CA, USA) [63]. All molecules were geometry optimized with the maxima and minima in the electrostatic potential surface (0.002 e/au isosurface) determined using positive point charge in the vacuum as a probe. The numbers indicate the Coulombic interaction energy (kJ/mol) between the positive point probe and the surface of the molecule at that particular point.

2.2. Hydrogen-Bond Energies (HBE)

The hydrogen-bond energies (HBE) were calculated using molecular electrostatic potentials (MEPs) combined with Hunter's parameters [28,29]. The hydrogen-bond parameters, α (hydrogen-bond donor) and β (hydrogen-bond acceptor) were determined using maxima and minima on the MEPS, respectively (Equations (1) and (2)), and the free energy of interaction is given by the product, $-\alpha\beta$ [28]. Only conventional hydrogen-bonds donors (pyrazole N-H, O-H) and acceptors (O=C, pyridine, pyrazole N) were included. In the HBE approach, both single-point interactions and two-point interactions (dimeric synthons) were considered.

$$\alpha = 0.0000162\, MEPmax^2 + 0.00962 MEP_{max} \tag{1}$$

$$\beta = 0.000146\, MEPmin^2 - 0.00930 MEP_{min} \tag{2}$$

$$E = -\sum_{ij} \alpha_i \beta_j \tag{3}$$

2.3. Hydrogen-Bond Propensity (HBP)

The hydrogen-bond propensity (HBP) is the probability of formation of an interaction based on defined functional groups and fitting data [44,46]. Possible values fall in the range of zero to one, where a value closer to zero indicates less likely occurrence and a value closer to one indicates a higher likely occurrence of an interaction in the given molecule.

2.4. Hydrogen-Bond Coordination (HBC)

The hydrogen-bond coordination (HBC) is the probability of observing a coordination number for any given hydrogen-bond donor/acceptor atom [45]. The coordination number (CN) is defined as the number of intermolecular hydrogen-bonds formed between a donor and an acceptor. =0, =1, =2, =3 denotes the number of times a functional group donates or accepts. The CN with the highest probability corresponds to the most optimal (likely) hydrogen-bond interaction. The numbers that are colored relate to the outcome present in the selected H-bonding network, if this is green it indicates that the outcome is optimal, whereas if it's red that indicates the outcome is sub-optimal.

3. Results

3.1. Molecular Electrostatics Potentials

MEPs values for each hydrogen-bond donor (pyrazole NH and amide NH) and hydrogen-bond acceptor (pyrazole N, C=O (amide)) for **P1–P8** (and additional pyridine N for **P9–P12**) are presented in Figures 1 and 2, respectively. In group 1 (**P1–P8**), pyrazole NH is the best donor and C=O (amide) is the best acceptor (as ranked by electrostatic potentials) followed by amide NH and pyrazole N, therefore we can postulate that B + C is the most likely synthon. In group 2 (**P9–P12**), pyrazole NH is the best donor and pyridine N is the best acceptor followed by amide NH and C=O (amide), therefore D + E is the most likely synthon.

3.2. Hydrogen-Bond Energies (HBE)

Hydrogen-bond energies for each combination of synthons is presented in Table 1, See Supporting Information (SI) for additional details. Based on hydrogen-bond energies, in **P1–P8**, both A + D and B + C have very similar energies, and thus they can be expected to have equal chances of appearing. Based on a similar analysis, in **P9–P12,** A + F, A + D, and C + E, are equally possible.

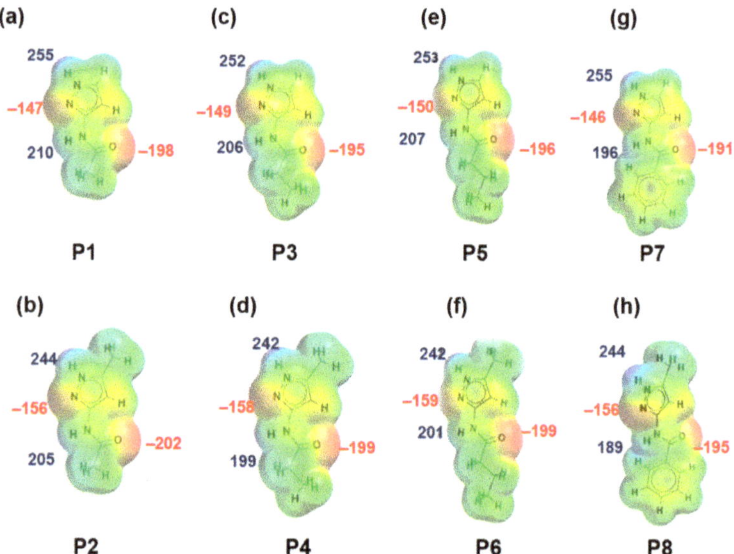

Figure 1. Electrostatic potential values (in kJ/mol); (**a**) **P1**, (**b**) **P2**, (**c**) **P3**, (**d**) **P4**, (**e**) **P5**, (**f**) **P6**, (**g**) **P7** and (**h**) **P8**. Red indicates MEPs on hydrogen-bond acceptors and blue indicates MEPs on hydrogen-bond donors.

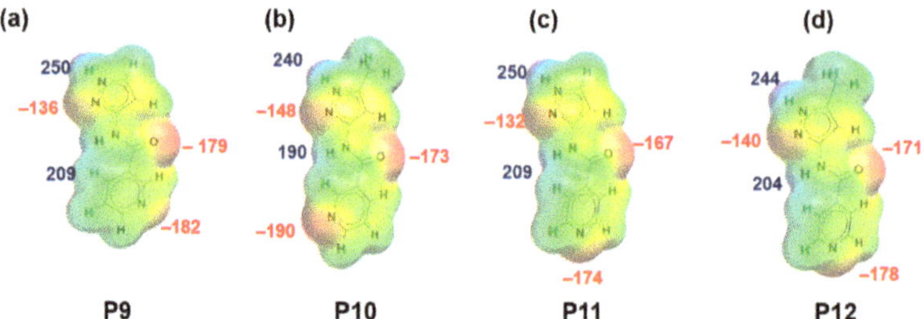

Figure 2. Electrostatic potential values (in kJ/mol); (**a**) **P9**, (**b**) **P10**, (**c**) **P11** and (**d**) **P12**. Red indicates MEPs on hydrogen-bond acceptors and blue indicates MEPs on hydrogen-bond donors.

Table 1. Hydrogen-bond energies (in kJ/mol) for each combination synthon for molecules **P1–P12**. Green shade indicates the most likely predicted synthons by this method.

	(A +() F)	(A + D)	(C + E)	(B + C)	(D + E)	(B + F)
AVG (P1–P8)	N/A	−52 ± 2	N/A	−50 ± 2	N/A	N/A
AVG (P9–P12)	−44 ± 2	−43 ± 1	−43 ± 2	−41 ± 1	−37 ± 2	−37 ± 2

3.3. Hydrogen-Bond Propensities (HBP)

The propensities calculations consider all possible interactions between two donors (pyrazole NH and amide NH) and two acceptors (pyrazole N and C=O) resulting in four propensity numbers for **P1–P8**. In molecules with an additional acceptor, **P9–P12**, six different combinations of propensities are obtained. The propensities of individual (see Supporting Information (SI) for details) and combination synthon are presented in Table 2. Based on combination approach in HBP, the most likely synthons to appear in **P1–P8** are A + D and B + C and for **P9–P12**, synthons A + F and B + F.

Table 2. Hydrogen-bond propensities (probability of interaction between a hydrogen-bond acceptor and donor) for combination synthons in molecules **P1–P12**. Combination synthon propensities are calculated by multiplying the individual synthon propensities. Green shade indicates the most likely predicted synthons by this method.

	Synthon (A + F)	Synthon (A + D)	Synthon (C + E)	Synthon (B + C)	Synthon (D + E)	Synthon (B + F)
P1	N/A	0.35	N/A	0.34	N/A	N/A
P2	N/A	0.35	N/A	0.30	N/A	N/A
P3	N/A	0.39	N/A	0.36	N/A	N/A
P4	N/A	0.39	N/A	0.37	N/A	N/A
P5	N/A	0.33	N/A	0.31	N/A	N/A
P6	N/A	0.35	N/A	0.32	N/A	N/A
P7	N/A	0.18	N/A	0.19	N/A	N/A
P8	N/A	0.16	N/A	0.17	N/A	N/A
P9	0.23	0.13	0.19	0.13	0.19	0.23
P10	0.24	0.12	0.19	0.13	0.18	0.23
P11	0.21	0.12	0.17	0.12	0.19	0.23
P12	0.22	0.12	0.18	0.12	0.19	0.23

3.4. Hydrogen-Bond Coordination (HBC)

HBC was calculated for each molecule, and the highest donor Coordination Number (CN) was matched with the highest acceptor CN in each molecule to determine the most likely synthon, Figure 3. Synthon (B + C) in **P1–P6**, (B + C) and (A + D) in **P7–P8**, (B + F) and (D + E) in **P9–P12** were predicted to be the most likely synthon.

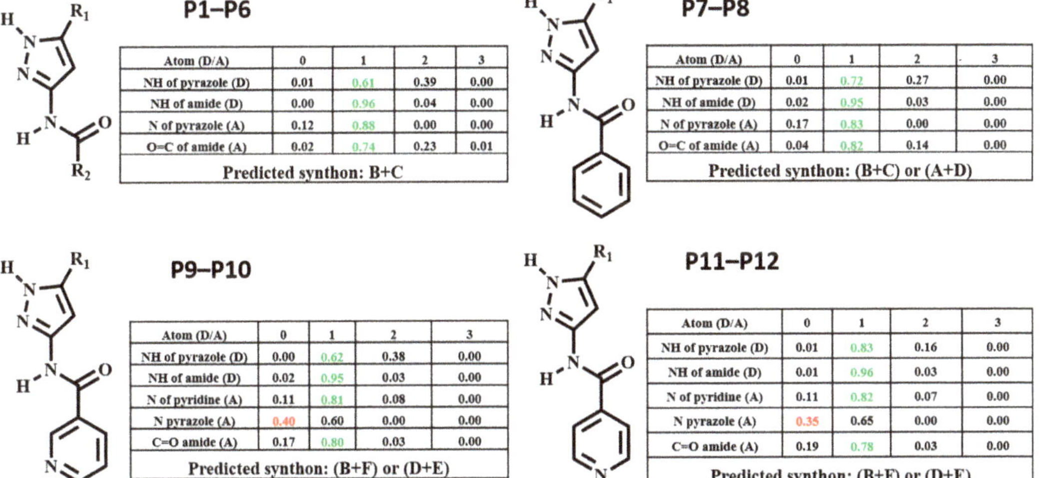

P1–P6

Atom (D/A)	0	1	2	3
NH of pyrazole (D)	0.01	0.61	0.39	0.00
NH of amide (D)	0.00	0.96	0.04	0.00
N of pyrazole (A)	0.12	0.88	0.00	0.00
O=C of amide (A)	0.02	0.74	0.23	0.01
Predicted synthon: B+C				

P7–P8

Atom (D/A)	0	1	2	3
NH of pyrazole (D)	0.01	0.72	0.27	0.00
NH of amide (D)	0.02	0.95	0.03	0.00
N of pyrazole (A)	0.17	0.83	0.00	0.00
O=C of amide (A)	0.04	0.82	0.14	0.00
Predicted synthon: (B+C) or (A+D)				

P9–P10

Atom (D/A)	0	1	2	3
NH of pyrazole (D)	0.00	0.62	0.38	0.00
NH of amide (D)	0.02	0.95	0.03	0.00
N of pyridine (A)	0.11	0.81	0.08	0.00
N pyrazole (A)	0.40	0.60	0.00	0.00
C=O amide (A)	0.17	0.80	0.03	0.00
Predicted synthon: (B+F) or (D+E)				

P11–P12

Atom (D/A)	0	1	2	3
NH of pyrazole (D)	0.01	0.83	0.16	0.00
NH of amide (D)	0.01	0.96	0.03	0.00
N of pyridine (A)	0.11	0.82	0.07	0.00
N pyrazole (A)	0.35	0.65	0.00	0.00
C=O amide (A)	0.19	0.78	0.03	0.00
Predicted synthon: (B+F) or (D+E)				

Figure 3. Predicted hydrogen-bond coordination **P1–P12** (D: hydrogen-bond donor, A: hydrogen-bond acceptor; green: optimal interactions, red: sub-optimal interactions).

3.5. Experimentally Observed Crystal Structures

Suitable crystals of **P1, P2, P3, P4, P7, P8, P10 and P11** were obtained by slow solvent evaporation method using methanol solvent. Few other solvents such as ethanol, THF, ethyl acetate etc. were tried to grow crystals for P5, P6, P9 and P12 with no positive hit. We obtained crystallographic data for eight (**P1, P2, P3, P4, P7, P8, P10 and P11**) out of twelve molecules. Our re-determination of the structure of **P2** was consistent with the reported structure in the CSD (ARAGUV) [64] and the crystal structure of **P8** is also reported in the CSD (PESQEK) [65], Figures 4 and 5.

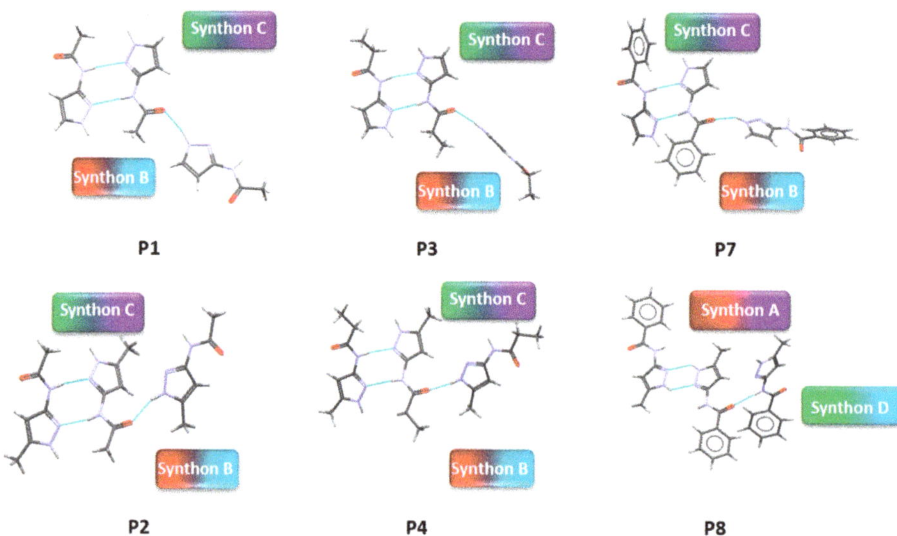

Figure 4. Crystal structures showing synthons (B + C) observed in **P1**, **P2**, **P3**, **P4**, **P7** and (A + D) observed in **P8**.

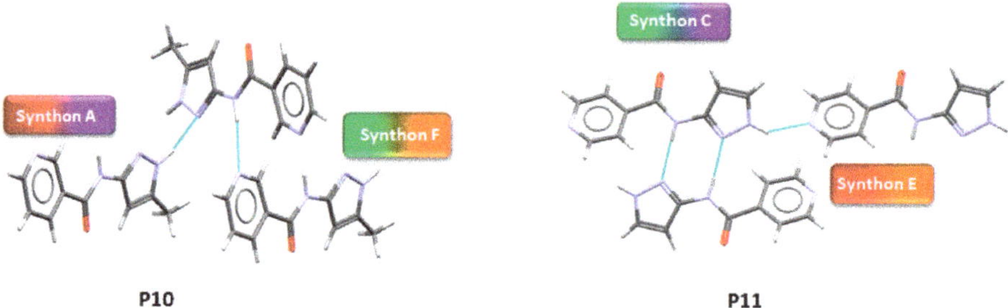

Figure 5. Crystal structures showing synthon (A + F) observed in **P10** and (C + E) in **P11**.

In group 1, six out of eight crystal structures (**P1–P4**, **P7** and **P8**) were obtained. In five of the six, B + C was observed (**P1–P4** and **P7**) and synthon (A + D) was found in **P8**. In group 2, when an extra acceptor group was added to the benzyl group, two crystal structures were obtained (**P10** and **P11**). Synthon (A + F) is observed in **P10** and (C + E) was observed in **P11**.

4. Discussion

4.1. Molecular Conformational Analysis

Molecular conformational analysis using DFT B3LYP (6-311++G** basis set in vacuum) shows that **P1–P12** with amide functionality occur as trans instead of cis isomer as the stable conformation, which was further confirmed based on a CSD search (Scheme 5, See Supporting Information (SI) for details). When a pyrazole group is added to the amide functionality, the most stable conformation is when pyrazole functional groups are cis to the amide NH group. When meta-substituted pyridine is added to this group, the more stable conformation is when pyridine N is trans to the amide NH group. However, the second conformation is only ~4 kJ/mol higher in energy and therefore is observed in **P10**. It is worth noting that the energy optimized conformations are not necessarily completely identical to those that may appear in the solid state, where a variety of close contacts and packing forces may distort some geometric parameters away from idealized gas phase values. However, these idealized conformations are likely to be most relevant in the solution phase at the point when target molecules begin to recognize and bind to each other during nucleation and growth (Scheme 5, See Supporting Information (SI) for additional details).

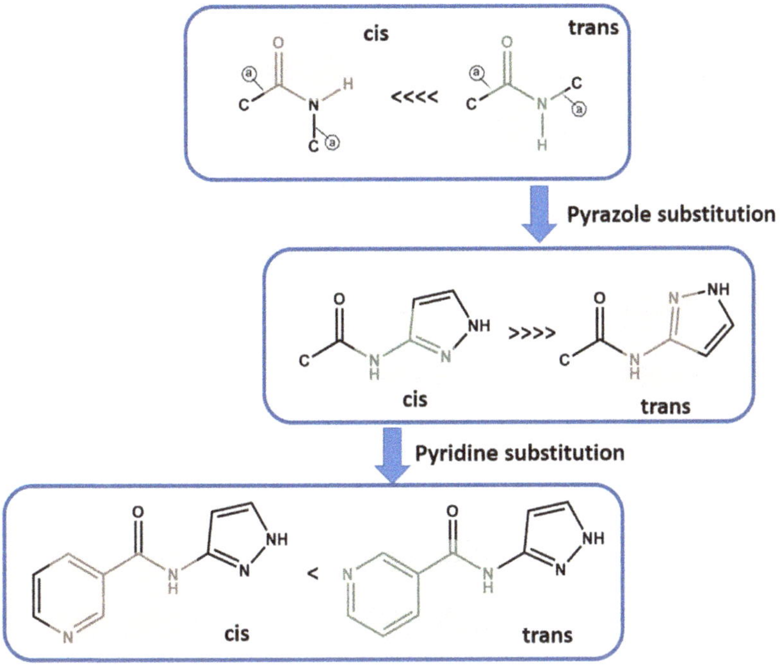

Scheme 5. Molecular conformation analysis of **P1–P12** using DFT B3LYP (6-311++G** basis set in vacuum). Green highlights the more stable conformation. Green highlights the likely conformation whereas red highlight the less likely conformation. ⓐ indicates acyclic moiety.

4.2. Prediction Analysis

Four different methods (MEPs, HBE, HBP, and HBC) were used to predict synthons in group 1 and group 2 molecules.

In group 1 (**P1–P8**), Synthon (B + C) was predicted by all four methods, however, synthon (A + D) was predicted by HBE and HBP method. In group 2 (**P9–P12**), A total of six synthons (A, B, C, D, E and F) and six different combination possibilities (A + F, A + D, C + E, B + C, D + E and B + F), made it complex to predict such synthons. In group 2, (D + E) was predicted by MEPs and HBC. (B + F) was predicted by HBP and HBC. Synthon (A + F) was predicted by both HBE and HBP. (C + E) and (A + D) were predicted by HBE only (Figure 6).

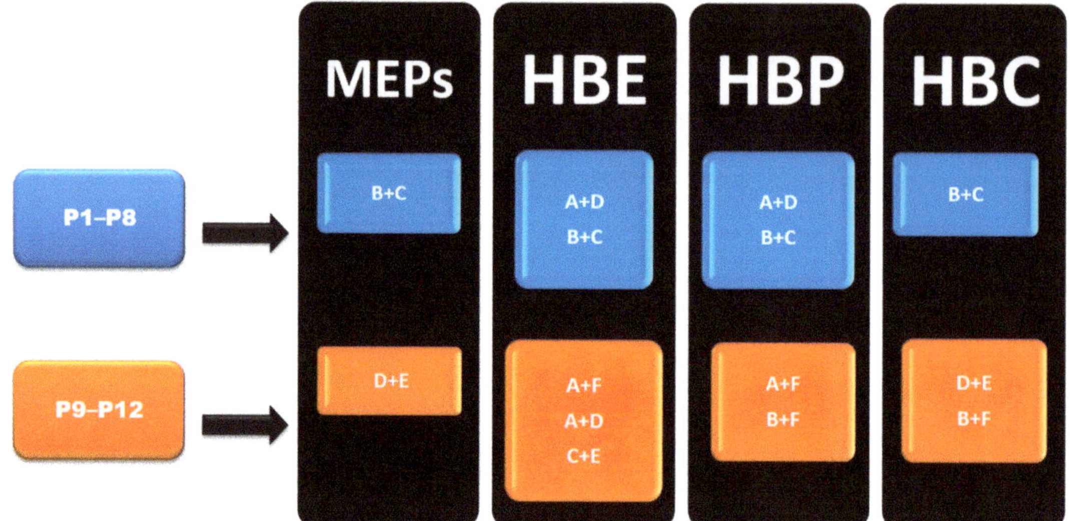

Figure 6. Summary of synthons predicted in **P1–P12**.

4.3. CSD Search-Molecular Geometric Complementarity

The molecular complementarity approach was used to determine whether synthon A or C is more likely to occur in pyrazole molecules with more than one binding site. The fragments used for CSD and resulting bond angles are listed in Figure 7. Based on this limited dataset, NH(amide) . . . .N(pyrazole) interaction is more linear compared to NH(pyrazole....N(pyrazole) indicating that synthon C has a geometric preference over synthon A. Moreover, a search performed with similar pyrazole binding pockets in the CSD gave 12 hits and in every case, synthon C was preferred over A.

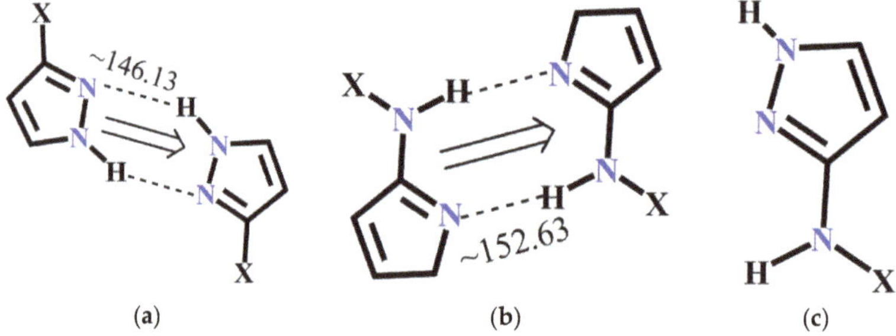

Figure 7. Pyrazole fragments used to perform CSD molecular complementarity search (**a**) Average bond angle for A (22 structures in CSD); (**b**) average bond angle for B (35 structures in CSD) and (**c**) pyrazole fragment for CSD search.

4.4. Validation Analysis

The comparison between the predictions from MEPs, HBE, HBP and HBC, and experimentally observed results is given in Table 3. In **P1–P8** with two donors and two acceptor sites, B + C was predicted to be the most likely synthon by all four methods which was also observed experimentally in five of the six compounds, **P1–P4**, and **P7**. In **P8**, A + D was present which was predicted by HBE and HBP as a possible synthon. By increasing the number of acceptor choices to three in **P9–P12** the challenge of getting the correct answer increased due to the enhanced possibility of synthon crossover and synthon polymorphism.

In **P10**, A + F was observed but it was only predicted correctly by HBE and HBP. In **P11**, C + E was observed experimentally, and this was only predicted correctly by HBE.

Table 3. Experimental vs. predicted synthons in the crystal structures of **P1–P14, P7–P8** and **P10-P11**.

	Experimental	MEPs	HBE	HBP	HBC
Group 1					
P1	Synthon (B + C)	Yes	Yes	Yes	Yes
P2	Synthon (B + C)	Yes	Yes	Yes	Yes
P3	Synthon (B + C)	Yes	Yes	Yes	Yes
P4	Synthon (B + C)	Yes	Yes	Yes	Yes
P7	Synthon (B + C)	Yes	Yes	Yes	Yes
P8	Synthon (A + D)	No	Yes	Yes	No
Group 2					
P10	Synthon (A + F)	No	Yes	Yes	No
P11	Synthon (C + E)	No	Yes	No	No
	Overall	62.5%	100%	87.5%	62.5%

4.5. Polymorph Assessment of Experimental Structures

A hydrogen-bond likelihood analysis was performed to understand the risk of synthon polymorphism in these molecules with multiple binding sites. Experimental structure was imported into the predicted hydrogen-bond coordination table to compare it against the predicted structures. Hypothetical structures in this tool are generated based on the combination of HBP and HBC parameters for each molecule. A correlation of the HB propensity vs. the mean HB coordination for all putative synthons possible in the structure of a given molecule is plotted. The most likely synthons should be found in the lower right corner of the plot. In **P1** (which is also a representative of **P2**, **P3**, **P4**, and **P7**), the experimental structure matched with the most likely synthon predicted by combined HBP an HBC. In **P8**, **P10** and **P11**, the experimental structures do not contain the most likely synthon predicted by a combination of HBP and HBC, highlighting the possibility of other structures with more optimal hydrogen-bond patterns, which indicates a reasonable risk of polymorphism in these three compounds. In **P8**, B + C was predicted by all four methods to be the most likely hydrogen bond, yet A + D was observed experimentally. Interestingly, this structure leads to a polymeric structure as reported by Daidone et al. [65], Figure 8.

In **P10**, multiple synthons were predicted by different methods, such as D + E, A + F, A + D, C + E and B + F. However A + F is in fact observed experimentally (and accurately predicted by HBE and HBP methods). These results indicate that even though A + F is geometrically constrained, as seen in CSD search, it can still form experimentally. Due to steric hindrance with meta-substituted pyridine, synthon A is observed as single-point interaction instead of dimeric synthon, resulting in a herringbone type arrangement, Figure 9.

In **P11**, experimentally observed synthon C + E was predicted correctly only by the HBE method, in addition, it was not deemed to be the most likely interaction by the hydrogen-bond likelihood analysis. This is less surprising as synthon C is the most likely synthon among all possibilities due to its linearity and dimeric chelate effect, leaving E as an alternate option because the pyridine nitrogen atom is a better acceptor than the C=O moiety, Figure 10.

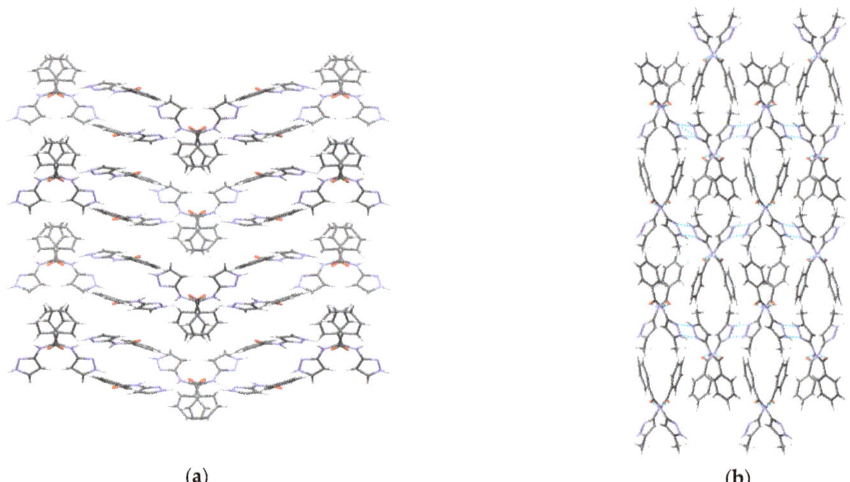

Figure 8. Hydrogen-bonding observed in the crystal structure of (**a**) **P7** and (**b**) **P8** along the crystallographic *a*-axis.

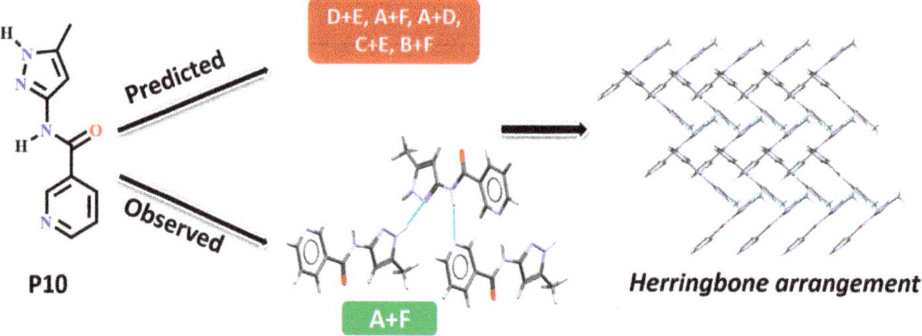

Figure 9. Synthon and packing observed in the crystal structure of **P10** leading to herringbone-like structure.

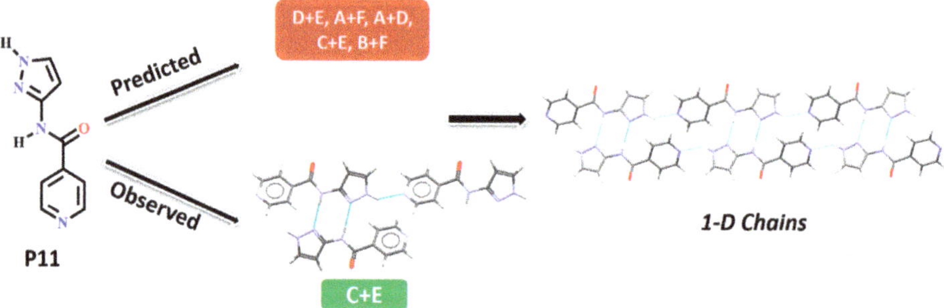

Figure 10. Synthon and packing observed in the crystal structure of **P11** leading to 1-D chains along the crystallographic *a*-axis.

Additionally, molecules for which crystal structure was known were screened through hydrogen-bond coordination likelihood analysis to understand the risk of synthon polymorphism. This tool suggested that three out of eight molecules (**P8**, **P10**, **P11**), where crystal structure is known, have a chance of synthon crossover and synthon polymorphism,

Figure 11. This tool is particularly useful as it can guide chemists to narrow down list of APIs that are at risk of delivering new structural polymorphs, which can cause havoc in late-stage formulation efforts.

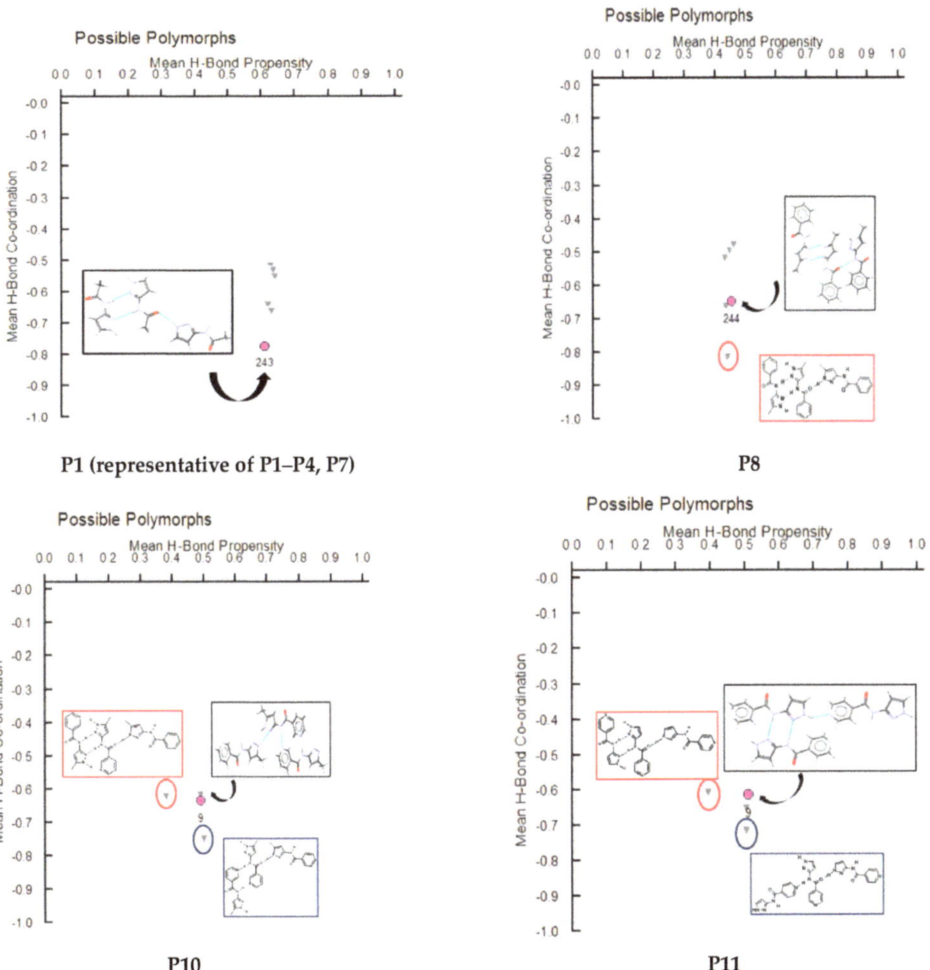

Figure 11. Polymorph assessment of **P1**, **P8**, **P10** and **P11** molecules showing the experimental structure and more stable hypothetical structure.

This work also highlights that small changes to a molecule can lead to profound changes in the hydrogen bonding, resulting in different packing arrangements in the crystal structure, Figure 12. For example, adding an electron donating methyl group to the pyrazole ring in **P7** changes the synthon from (B + C) in **P7** to (A + D) in **P8**. Comparing the electrostatic potential of **P7** and **P8**, adding a methyl group decreases the charges on both H-bond donors (amide NH and pyrazole NH) whereas increase the charges on the H-bond acceptor (pyrazole N), highlighting these two molecules as a classic example of synthon crossover. The addition of pyridine (para or meta) group changes the ranking of H-bond acceptors in **P10** and **P11**. Therefore, pyrazole NH and amide NH binds to pyridine N in **P11** and **P10** instead of C=O (as was observed in **P7** and **P8**) respectively. The position of a substituent (meta vs. para) as well as addition of methyl group on the pyrazole ring also leads to different crystal packing. For example, changing the pyridine N from meta to para position and adding methyl group on the pyrazole ring changes the synthon from (A + F) in **P10** to (C + E) in **P11**, resulting in varying packing arrangements.

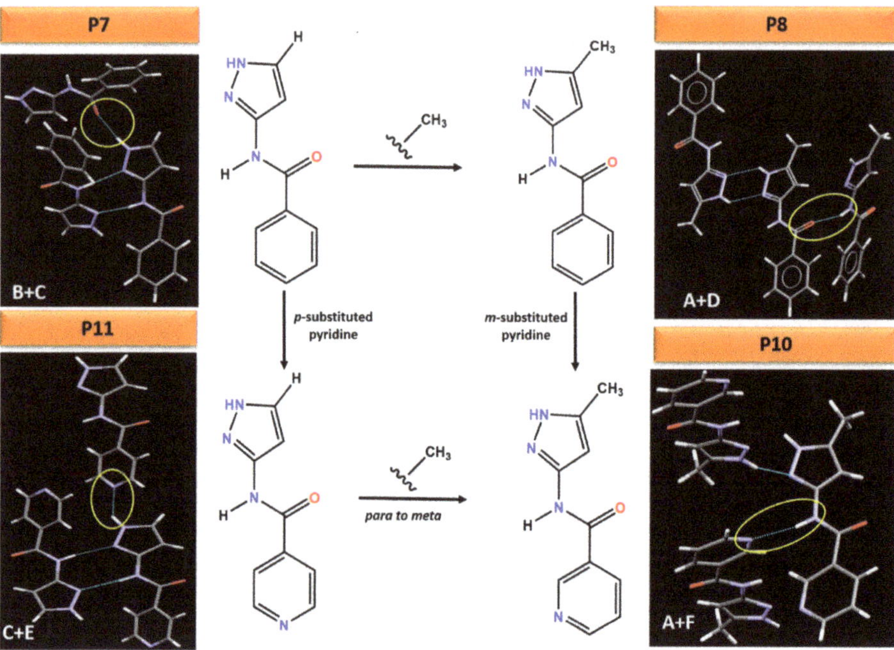

Figure 12. Comparison of subtle variations on the molecular fragments leading to different hydrogen bonding interactions in crystal structures of **P7**, **P8**, **P10** and **P11**.

Another key observation is the ability of hydrogen-bond propensity model to incorporate effect of aromaticity on the probability of hydrogen–bond interaction. For example, presence of aromatic groups such as phenyl and pyridine rings in molecules **P8–P12** leads to higher values of donor and acceptor aromaticity (0.56) in the logistic regression model compared to molecules **P1–P6** (0.27). This further impacts the final probability values where **P1–P6** has higher HBP values for synthon B and D compared to **P7** and **P8**, see Table 2 and Table S5 in Supporting Information for more details.

This study didn't include extensive crystallization experiments to grow suitable crystals for **P5**, **P6**, **P9** and **P12** but it is possible that with further crystallization work, crystals of these molecules can be achieved. One of the reasons behind why **P5** and **P6** might be difficult to crystallize compared to **P1–P4** is due to the presence of butyramide side chain which increases the number of rotatable bonds and hence number of conformations in the solution state. The possibility of multiple conformations very close in energy to each other can cause difficulties in crystallizing molecules by hindering the nucleation and crystal growth phase. Additionally, further crystallization work can also be done for molecules with risk of synthon polymorphism such as **P8**, **P10** and **P11** to validate the computational studies.

In this study, we examined two energy based, and two informatics-based protocols for predicting hydrogen—bond based synthons in crystal structures of a family of pyrazoles. The overall outcome gave the following ranking in terms pf predictive quality: HBE (100%) > HBP (87.5%) > HBC = MEPs (62.5%), Figure 13. Even though HBE is a step forward of MEPs method, it worked better than the latter in pyrazole based molecules because of inclusion of energies of dimeric interactions due to a chelating effect. It is important to note that even though pyrazole molecules studied in this work with high predictive success rate have with very limited structural diversity (<250 gmol^{-1} molecular weight, <3 rotatable bonds, limited functional groups), such models can be applicable to diverse and more flexible drug-like molecules with similar functional groups for prediction analysis. We encourage scientific community to test these CCDC provided tools on actual drug molecules to further understand their applicability and range of predictive ability of such tools.

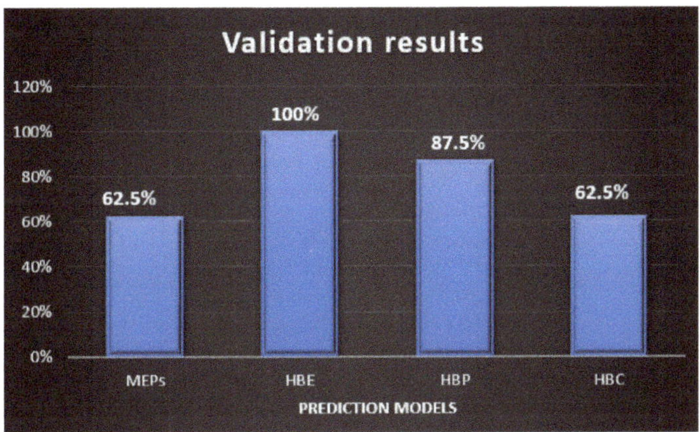

Figure 13. Validation results of supermolecular synthons observed in the pyrazole analogues using four different prediction models.

5. Conclusions

One useful consequence of predicting a correct supramolecular synthon is the ability to predict the right crystal structure. Four prediction models based on molecular electrostatic potentials (MEPs), hydrogen-bond energies (HBE), hydrogen-bond propensity (HBP) and hydrogen-bond coordination (HBC) were studied for their ability to predict synthons in small pyrazole based targets, **P1–P12**. Molecules were grouped into categories based on number of conventional hydrogen-bond donors and acceptors, group 1 (**P1–P8**) and group 2 (**P9–P12**). In group 1, both HBE and HBP gave 100% success rate as predicted synthon matched with the experimental results for molecules **P1–P4** and **P7**. However, when a strong acceptor group such as pyridine nitrogen was added as in **P9–P12**, synthon prediction became complex, so did the experimentally observed synthons. Only two crystal structures (**P10** and **P11**) were obtained experimentally. HBE predicted the synthons correctly for both molecules whereas HBP predicted it correctly for **P10**. Additionally, hydrogen-bond coordination likelihood analysis suggested that **P8**, **P10** and **P11** are at risk of synthon crossover and synthon polymorphism based on the net putative interaction likelihoods. Methodologies used in this study are a valuable tool to determine which synthon is likely to form in the crystal structure of a molecule and if the molecule is at a risk of synthon polymorphism. Therefore, a simple health check on these molecules using structural informatics tools such as MEPs, HBE, HBP and HBC for mapping out the structural landscape of these types of molecules will have significant practical applications in various fields.

Supplementary Materials: The following are available online at https://www.mdpi.com/article/10.3390/chemistry3020043/s1, Figure S1. (a) *Cis* and *trans* amide functionality (both bonds are acyclic representing using symbol @) used to perform the torsion angle search. (b) Pie chart indicating number of structures with torsions for *cis* (yellow, ~32 structures, 0.5%) and *trans* (red, ~6303 Structures, 99.5%) conformations; Figure S2. (a)Propensity-coordination chart of **P1-P6** molecules and (b) coordination of each functional group in all predicted motifs; Figure S3. Propensity-coordination chart of **P7-P8** molecules and the coordination of each functional group in all predicted motifs; Figure S4. Propensity-coordination chart of **P9-P10** molecules and the coordination of each functional group in all predicted motifs; Figure S5. Propensity-coordination chart of **P11-P12** molecules and the coordination of each functional group in all predicted motifs; Figure S6. ^{1}H NMR of 3-acetamido-1H-pyrazole, **P1**; Figure S7. ^{1}H NMR of 3-acetamido-5methyl-1H-pyrazole, **P2**.; Figure S8. ^{1}H NMR of 3-propamido-1H-pyrazole, **P3**; Figure S9. ^{1}H NMR of 3-propamido-5methyl-1H-pyrazole, **P4**; Figure S10. 1H NMR spectrum of 3-butyramido pyrazole, **P5**; Figure S11. NMR spectrum of 3-butyramido 5-methyl pyrazole, **P6**; Figure S12. ^{1}H NMR of 3-benzamido-1H-pyrazole, **P7**; Figure S13. ^{1}H NMR of 3-benzamido-5methyl-1H-pyrazole, **P8**; Figure S14. ^{1}H NMR of N-(pyrazole-2-yl)nicotinamide, **P9**; Figure S15. NMR spectrum of N-(5-methylpyrazol-2-yl)nicotinamide, **P10**; Figure S16. NMR spectrum of N-(pyrazole-2-yl)isonicotinamide, **P11**; Figure S17. NMR spectrum of N-(5-methylpyrazol-2-yl)isonicotinamide, **P12**; Table S1. Energies of each *trans* amide conformation relative to most stable *trans* conformation is shown below in kJ/mol. The conformations with duplicate energies were ignored. Note: methyl-based target molecule conformations are not shown here; Table S2. Hydrogen-bond energies (in kJ/mol) for each individual synthon for molecules **P1-P12**. Synthon A and C are dimeric synthons; therefore, energies are presented for pairs of molecules; Table S3. Hydrogen-bond energies (in kJ/mol) for each combination synthon for molecules **P1-P12**; Table S4. Functional groups used to determine the hydrogen-bond propensities for the **P1-P12** target molecules. The labels in the figures can be explained as follows: Tn = atom makes n bonds, c = atom is cyclic, ⓔ = bond is acyclic, and Hn = n bonded hydrogen atoms; Table S5. Hydrogen-bond propensities for each individual synthon possible in molecules **P1-P12**; Table S6. Hydrogen-bond propensities for combination synthons possible in molecules **P1-P12**. Combination synthon propensities are calculated by multiplying the individual synthon propensities; Table S7. Experimental details of crystals obtained in this study.

Author Contributions: B.S. and C.B.A. conceived and designed the experiments; B.S. and A.M. performed the experiments; A.S.S. and J.D. performed the single-crystal X-ray crystallography; B.S. and C.B.A. analyzed the data and wrote the paper. All authors have read and agreed to the published version of the manuscript.

Funding: This research was funded by the Johnson Cancer Research Center, Kansas State University.

Data Availability Statement: Details of chemical synthesis, calculations, and NMR and IR spectra are available online. The crystallographic data for this paper (CCDC 2072631-2072636) can be obtained free of charge via www.ccdc.cam.ac.uk/data_request/cif, or by emailing data_request@ccdc.cam.ac.uk, or by contacting The Cambridge Crystallographic Data Centre, 12 Union Road, Cambridge CB2 1EZ, UK; Fax: +44-1223-336033.

Conflicts of Interest: The authors declare no conflict of interest. The funders had no role in the design of the study; in the collection, analyses, or interpretation of data; in the writing of the manuscript, or in the decision to publish the results.

References

1. Price, S.L. Predicting crystal structures of organic compounds. *Chem. Soc. Rev.* **2014**, *43*, 2098–2111. [CrossRef]
2. Desiraju, G.R. Supramolecular Synthons in Crystal Engineering—A New Organic Synthesis. *Angew. Chem. Int. Ed. Engl.* **1995**, *34*, 2311–2327. [CrossRef]
3. Fedyanin, I.V.; Karnoukhova, V.A.; Lyssenko, K.A. Conformational analysis of a supramolecular synthon: A case study of 8-hydroxyquinoline. *CrystEngComm* **2018**, *20*, 652–660. [CrossRef]
4. Thalladi, V.R.; Goud, B.S.; Hoy, V.J.; Allen, F.H.; Howard, J.A.K.; Desiraju, G.R. Supramolecular synthons in crystal engineering. Structure simplification, synthon robustness and supramolecular retrosynthesis. *Chem. Commun.* **1996**, *3*, 401–402. [CrossRef]
5. Nangia, A.; Desiraju, G.R. Supramolecular Synthons and Pattern Recognition. In *Design of Organic Solids*; Weber, E., Aoyama, Y., Caira, M.R., Desiraju, G.R., Glusker, J.P., Hamilton, A.D., Meléndez, R.E., Nangia, A., Eds.; Springer: Berlin/Heidelberg, Germany, 1998; pp. 57–95. [CrossRef]
6. Hofmann, D.W.M.; Kuleshova, L.N.; Antipin, M.Y. Supramolecular Synthons and Crystal Structure Prediction of Organic Compounds. *Cryst. Growth Des.* **2004**, *4*, 1395–1402. [CrossRef]
7. Mukherjee, A.; Dixit, K.; Sarma, S.P.; Desiraju, G.R. Aniline-phenol recognition: From solution through supramolecular synthons to cocrystals. *IUCrJ* **2014**, *1*, 228–239. [CrossRef]

8. Reddy, D.S.; Ovchinnikov, Y.E.; Shishkin, O.V.; Struchkov, Y.T.; Desiraju, G.R. Supramolecular Synthons in Crystal Engineering. 3. Solid State Architecture and Synthon Robustness in Some 2,3-Dicyano-5,6-dichloro-1,4-dialkoxybenzenes1. *J. Am. Chem. Soc.* **1996**, *118*, 4085–4089. [CrossRef]
9. Desiraju, G.R. Crystal Engineering: A Holistic View. *Angew. Chem. Int. Ed.* **2007**, *46*, 8342–8356. [CrossRef] [PubMed]
10. Desiraju, G.R. Designer crystals: Intermolecular interactions, network structures and supramolecular synthons. *Chem. Commun.* **1997**, *16*, 1475–1482. [CrossRef]
11. Merz, K.; Vasylyeva, V. Development and boundaries in the field of supramolecular synthons. *CrystEngComm* **2010**, *12*, 3989–4002. [CrossRef]
12. Desiraju, G.R. Crystal engineering: Solid state supramolecular synthesis. *Curr. Opin. Solid State Mater. Sci.* **1997**, *2*, 451–454. [CrossRef]
13. Lehn, J.-M. Supramolecular Chemistry and Chemical Synthesis. In *Chemical Synthesis: Gnosis to Prognosis*; Chatgilialoglu, C., Snieckus, V., Eds.; Springer: Dordrecht, The Netherlands, 1996; pp. 511–524. [CrossRef]
14. Desiraju, G.R. Crystal Engineering: From Molecule to Crystal. *J. Am. Chem. Soc.* **2013**, *135*, 9952–9967. [CrossRef] [PubMed]
15. Lehn, J.-M. Supramolecular Chemistry—Scope and Perspectives Molecules, Supermolecules, and Molecular Devices (Nobel Lecture). *Angew. Chem. Int. Ed. Engl.* **1988**, *27*, 89–112. [CrossRef]
16. Dunitz, J.D. Introduction. In *X-ray Analysis and the Structure of Organic Molecules*; © Verlag Helvetica Chimica Acta: Zürich, Switzerland, 1995; pp. 17–21. [CrossRef]
17. Dunitz, J.D. Diffraction of X-Rays by Crystals. In *X-ray Analysis and the Structure of Organic Molecules*; © Verlag Helvetica Chimica Acta: Zürich, Switzerland, 1995; pp. 23–72. [CrossRef]
18. Dunitz, J.D. Internal Symmetry of Crystals. In *X-ray Analysis and the Structure of Organic Molecules*; © Verlag Helvetica Chimica Acta: Zürich, Switzerland, 1995; pp. 73–111. [CrossRef]
19. Dey, A.; Pati, N.N.; Desiraju, G.R. Crystal structure prediction with the supramolecular synthon approach: Experimental structures of 2-amino-4-ethylphenol and 3-amino-2-naphthol and comparison with prediction. *CrystEngComm* **2006**, *8*, 751–755. [CrossRef]
20. Sarma, J.A.R.P.; Desiraju, G.R. The Supramolecular Synthon Approach to Crystal Structure Prediction. *Cryst. Growth Des.* **2002**, *2*, 93–100. [CrossRef]
21. Aakeröy, C.B.; Wijethunga, T.K.; Desper, J. Molecular electrostatic potential dependent selectivity of hydrogen bonding. *New J. Chem.* **2015**, *39*, 822–828. [CrossRef]
22. Etter, M.C. Encoding and decoding hydrogen-bond patterns of organic compounds. *Acc. Chem. Res.* **1990**, *23*, 120–126. [CrossRef]
23. Sandhu, B.; McLean, A.; Sinha, A.S.; Desper, J.; Sarjeant, A.A.; Vyas, S.; Reutzel-Edens, S.M.; Aakeröy, C.B. Evaluating Competing Intermolecular Interactions through Molecular Electrostatic Potentials and Hydrogen-Bond Propensities. *Cryst. Growth Des.* **2018**, *18*, 466–478. [CrossRef]
24. Singh, M.K. Predicting lattice energy and structure of molecular crystals by first-principles method: Role of dispersive interactions. *J. Cryst. Growth* **2014**, *396*, 14–23. [CrossRef]
25. Atahan-Evrenk, S.; Aspuru-Guzik, A. *Prediction and Calculation of Crystal Structures: Methods and Applications*; Springer: Berlin/Heidelberg, Germany, 2014.
26. Gharagheizi, F.; Sattari, M.; Tirandazi, B. Prediction of Crystal Lattice Energy Using Enthalpy of Sublimation: A Group Contribution-Based Model. *Ind. Eng. Chem. Res.* **2011**, *50*, 2482–2486. [CrossRef]
27. Day, G.M. Current approaches to predicting molecular organic crystal structures. *Crystallogr. Rev.* **2011**, *17*, 3–52. [CrossRef]
28. Musumeci, D.; Hunter, C.A.; Prohens, R.; Scuderi, S.; McCabe, J.F. Virtual cocrystal screening. *Chem. Sci.* **2011**, *2*, 883–890. [CrossRef]
29. Hunter, C.A. Quantifying Intermolecular Interactions: Guidelines for the Molecular Recognition Toolbox. *Angew. Chem. Int. Ed.* **2004**, *43*, 5310–5324. [CrossRef] [PubMed]
30. Scheiner, S. Comparison of Bifurcated Halogen with Hydrogen Bonds. *Molecules* **2021**, *26*, 350. [CrossRef]
31. Hou, L.; Gao, L.; Zhang, W.; Yang, X.-J.; Wu, B. Quaternary Cocrystals Based on Halide-Binding Foldamers through Both Hydrogen and Halogen Bonding. *Cryst. Growth Design* **2021**. [CrossRef]
32. Aakeröy, C.B.; Chopade, P.D.; Desper, J. Avoiding "Synthon Crossover" in Crystal Engineering with Halogen Bonds and Hydrogen Bonds. *Cryst. Growth Des.* **2011**, *11*, 5333–5336. [CrossRef]
33. Aakeröy, C.B.; Fasulo, M.; Schultheiss, N.; Desper, J.; Moore, C. Structural Competition between Hydrogen Bonds and Halogen Bonds. *J. Am. Chem. Soc.* **2007**, *129*, 13772–13773. [CrossRef]
34. Zhang, Y.; Zhu, B.; Ji, W.-J.; Guo, C.-Y.; Hong, M.; Qi, M.-H.; Ren, G.-B. Insight into the Formation of Cocrystals of Flavonoids and 4,4'-Vinylenedipyridine: Heteromolecular Hydrogen Bonds, Molar Ratio, and Structural Analysis. *Cryst. Growth Design* **2021**. [CrossRef]
35. Aakeroy, C.B.; Spartz, C.L.; Dembowski, S.; Dwyre, S.; Desper, J. A systematic structural study of halogen bonding versus hydrogen bonding within competitive supramolecular systems. *IUCrJ* **2015**, *2*, 498–510. [CrossRef] [PubMed]
36. Perera, M.D.; Desper, J.; Sinha, A.S.; Aakeröy, C.B. Impact and importance of electrostatic potential calculations for predicting structural patterns of hydrogen and halogen bonding. *CrystEngComm* **2016**, *18*, 8631–8636. [CrossRef]
37. Zheng, Q.; Unruh, D.K.; Hutchins, K.M. Cocrystallization of Trimethoprim and Solubility Enhancement via Salt Formation. *Cryst. Growth Design* **2021**, *21*, 1507–1517. [CrossRef]

38. Aakeröy, C.B.; Wijethunga, T.K.; Desper, J.; Đaković, M. Electrostatic Potential Differences and Halogen-Bond Selectivity. *Cryst. Growth Des.* **2016**, *16*, 2662–2670. [CrossRef]

39. Wang, L.; Liu, S.; Chen, J.-m.; Wang, Y.-x.; Sun, C.C. Novel Salt-Cocrystals of Berberine Hydrochloride with Aliphatic Dicarboxylic Acids: Odd–Even Alternation in Physicochemical Properties. *Mol. Pharm.* **2021**, *18*, 1758–1767. [CrossRef]

40. Gunawardana, C.A.; Desper, J.; Sinha, A.S.; Đaković, M.; Aakeröy, C.B. Competition and selectivity in supramolecular synthesis: Structural landscape around 1-(pyridylmethyl)-2,2′-biimidazoles. *Faraday Discuss.* **2017**, *203*, 371–388. [CrossRef] [PubMed]

41. Aakeröy, C.B.; Beatty, A.M.; Helfrich, B.A. "Total Synthesis" Supramolecular Style: Design and Hydrogen-Bond-Directed Assembly of Ternary Supermolecules. *Angew. Chem. Int. Ed.* **2001**, *40*, 3240–3242. [CrossRef]

42. Bennion, J.C.; Matzger, A.J. Development and Evolution of Energetic Cocrystals. *Acc. Chem. Res.* **2021**, *54*, 1699–1710. [CrossRef] [PubMed]

43. Groom, C.R.; Allen, F.H. The Cambridge Structural Database in Retrospect and Prospect. *Angew. Chem. Int. Ed.* **2014**, *53*, 662–671. [CrossRef] [PubMed]

44. Wood, P.A.; Feeder, N.; Furlow, M.; Galek, P.T.A.; Groom, C.R.; Pidcock, E. Knowledge-based approaches to co-crystal design. *CrystEngComm* **2014**, *16*, 5839–5848. [CrossRef]

45. Galek, P.T.A.; Chisholm, J.A.; Pidcock, E.; Wood, P.A. Hydrogen-bond coordination in organic crystal structures: Statistics, predictions and applications. *Acta Crystallogr. Sect. B* **2014**, *70*, 91–105. [CrossRef] [PubMed]

46. Nauha, E.; Bernstein, J. "Predicting" Crystal Forms of Pharmaceuticals Using Hydrogen Bond Propensities: Two Test Cases. *Cryst. Growth Des.* **2014**, *14*, 4364–4370. [CrossRef]

47. Sarkar, N.; Aakeröy, C.B. Evaluating hydrogen-bond propensity, hydrogen-bond coordination and hydrogen-bond energy as tools for predicting the outcome of attempted co-crystallisations. *Supramol. Chem.* **2020**, *32*, 81–90. [CrossRef]

48. Sarkar, N.; Sinha, A.S.; Aakeröy, C.B. Systematic investigation of hydrogen-bond propensities for informing co-crystal design and assembly. *CrystEngComm* **2019**, *21*, 6048–6055. [CrossRef]

49. Naim, M.; Alam, O.; Nawaz, F.; Alam, M.; Alam, P. Current status of pyrazole and its biological activities. *J. Pharm. Bioallied Sci.* **2016**, *8*, 2–17. [CrossRef]

50. Ansari, A.; Ali, A.; Asif, M.; Shamsuzzaman. Review: Biologically active pyrazole derivatives. *New J. Chem.* **2017**, *41*, 16–41. [CrossRef]

51. Karrouchi, K.; Radi, S.; Ramli, Y.; Taoufik, J.; Mabkhot, Y.N.; Al-Aizari, F.A.; Ansar, M.h. Synthesis and Pharmacological Activities of Pyrazole Derivatives: A Review. *Molecules* **2018**, *23*, 134. [CrossRef] [PubMed]

52. Zhang, J.; Tan, D.-J.; Wang, T.; Jing, S.-S.; Kang, Y.; Zhang, Z.-T. Synthesis, crystal structure, characterization and antifungal activity of 3,4-diaryl-1H-Pyrazoles derivatives. *J. Mol. Struct.* **2017**, *1149*, 235. [CrossRef]

53. El Shehry, M.F.; Ghorab, M.M.; Abbas, S.Y.; Fayed, E.A.; Shedid, S.A.; Ammar, Y.A. Quinoline derivatives bearing pyrazole moiety: Synthesis and biological evaluation as possible antibacterial and antifungal agents. *Eur. J. Med. Chem.* **2018**, *143*, 1463–1473. [CrossRef] [PubMed]

54. Du, S.; Tian, Z.; Yang, D.; Li, X.; Li, H.; Jia, C.; Che, C.; Wang, M.; Qin, Z. Synthesis, Antifungal Activity and Structure-Activity Relationships of Novel 3-(Difluoromethyl)-1-methyl-1H-pyrazole-4-carboxylic Acid Amides. *Molecules* **2015**, *20*, 8395–8408. [CrossRef]

55. Meta, E.; Brullo, C.; Tonelli, M.; Franzblau, S.G.; Wang, Y.; Ma, R.; Baojie, W.; Orena, B.S.; Pasca, M.R.; Bruno, O. Pyrazole and imidazo[1,2-b]pyrazole Derivatives as New Potential Antituberculosis Agents. *Med. Chem.* **2019**, *15*, 17–27. [CrossRef]

56. Sun, J.; Zhou, Y. Synthesis and antifungal activity of the derivatives of novel pyrazole carboxamide and isoxazolol pyrazole carboxylate. *Molecules* **2015**, *20*, 4383–4394. [CrossRef]

57. Liu, J.J.; Zhao, M.Y.; Zhang, X.; Zhao, X.; Zhu, H.L. Pyrazole derivatives as antitumor, anti-inflammatory and antibacterial agents. *Mini Rev. Med. Chem.* **2013**, *13*, 1957–1966. [CrossRef] [PubMed]

58. Wu, J.; Wang, J.; Hu, D.; He, M.; Jin, L.; Song, B. Synthesis and antifungal activity of novel pyrazolecarboxamide derivatives containing a hydrazone moiety. *Chem. Cent. J.* **2012**, *6*, 51. [CrossRef]

59. Taylor, R.; Macrae, C.F. Rules governing the crystal packing of mono- and dialcohols. *Acta Crystallogr. Sect. B* **2001**, *57*, 815–827. [CrossRef] [PubMed]

60. Bruno, I.J.; Cole, J.C.; Edgington, P.R.; Kessler, M.; Macrae, C.F.; McCabe, P.; Pearson, J.; Taylor, R. New software for searching the Cambridge Structural Database and visualizing crystal structures. *Acta Crystallogr. Sect. B* **2002**, *58*, 389–397. [CrossRef] [PubMed]

61. Macrae, C.F.; Edgington, P.R.; McCabe, P.; Pidcock, E.; Shields, G.P.; Taylor, R.; Towler, M.; van de Streek, J. Mercury: Visualization and analysis of crystal structures. *J. Appl. Crystallogr.* **2006**, *39*, 453–457. [CrossRef]

62. Macrae, C.F.; Bruno, I.J.; Chisholm, J.A.; Edgington, P.R.; McCabe, P.; Pidcock, E.; Rodriguez-Monge, L.; Taylor, R.; van de Streek, J.; Wood, P.A. Mercury CSD 2.0—New features for the visualization and investigation of crystal structures. *J. Appl. Crystallogr.* **2008**, *41*, 466–470. [CrossRef]

63. Shao, Y.; Molnar, L.F.; Jung, Y.; Kussmann, J.; Ochsenfeld, C.; Brown, S.T.; Gilbert, A.T.B.; Slipchenko, L.V.; Levchenko, S.V.; O'Neill, D.P.; et al. Advances in methods and algorithms in a modern quantum chemistry program package. *Phys. Chem. Chem. Phys.* **2006**, *8*, 3172–3191. [CrossRef] [PubMed]

64. Lu, Y.; Kraatz, H.B. Metal complexes of 3-acetamido-5-methylpyrazole. *Inorg. Chim. Acta* **2004**, *357*, 159–166. [CrossRef]

65. Daidone, G.; Maggio, B.; Raimondi, M.V.; Bombieri, G.; Marchini, N.; Artali, R. Comparative Structural Studies of 4-Diazopyrazole Derivatives by X-Ray Diffraction and Theoretical Investigation. *Heterocycles* **2005**, *65*, 2753. [CrossRef]

 *chemistry*

Review

The Crystal Chemistry of Inorganic Hydroborates [†]

Radovan Černý *, Matteo Brighi and Fabrizio Murgia

Laboratory of Crystallography, DQMP, University of Geneva, 24 quai E. Ansermet, CH-1211 Geneva, Switzerland; Matteo.Brighi@unige.ch (M.B.); Fabrizio.Murgia@unige.ch (F.M.)
* Correspondence: Radovan.Cerny@unige.ch
† Dedicated to Dr. Howard Flack (1943–2017).

Academic Editor: Catherine Housecroft
Received: 28 August 2020; Accepted: 23 September 2020; Published: 29 September 2020

Abstract: The crystal structures of inorganic hydroborates (salts and coordination compounds with anions containing hydrogen bonded to boron) except for the simplest anion, borohydride BH_4^-, are analyzed regarding their structural prototypes found in the inorganic databases such as Pearson's Crystal Data [Villars and Cenzual (2015), Pearson's Crystal Data. Crystal Structure Database for Inorganic Compounds, Release 2019/2020, ASM International, Materials Park, Ohio, USA]. Only the compounds with hydroborate as the only type of anion are reviewed, although including compounds gathering more than one different hydroborate (mixed anion). Carbaborane anions and partly halogenated hydroborates are included. Hydroborates containing anions other than hydroborate or neutral molecules such as NH_3 are not discussed. The coordination polyhedra around the cations, including complex cations, and the hydroborate anions are determined and constitute the basis of the structural systematics underlying hydroborates chemistry in various variants of anionic packing. The latter is determined from anion–anion coordination with the help of topology analysis using the program TOPOS [Blatov (2006), IUCr CompComm. Newsl. 7, 4–38]. The Pauling rules for ionic crystals apply only to smaller cations with the observed coordination number within 2–4. For bigger cations, the predictive power of the first Pauling rule is very poor. All non-molecular hydroborate crystal structures can be derived by simple deformation of the close-packed anionic lattices, i.e., cubic close packing (*ccp*) and hexagonal close packing (*hcp*), or body-centered cubic (*bcc*), by filling tetrahedral or octahedral sites. This review on the crystal chemistry of hydroborates is a contribution that should serve as a roadmap for materials engineers to design new materials, synthetic chemists in their search for promising compounds to be prepared, and materials scientists in understanding the properties of novel materials.

Keywords: hydroborate; anions packing; crystal structure

---

1. Introduction

Hydroborates are anions containing hydrogen bonded to boron. They are also sometimes referred to as "boranes"; this term is, however, used for neutral molecules B_xH_y according to IUPAC [1]. Inorganic hydroborates are salts or coordination compounds where one of the ligands is the hydroborate. The bonding in boron clusters of boranes and hydroborates was explained by Lipscomb [2] using the concept of a three-electron-two-center bond. The concept was further developed in the Polyhedral Skeletal Electron Pair theory (PSEPT) also known as Wade–Mingos rules based on a molecular orbital treatment of the bonding [3,4]. The naming of the boron and hydroborate clusters, used in this review, follows the Wade-Mingos rules: (i) a number of vertices in the polyhedral boron cluster, i.e., dodecaborate for $B_{12}H_{12}^{2-}$ anion, and (ii) the name that describes the topology of the polyhedral cluster, i.e., *closo-*, *nido-* and *arachno*-borates for all polyhedron vertices occupied by boron, and one or two boron atoms missing, respectively.

Inorganic hydroborates were studied as fuels for military applications [5], reducing agents in organic syntheses [6], weakly coordinating anions in catalysis [7,8], for the delivery of ^{10}B for Boron Neutron Capture Therapy (BNCT) [9], as nanocarriers for the delivery of various chemotherapy drugs [10], magnetic resonance imaging (MRI) agents [11], liquid electrolytes [12], and more recently as solid ionic conductors [13,14]. It is exactly the increase of interest in hydroborates as solid-state electrolytes for application in all-solid-state batteries that motivated us to assemble the structural characteristic of known compounds, analyze the coordination of cations and anions, study the type of anion packing, and find the structural aristotypes. The structural systematics discussed in this review attempt to provide a guide to functional inorganic hydroborates design. The hydroborates containing the simplest anion, borohydride BH_4^-, were recently reviewed by us [15] and are not included herein. We will limit our review only to the compounds containing the hydroborate as the only type of anion, although including compounds gathering more than one different hydroborate (mixed anion). We will include in our review also the carbaborane anions where one or more boron atoms are replaced by carbon and partly halogenated hydroborates, i.e., where there is the partial replacement of hydrogen by a halogen. Hydroborates containing anions other than hydroborate or neutral molecules such as NH_3 will not be discussed. Only the compounds in which the crystal structure has been experimentally fully characterized are included; the structures predicted by ab initio calculations or any other method of prediction, but not experimentally confirmed, are excluded. According to our analysis, the crystal chemistry of inorganic large hydroborates is much simpler compared to that of containing small BH_4^- [15] reducing the anion arrangement, for most of the presented compounds, to the packing of hard spheres. We will also show that the first Pauling rule on preferred cation coordination in these (mostly) ionic compounds is limited to small cations, with the coordination number within 2–4 [16].

2. Anions Packing

The hydroborates that form the compounds mostly used in various applications are *closo*-dodecaborates $B_{12}H_{12}^{2-}$ and *closo*-decaborates $B_{10}H_{10}^{2-}$ (Figure 1). While the boron cluster in the first has the form of an ideal icosahedron with the symmetry I_h, the latter is the gyroelongated square bipyramid (bi-capped square antiprism) with the symmetry D_{4d}. The non-crystallographic symmetry of both clusters is at the origin of their orientation disorder in the *ht* phases of many hydroborates.

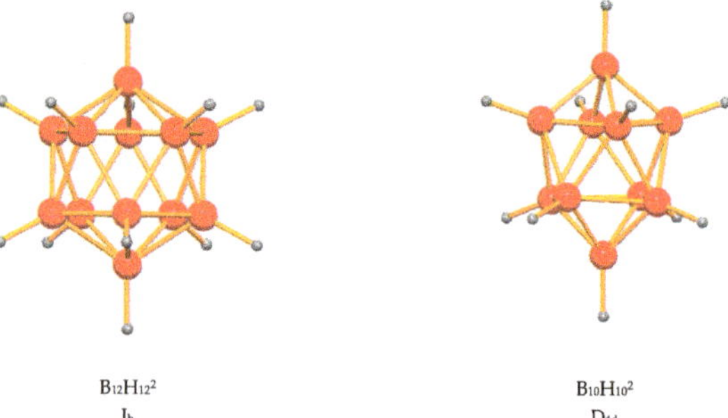

Figure 1. The two frequently used hydroborate anions: *closo*-dodecaborate (**left**) and *closo*-decaborate (**right**) and their point group symmetry.

The high symmetry of the two anions is also responsible for the building principle of corresponding crystal structures based on the packing of hard spheres. Inorganic hydroborates usually contain three types of anion packing: two close packings, cubic close packing (*ccp*) and hexagonal close packing (*hcp*), and one less dense packing in a body-centered cubic cell (*bcc*). However, the poly-anion nature of hydroborates complicates the detections of the packing type, which usually deviates from the ideal packing. For instance, the hexagonal layers of *ccp* or *hcp* are not always parallel to simple crystallographic planes. In some works, the anion packing was analyzed with the algorithms developed for molecular dynamics (MD) simulation [17], which gives the frequency with which each basic packing maps locally the analyzed structure. This is justified when studying with MD the temporal evolution of some microscopic aspect (such as the distribution of energy barriers for diffusion), and necessarily the atomic positions are given as the atoms' distribution, hence anions packing distribution. However, it makes no sense to use such an algorithm for the analysis of a periodic structure, which gives the averaged description of the structure and where the atoms in the unit cell are located on Wyckoff sites even when the structure is disordered. Instead, the easiest method is to analyze the shape of the first and second coordination sphere of anion–anion interaction, i.e., considering the poly-anions centers: in both close packings, the coordination number of the first sphere is 12, and the polyhedron shape is cuboctahedron for *ccp* (Figure 2a) and anti-cuboctahedron for *hcp* (Figure 2b). In *bcc* packing, the first coordination sphere has a cubic shape, since there are eight nearest neighbors, while the second coordination sphere contains six anions forming an octahedron. Together, the first and second coordination spheres form a rhombic dodecahedron in *bcc* (Figure 2c). We will allocate a type of packing to a given hydroborate, based on the anion–anion coordination polyhedron (which may be deformed) and on the number, type, and connectivity of the interstitial sites, which must not change with respect to the ideal packing.

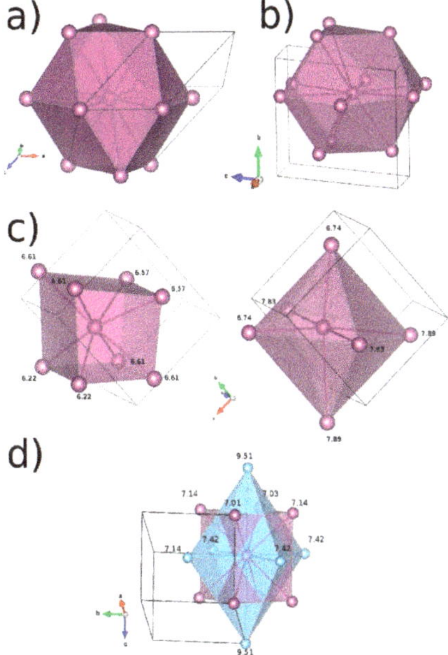

Figure 2. Anion–anion coordination for (**a**) monoclinic *rt*-Na$_2$B$_{12}$H$_{12}$ with cubic close packing (*ccp*) of anions. The first coordination sphere contains 12 nearest neighbors at the distance 7.01–7.42 Å forming a cuboctahedron; (**b**) monoclinic *rt*-K$_2$B$_{10}$H$_{10}$ with hexagonal close packing (*hcp*) of anions. The first coordination sphere forms an anti-cuboctahedron with a bond distance 6.67–7.68 Å; (**c**) orthorhombic *rt*-NaCB$_9$H$_{10}$ with body-centered cubic (*bcc*) packing of anions. The first coordination shell is shown on the left, while the second shell is shown on the right (cube and octahedron, respectively). (**d**) Wrong determination of coordination polyhedron for the *rt*-Na$_2$B$_{12}$H$_{12}$ leading to the false determination of a *bcc* sublattice. The apical atoms in the octahedron are indeed at the distance of 9.51 Å, while all the others fall in the range 7.0–7.4 Å.

3. Controlling Anions Packing

The type of anion packing is an important parameter controlling hydroborates properties such as cation mobility [18]. While the packing type for monoatomic structures such as metals is simply driven by space-filling efficiency, it becomes more complex for poly-atomic structures such as ionic compounds. The ability of smaller cations to fill available interstitial sites in the packing of larger anions may control the packing type of the latter [16]. In the case of metal hydroborates, the polyanions are only approximately spherical, and their orientation is controlled by directional metal–hydrogen bonding. It means that in addition to the space-filling and cation/anion size ratio, the cation–anion directional bonding is a third parameter controlling the anion packing [19] at a given temperature and pressure.

Cation distribution on interstitial sites and cation–anion directional bonding are at the basis of the anion packing design and engineering: (1) cation mixing where the larger cation fills the interstitial sites left empty by the smaller (mobile) cation [20,21]; (2) anion mixing, which perturbates the directional cation–anion bonding [22,23], and with the same effect, (3) anion modification such as partial halogenation [21]. In all cases, the volume change and its effect on the total crystal energy cannot be excluded [21].

Anions packing varies also when a neutral molecule such as NH_3 is introduced into the structure of the hydroborate salt [24]. In that case, when the molecule coordinates to the cation, the mobility of the latter is strongly decreased [25].

From the analysis of the anion packing presented here below, it seems that increasing anion anisotropy, i.e., partial halogenation of $B_{12}H_{12}$, use of elongated $B_{10}H_{10}$, or by *carba*-borate and *nido*-borate, increases the probability of having an *hcp* lattice.

4. Preferred Cation Sites

The first Pauling rule predicts the preferred cation coordination in iono-covalent compounds from the ratio of cation and anion radii [16]. We have analyzed all single metal hydroborates presented in this review according to the first Puling rule. The cation radii are according to Shannon [26]. We have used an ionic radius for a given oxidation state contrary to Pauling's original work, which is based on univalent radii. As only very few estimations of hydroborate radii are available in the literature (for BH_4 see [15]), we have used the radii from [27], where the diameter of the anion was calculated as the maximal distance between two terminal hydrogens. Clearly, these radii correspond to the lower limit of the estimation. The calculated cation/anion radius ratios are plotted in Supplementary Materials Figure S1 together with regions corresponding to different cation coordination polyhedra. The agreement of the observed coordination (color of the data point) with the predicted region is relatively good for smaller cations with the observed coordination number within 2–4. For larger cations, the predictive power of the first Pauling rule is very poor, which is in agreement with the recent analysis of Pauling rules validity for oxides [28]. For example, we observe the tetragonal coordination in the region where the octahedral coordination should exist. This discrepancy would be removed if larger anion radii were used, such as estimated from solid crystal structures: 3.18 Å for $B_{10}H_{10}$ [29] or 3.28 Å for $B_{12}H_{12}$ [30] (3.46–3.5 Å from hard spheres approximation) or even larger from the calculation of electrostatic potential surface maps [31]. With larger estimation of the anion size, the predicted region of existence would be shifted to smaller coordination numbers.

5. Anions and Cations Dynamics

The complex anions in the hydroborates can participate in the fast reorientation (rotational) motion. This important dynamical feature contributes to the entropic part of the free energy balance determining thermodynamic stability. Therefore, the information on the anion reorientation dynamics is important for understanding the fundamental properties of hydroborate salts such as the thermal stability and symmetry of the crystal structure. When rotating, the anions approach the spherical symmetry, averaging off the local hydrogen–cation interactions and following more precisely the ideal packing of hard spheres, which leads to stabilizing the higher crystal symmetries in *ht* phases. In addition to the localized reorientation motion of the anions, the long-range motion of the cations is observed in the compounds with high cation conductivity. These two types of atomic motion may be related, i.e., the fast cations diffusion is accompanied by the fast reorientation motion of the anions [32], as was also shown by ab initio molecular dynamics calculations [33]. The anion and cation dynamics in hydroborates are studied by nuclear magnetic resonance (NMR) [34] and quasi-elastic neutron scattering (QENS) [35]. For more details on the experimental and theoretical studies of the ions dynamics in hydroborates, we refer to the recent review [34].

The ions dynamic is practically invisible in the diffraction experiment using Bragg scattering, which corresponds to the space and time average of atomic positions. The anions reorientation motion and cations long-range diffusion can be only indirectly judged from the observed disorder in atomic positions. If both dynamic and static disorders are present, they may be separated by a temperature-dependent diffraction. The static disorder, if present, persists down to low temperatures where the dynamic one disappears. Some details of the local anion–anion correlation can be obtained by modeling the total scattering (Bragg and diffuse scattering), for example using the Pair Distribution Function analysis. Such experiments have not yet been done on inorganic hydroborates except for salts with a BH_4^- anion [36,37]. The disorder of atomic positions observed from Bragg scattering needs a correct understanding of the superposition of anion point symmetry (Figure 1) with the point symmetry of Wyckoff positions in the crystal. The high non-crystallographic symmetry of the hydroborate anion can fit only partly to a Wyckoff position, and the anion needs to be oriented in such a way that the symmetry elements of a given site match the corresponding symmetry elements of the anions point group. Only in this case do the atomic positions of boron and hydrogen appear fully ordered. If in such ordered structure the anions reorientation happens between orientations that are equivalent by the site symmetry, Bragg scattering cannot detect it. However, if the reorientation occurs between non-equivalent orientations, the structure appears disordered in atomic positions. If this is the case, the Bragg scattering cannot still say whether the disorder is dynamic (i.e., the reorientation of the anion) or whether it is of static nature (i.e., a long-range disorder of anions orientation) [38].

We can demonstrate the interaction between the anions point symmetry and Wyckoff site symmetry on two examples: $Cs_2B_{12}H_{12}$ crystallizes at *rt* in an ordered anti-CaF_2 structure with the crystal symmetry reduced from *Fm-$\bar{3}$m* to *Fm-$\bar{3}$* due to the icosahedral symmetry of the *closo*-anion localized in one orientation (light gray, Figure 3 left). At 529 K, it undergoes a second-order transition of order–disorder type disordering the *closo*-anion by {110} mirror planes (absent in the point symmetry of the anion) increasing the crystal symmetry to *Fm-$\bar{3}$m*. Then, the observed disordered anion is a superposition of the anion in two orientations (light and dark gray, Figure 3 right) related by the {110} mirror plane or by the rotation around the anions axis C_3 by 45° (which is equivalent), as shown in Figure 3. This disorder is static as shown by QENS studies where different reorientation jumps about two molecular axes, C_3 by 120° and C_5 by 72°, were suggested [39]. Both jumps exist in *lt* ordered and *ht* disordered phases. They operate independently at a lower temperature, but combine to two-axial jumps at higher temperatures, as shown by ^1H and ^{11}B NMR [40].

$Li_2B_{12}H_{12}$ crystallizes at *rt* in an ordered *anti*-CaF_2 structure with its symmetry reduced from *Fm-3m* to *Pa-3* resulting in two *closo*-anions in the unit cell in different orientations but related by the *a*-glide plane (light and dark gray, Figure 3 left). $Li_2B_{12}H_{12}$ undergoes at 628 K a first-order phase transition into the *s.g. Fm-3m* where the disordered anion may be explained as a superposition of the two orientations from the *rt*-phase. As the reorientation dynamics in $Li_2B_{12}H_{12}$ has not yet been studied, we cannot conclude whether the disorder in the *ht*-$Li_2B_{12}H_{12}$ is of dynamic or static nature. The phase transition in $Li_2B_{12}H_{12}$ may be compared with that in C60, which crystallizes in an ordered *Pa-3* structure below 250 K and transforms with first-order order–disorder phase transition connected with the rotation of C60 molecules into an *Fm-3m* structure (see Figure S2) [41].

The understanding of cations disorder is easier as there is no interaction between the cations symmetry (single atom) and the symmetry of a Wyckoff site. The observed disorder is always a position disorder, but its nature (dynamic or static) cannot be concluded from Bragg intensities. For example, it is easily identified by measuring the electrical conductivity based on the given cation. As local correlations between moving cations are suggested by molecular dynamic studies [33], the total scattering studies on ionic conductors are of high interest.

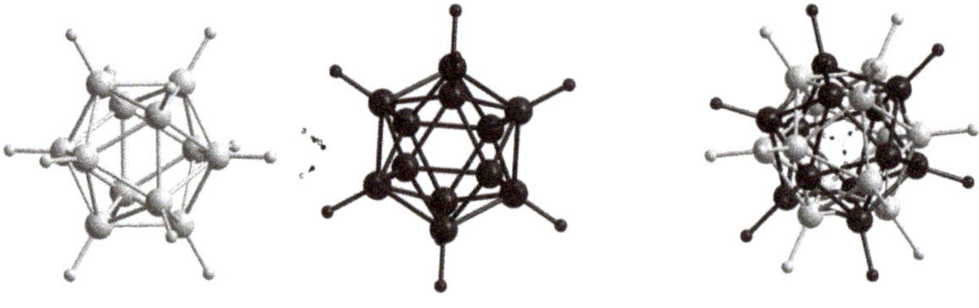

rt-Li$_2$B$_{12}$H$_{12}$ and *rt*-Cs$_2$B$_{12}$H$_{12}$ *ht*-Li$_2$B$_{12}$H$_{12}$ and *ht*-Cs$_2$B$_{12}$H$_{12}$

Figure 3. Two orientations, dark and light gray, of the *closo*-anion in the *rt* ordered (s.g. *Pa*-3) and in the *ht* disordered Li$_2$B$_{12}$H$_{12}$ (s.g. *Fm*-3) viewed along the C$_3$ rotation axis of the *closo*-anion. The *rt*-Cs$_2$B$_{12}$H$_{12}$ (s.g. *Fm*-3) contains one anion in the orientation identical to the light gray anion in *rt*-Li$_2$B$_{12}$H$_{12}$, while the *ht*-Cs$_2$B$_{12}$H$_{12}$ (s.g. *Fm-3m*) has an identical orientation disorder (rotation around C$_3$ by 45°) as *ht*-Li$_2$B$_{12}$H$_{12}$.

6. Anions Chemistry

According to the aforementioned PSEPT, hydroborates can be thought of as formally made from simple electron donors, i.e., B-H groups, but also C-H in the case of carbaboranes, providing electron density to build up a cluster. The main building block of hydroborates, the B-H unit, accounts for four valence electrons. Excluding the electron pair involved in the boron–hydrogen bond, the remaining two electrons are available to link the boron atoms in non-classical connectivity, where on average two electrons are shared between three boron atoms (three-centers two-electrons). This multicenter interaction is typical of boron-based compounds, due to the intrinsic boron electron deficiency and relatively low electronegativity. As a direct consequence, the possibility of arranging several B-H groups is limited (with fewer exceptions in some open-cage hydroborates [42]), according to 4n's Wade–Mingos rules, to a deltahedron, i.e., a polyhedron with all faces as equilateral triangles. From the point of view of molecular orbitals (MO), each B-H group participates in the formation of three MOs for the cluster bonding, where two are tangent to the cage surface, giving rise to σ superposition of p orbitals, while the remaining is oriented inside the cluster and accounts for a further MO. Therefore, n B-H building blocks lead to n+1 total MOs in the framework. Such non-classical electron distribution implies electron resonance all along the σ-bonding in the cage, which, in analogy with carbon chemistry, is often referred to as "superaromaticity". This resonance energy (that accounts for the superior stability of hydroborates compared to other compounds) has been determined for several *closo*-boranes and carba*closo*-boranes up to 12 vertices [43,44]. It is worth noting that the greater the number of B-H units, the higher the stability due to the resonance effect, with a special stabilization shown by 6- and 12-vertex polyhedra. However, in the mono and bi-substituted carba*closo* series, the difference in energy decreases progressively [44], which is likely because the greater electronegativity of the carbon atom(s) withdraws electronic density and limits the resonance along the cage [45]. The progressive removal of one or two boron vertices from a *closo* cage leads to stable open-cage structures, namely *nido*- and *arachno*-boranes. Indeed, retrieving one or two vertices does not change the number of MOs involved in the skeletal binding, since the missing electrons are provided by the addition of hydrogen atoms. However, the loss of quasi-spherical geometry and the consequent reduction of electronic delocalization, besides the more hydridic character of apical hydrogens and their ability to withdraw electrical density from the cage [46], results in the lower chemical stability of the *nido*- and *arachno*-frameworks, with respect to their parent *closo* compounds [47].

7. Classification of Inorganic Hydroborates

Contrary to metal borohydrides [15], the known hydroborates containing larger poly-anions are mostly limited to alkali-metals, alkali-earth, and 3d transition metals. No double-cation hydroborate with a large electronegativity difference between the cations is known. Contrary to the borohydrides, many hydroborate structures with larger hydroborate clusters can be derived not only from a close packed anion lattice (*ccp* or *hcp*) by filling the interstitial tetrahedral (T) and octahedral (O) sites with cations but also from the *bcc* anion packing. The O and T sites are deformed and interpenetrate in *bcc* packing, which makes their occupation by cations less favorable, and neighboring sites cannot be simultaneously occupied. Only very few ionic compounds are based on *bcc* anion packing, contrary to monoatomic structures such as metals. Well-known examples are the *ht* phases of AgI and Ag chalcogenides (S-Te) with Ag_2S being a structural prototype for the *ht* phase of $Na_2B_{12}H_{12}$ (see Figure S2). A common feature of all three *ht* phases, AgI, Ag_2S, and $Na_2B_{12}H_{12}$, is being a superionic conductor with *bcc* anions packing. It is interesting to note that while not a single metal borohydride is based on a *bcc* packing of anions, this anion packing is very often observed among the here-discussed hydroborates with poly-anions, especially at higher temperatures. This is probably a consequence of the weaker cation–hydrogen interaction compared to borohydride, since in hydroborates, the hydridic nature of hydrogen is depleted by the strong electronic density withdrawing toward the boron cage, which is due to the high delocalization among the boron–boron σ bonds in the cage [31,48].

The hydroborates discussed in this review are listed in the following tables: Table 1 contains single cation hydroborates, Table 2 contains double cation hydroborates, and Table 3 contains the hydroborates with inorganic poly-cation. The structural prototypes given in the tables can be found in different crystallographic databases of inorganic compounds such as the Inorganic Crystal Structure Database [49] or Pearson's Crystal Data [50], and in the following, we will not be providing the references for each structural prototype. The occurrence of structural prototypes among the hydroborates is shown in Figure 4, and it is compared with the occurrence among all inorganic compounds as extracted from the Linus Pauling File (LPF) [51].

While the inorganic compounds are dominated by four structure prototypes, NaCl, perovskite, sphalerite, and CsCl, the hydroborates prefer CaF_2, Ni_2In, perovskite (including Ag_3SI), NaCl, and wurtzite (anti-types are included). As already discussed, the prototypes based on anionic *bcc* packing (AgI, Ag_2S, and perovskites) are very common among the hydroborate salts. Other information on the chemistry, synthesis, and application of inorganic hydroborates can be found in recent reviews [15,27,52–58].

The content of the tables is organized in the following way: *First column*: Hydroborates are ordered according to the order of cations in the periodic table. The name describing the topology of the polyhedral cluster is given, too. *Second column*: The symbol of the phase is given as a prefix of the chemical formula only if defined in the original publication. *Third column*: The space group symbol as used in the original publication. If different from the standard setting, the latter is given in parentheses. *Fourth column*: The structural prototype among inorganic compounds. The choice of the prototype is not always in the sense of an *aristotype*, i.e., the most symmetrical structure in the Bärnighausen tree of the given structure (see for example [59]), but rather as a closest structure from which the hydroborate may be derived by replacing anions (i.e., oxides, halides) with hydroborates. The naming of the structural prototypes is according to LPF, i.e., name, Pearson symbol, and space group number [51]. The space group of the prototype or its mineral name is given only if more than one polymorph of the prototype exists. *Fifth column*: Cation coordination. The denticity of the hydroborate coordination, i.e., the number of hydrogen atoms of one hydroborate group coordinating to a given cation, is listed only if the crystal structure was determined by neutron diffraction, single crystal X-ray diffraction, or ab initio solid-state calculations. *Sixth column*: Anion coordination. *Seventh column*: Type of anions packing. If different from three basic packings, the symbol of the anion's net topology is given. The topological analysis was performed with the program TOPOS [60]. The type of molecules packing is given for molecular compounds. The nomenclature used to represent the different nets has been chosen according to the symbolism used in the TOPOS Topological Database (TTD) [61]. *Eighth column*: The given references relate to the most reliable crystal structure determination. If available, one reference for the X-ray and one for the neutron diffraction experiment are given as well as one for the validation of the structure by ab initio solid-state calculations.

76

Table 1. Structural classification of single cation metal hydroborates.

Cation	Compound	Space Group	Structural Prototype	Cation Coordination by Anions	Cation Coordination by Hydrogen and Halogen	Anion Coordination	Aristotype of Anion Packing	Ref.
Li$^+$ closo-	rt-α-Li$_2$B$_{12}$H$_{12}$	$Pa\bar{3}$	anti-CaF$_2$, cF12, 225	Tri-bi	Oct	Oct	ccp	[62]
closo-	ht-β-Li$_2$B$_{12}$H$_{12}$	$Fm\bar{3}$	anti-CaF$_2$, cF12, 225	Tri			ccp	[63,64]
closo-	rt-Li$_2$B$_{10}$H$_{10}$	$P6_2$22	anti-CaF$_2$, cF12, 225	Tri-bi	Oct	6-fold	{$3^{36}.4^{46}.5^9$} 14-c net; tcc-x	[29]
closo-	ht-Li$_2$B$_{10}$H$_{10}$							[29]
closo-	rt-LiCB$_{11}$H$_{12}$	$Pca2_1$	NaCl, cF8, 225	Tri-bi	Oct	Tri	ccp	[65]
closo-	ht-LiCB$_{11}$H$_{12}$	$Fm\bar{3}m$					ccp	[65]
closo-	rt-α-LiCB$_{11}$H$_6$Cl$_6$	$Pnma$	NaCl, cF8, 225	Lin-tri	Oct	NLin	ccp	[66]
closo-	ht-γ-LiCB$_{11}$H$_6$Cl$_6$	$P6_3/mmc$		Lin, Tri			hcp	[66]
closo-	rt-α-LiCB$_{11}$H$_6$Br$_6$	$Pnma$	NaCl, cF8, 225	Lin-tri	Oct	NLin	ccp	[66]
closo-	ht-γ-LiCB$_{11}$H$_6$Br$_6$	$P6_3/mmc$		Lin, Tri			hcp	[66]
closo-	rt-LiCB$_9$H$_{10}$							[48]
closo-	ht1-LiCB$_9$H$_{10}$	$Cmc2_1$	wurtzite, hP4, 186				hcp	[48]
closo-	ht2-LiCB$_9$H$_{10}$	$P3_1c$	wurtzite, hP4, 186				hcp	[48]
nido-	rt-LiB$_{11}$H$_{14}$	$Pbca$	sphalerite, cF8, 216	Tri-mon	Oct	NTri	ccp	[67]
nido-	ht-LiB$_{11}$H$_{14}$	$Fm\bar{3}$	sphalerite, cF8, 216	Tri			ccp	[67]
closo-	rt1-Li$_2$(CB$_9$H$_{10}$) (CB$_{11}$H$_{12}$)	$P31c$	wurtzite, hP4, 186				hcp	[68]
closo-	rt2-Li$_2$(CB$_9$H$_{10}$) (CB$_{11}$H$_{12}$)	$Fm\bar{3}m$					ccp	[68]
nido, closo-	Li(B$_{11}$H$_{14}$)$_x$ (CB$_{11}$H$_{12}$)$_{1-x}$ (x = 1/2, 1/3, 2/3)	$Pca2_1$	NaCl, cF8, 225	Tri		Tri	ccp	[67]
nido, closo-	Li$_{2-x}$(B$_{11}$H$_{14}$)$_x$(B$_{12}$H$_{12}$)$_{1-x}$ (x = 1/2)	$Pa\bar{3}$	anti-CaF$_2$, cF12, 225	Tri		Tri	ccp	[67]
nido, closo-	Li(B$_{11}$H$_{14}$)$_x$ (CB$_9$H$_{10}$)$_{1-x}$ (x = 1/2)	$P31c$	wurtzite, hP4, 186	Tet		Tet	hcp	[67]
Na$^+$ closo-	α-rt-Na$_2$B$_{12}$H$_{12}$	$P2_1/n$ ($P2_1/c$)	anti-CaF$_2$, cF12, 225	Tet-mon, bi, tri	10-fold	Cub	ccp	[21,30]
closo-	β-ht1-Na$_2$B$_{12}$H$_{12}$	$Pm\bar{3}n$	Ag$_2$S, cI20, 229	Tet			bcc	[21,30]
closo-	γ-ht2-Na$_2$B$_{12}$H$_{12}$	$Im\bar{3}m$	Ag$_2$S, cI20, 229	Tet			bcc	[21,30]
closo-	δ-ht3-Na$_2$B$_{12}$H$_{12}$	$Fm\bar{3}$	anti-CaF$_2$, cF12, 225	Tet		Cub	ccp	[21]

Table 1. *Cont.*

Cation	Compound	Space Group	Structural Prototype	Cation Coordination by Anions	Cation Coordination by Hydrogen and Halogen	Anion Coordination	Aristotype of Anion Packing	Ref.
closo-	ε-*hp*1-$Na_2B_{12}H_{12}$	*Pbca*	anti-CaF_2, cF12, 225	Tri-bi, tri	7-fold	Oct	*ccp*	[69]
closo-	ζ-*hp*2-$Na_2B_{12}H_{12}$	*Pnnm*	anti-CaF_2, cF12, 225	Tet-mon, bi, tri	9-fold	Cub	*ccp*	[69]
closo-	*m*-$Na_2B_{12}H_{12-x}I_x$	*Pc*	Ni_2In ($Co_{1.75}Ge$), hP6, 194	Tet, Tri		6-fold	*hcp*	[21]
closo-	*h*-$Na_2B_{12}H_{12-x}I_x$	*P6_3mc*	Ni_2In ($Co_{1.75}Ge$), hP6, 194	Tet, Tri		Tet	*hcp*	[21]
closo-	*rt*-$Na_2B_{10}H_{10}$	*P2_1/c*	anti-CaF_2, cF12, 225	Tet-bi, Tet-mon, bi, tri	8-fold, 7-fold	Cub	*ccp*	[70]
closo-	*ht*-$Na_2B_{10}H_{10}$	*Fm$\bar{3}$m*	anti-CaF_2, cF12, 225	Tet, Oct			*ccp*	[70]
arachno-	NaB_3H_8	*Pmn2_1*	NaCl, cF8, 225	Oct-mon, bi	11-fold	Oct	*ccp*	[71]
closo-	*rt*-$NaCB_{11}H_{12}$	*Pca2_1*	NaCl, cF8, 225	Tri-bi	Oct	Tri	*ccp*	[65]
closo-	*ht*-$NaCB_{11}H_{12}$	*Fm$\bar{3}$m*					*ccp*	[65]
closo-	*rt*-α-$NaCB_{11}H_6Cl_6$	*Pbca*	sphalerite, cF8, 216	Tri-bi, tri	PentBiPyr	NTri	*ccp*	[66]
closo-	*ht*-γ-$NaCB_{11}H_6Cl_6$	*P6_3/mmc*		Lin, Tri			*hcp*	[66]
closo-	*ht*-γ-$NaCB_{11}H_6Br_6$	*P6_3/mmc*		Lin, Tri			*hcp*	[66]
closo-	*lt*-$NaCB_9H_{10}$	*P2_1/c*	AgI, cI38, 229	Tet-bi	8-fold	Tet	*bcc*	[72]
closo-	*rt*-$NaCB_9H_{10}$	*Pna2_1*	AgI, cI38, 229	Tet-bi	8-fold	Tet	*bcc*	[72]
closo-	*ht*-$NaCB_9H_{10}$	*P3_1c*	wurtzite, hP4, 186				*hcp*	[48]
nido-	*rt*-$NaB_{11}H_{14}$	*Pna2_1*	wurtzite, hP4, 186	Tet, Oct		Tet	*hcp*	[47,73]
nido-	*lt*-$NaB_{11}H_{14}$	*Pna2_1*	wurtzite, hP4, 186	Tet-bi, tri	9-fold	Tet	*hcp*	[47]
nido-	*ht*-$NaB_{11}H_{14}$	*I$\bar{4}$3d*					*bcc*	[73]
nido-	*rt*-Na-7-$CB_{10}H_{13}$	*Pna2_1*	wurtzite, hP4, 186	Tet-mon, bi, tri	SqAntPris	4-fold	*hcp*	[73]
nido-	*ht*-Na-7-$CB_{10}H_{13}$	*Fm$\bar{3}$m*					*ccp*	[73]
nido-	*rt*-Na-7,8-$C_2B_9H_{12}$	*P2_1*						[73]
nido-	*ht*-Na-7,8-$C_2B_9H_{12}$	*P31c*					*hcp*	[73]
nido-	*rt*-Na-7,9-$C_2B_9H_{12}$	*Pna2_1*	wurtzite, hP4, 186	Tet-mon, bi, tri	1cap-SqAntPris	4-fold	*hcp*	[73]
nido-	*ht*-Na-7,9-$C_2B_9H_{12}$	*Fm$\bar{3}$m*					*ccp*	[73]
closo-	$Na_2(B_{12}H_{12})_{0.75}(B_{10}H_{10})_{0.25}$						*bcc*	[74]

Table 1. *Cont.*

Cation	Compound	Space Group	Structural Prototype	Cation Coordination by Anions	Cation Coordination by Hydrogen and Halogen	Anion Coordination	Aristotype of Anion Packing	Ref.
closo-	$Na_2(B_{12}H_{12})_{0.50}(B_{10}H_{10})_{0.50}$	$Fm\bar{3}m$	anti-CaF$_2$, cF12, 225	Tet, Oct			*ccp*	[75]
closo-	*rt*-$Na_{2-x}(CB_{11}H_{12})_x(B_{12}H_{12})_{1-x}$ (x = 1/2, 1/3, 2/3)	*I*23	Ag$_3$SI, (anti-perovskite) cP5, 221	Tet			*bcc*	[22]
closo-	*ht1*-$Na_{2-x}(CB_{11}H_{12})_x(B_{12}H_{12})_{1-x}$ (x = 1/3)	$Pm\bar{3}n$	Ag$_3$SI, (anti-perovskite) cP5, 221	Tet			*bcc*	[22]
closo-	*ht2*-$Na_{2-x}(CB_{11}H_{12})_x(B_{12}H_{12})_{1-x}$ (x = 1/3)	$Im\bar{3}m$	Ag$_3$SI, (anti-perovskite) cP5, 221	Tet			*bcc*	[22]
closo-	$Na_3(CB_9H_{10})$ $(B_{12}H_{12})$	$P\bar{4}2_1c$	Ag$_3$SI, (anti-perovskite) cP5, 221	Tet			*bcc*	[18]
closo-	$Na_3(CB_{11}H_{12})$ $(B_{10}H_{10})$	$Pnnm$	anti-CaF$_2$, cF12, 225	Tet, Tri			*ccp*	[18]
closo-	*rt1*-$Na_2(CB_9H_{10})$ $(CB_{11}H_{12})$	$P3_1c$	wurtzite, hP4, 186	Tri			*hcp*	[18,68]
closo-	*rt2*-$Na_2(CB_9H_{10})$ $(CB_{11}H_{12})$	$Fm\bar{3}m$					*ccp*	[68]
nido, closo-	$Na_{2-x}(B_{11}H_{14})_x(B_{12}H_{12})_{1-x}$ (x = 1/2, 2/3)	$Im\bar{3}m$	Ag$_3$SI, (anti-perovskite) cP5, 221	Tet			*bcc*	[47]
nido, closo-	*rt*-$Na_{2-x}(B_{11}H_{14})_x(B_{12}H_{12})_{1-x}$ (x = 1/3)	$Pm\bar{3}n$	Ag$_3$SI, (anti-perovskite) cP5, 221	Tet			*bcc*	[47]
nido, closo-	*ht*-$Na_{2-x}(B_{11}H_{14})_x(B_{12}H_{12})_{1-x}$ (x = 1/3)	$Im\bar{3}m$	Ag$_3$SI, (anti-perovskite) cP5, 221	Tet			*bcc*	[47]
nido, closo-	$Na_3BH_4B_{12}H_{12}$	$Cmc2_1$	novel	Tet			*CrB*	[23]
K$^+$ *closo-*	$K_2B_{12}H_{12}$	$Fm\bar{3}$	anti-CaF$_2$, cF12, 225	Tet-tri	Cuboct	Cub	*ccp*	[76]
closo-	$K_2B_{10}H_{10}$	$P2_1/c$	Ni$_2$In (Co$_{1.75}$Ge), hP6, 194	Tet-tri, Oct-mon, bi, tri, tetra	12-fold, 14-fold	2cap-SqAntPris	*hcp*	[77]
arachno-	α-KB_3H_8	$P2_1/m$	NaCl, cF8, 225	Oct-tri, tetra, penta	24-fold	Oct	*ccp*	[78]
arachno-	α′-KB_3H_8	$Cmcm$	NaCl, cF8, 225	Oct-bi, tri, tetra	18-fold	Oct	*ccp*	[78]
arachno-	β-KB_3H_8	$Fm\bar{3}m$	NaCl, cF8, 225	Oct		Oct	*ccp*	[78]
closo-	$K_2B_6H_6$	$Fm\bar{3}m$	anti-CaF$_2$, cF12, 225	Tet-tri	Cuboct	Cub	*ccp*	[79]
closo-	*rt*-$KCB_{11}H_{12}$	$P2_1/c$	NaCl, cF8, 225	Oct-mon, bi, tri	12-fold, 10-fold	Oct	*ccp*	[80]
closo-	*ht*-$KCB_{11}H_{12}$	$Fm\bar{3}m$	NaCl, cF8, 225	Oct		Oct	*ccp*	[80]

Table 1. *Cont.*

Cation	Compound	Space Group	Structural Prototype	Cation Coordination by Anions	Cation Coordination by Hydrogen and Halogen	Anion Coordination	Aristotype of Anion Packing	Ref.
closo, *closo-*	$KB_{21}H_{18}$	$C2$	NiAs, hP4, 194	TriAntPris-mon, tri	14-fold	TriPris	*hcp*	[81]
nido, *closo-*	m-$K_3BH_4B_{12}H_{12}$	$P2/c$	Ag$_3$SI, (anti-perovskite) cP5, 221	Oct-bi, tri	14-fold	BH_4-Oct, $B_{12}H_{12}$-Cuboct	*bcc*	[82]
nido, *closo-*	r-$K_3BH_4B_{12}H_{12}$	$R\bar{3}m$	Ag$_3$SI, (anti-perovskite) cP5, 221	Oct		BH_4-Oct, $B_{12}H_{12}$-Cuboct	*bcc*	[82]
nido, *closo-*	c-$K_3BH_4B_{12}H_{12}$	$P23$	Ag$_3$SI, (anti-perovskite) cP5, 221	Oct		BH_4-Oct, $B_{12}H_{12}$-Cuboct	*bcc*	[82]
Rb+ *closo-*	$Rb_2B_{12}H_{12}$	$Fm\bar{3}$	anti-CaF$_2$, cF12, 225	Tet-tri	Cuboct	Cub	*ccp*	[83]
closo-	rt-$Rb_2B_{10}H_{10}$	$P2_1/c$	Ni$_2$In (Co$_{1.75}$Ge), hP6, 194	Tet-bi, tri, tetra Oct-mon, bi, tri, tetra	12-fold, 17-fold	3cap-SqAntPris	*hcp*	[77,84]
closo-	lt-$Rb_2B_{10}H_{10}$	$P\bar{1}$	Ni$_2$In (Co$_{1.75}$Ge), hP6, 194	Tet-bi, tri, tetra Oct-mon, bi, tri, tetra	12-fold, 17-fold	3cap-SqAntPris	*hcp*	[84]
closo-	$Rb_2B_9H_9$	$P4/nmm$	anti-BiF$_3$, cF16, 225	Tet-tri, SqPyr-bi, tri	Cuboct, 14-fold	4cap-TriPris	*ccp*	[85]
closo, *closo-*	$Rb_2B_{20}H_{18}$	$Pna2_1$	anti-MoSi$_2$, tI6, 139	SqPyr-mon, tri	13-fold	2cap-Cub	Mo in MoSi$_2$ {$3^{48}.4^{66}.5^6$} 16-c net	[86]
nido, *closo-*	$Rb_3BH_4B_{12}H_{12}$	$P23$	Ag$_3$SI, (anti-perovskite) cP5, 221	Oct		BH_4-Oct, $B_{12}H_{12}$-Cuboct	*bcc*	[87]
Cs$^+$ *closo-*	rt-$Cs_2B_{12}H_{12}$	$Fm\bar{3}$	anti-CaF$_2$, cF12, 225	Tet-tri	Cuboct	Cub	*ccp*	[88]
closo-	ht-$Cs_2B_{12}H_{12}$	$Fm\bar{3}m$	anti-CaF$_2$, cF12, 225	Tet		Cub	*ccp*	[89]
closo-	$Cs_2B_6H_6$	$Fm\bar{3}m$	anti-CaF$_2$, cF12, 225	Tet-tri	Cuboct	Cub	*ccp*	[90]
arachno-	CsB_3H_8	$Ama2$	NaCl, cF8, 225	Oct-bi, tri, tetra	16-fold	Oct	*ccp*	[90]
closo-	$Cs_2B_8H_8$	$Pnma$	Ni$_2$In (Co$_{1.75}$Ge), hP6, 194	Tet-tri, SqPyr-bi, tri	12-fold	1cap-SqAntPris	*hcp*	[91]
nido-	CsB_9H_{14}	$P\bar{1}$	CsCl, cP2, 221	Cub		Cub	*aa*	[92]
closo, *closo-*	$CsB_{21}H_{18}$	$P2_1/c$	NaCl, cF8, 225	TriAntPris-mon, bi, tri	15-fold	TriAntPris	*ccp*	[93]
closo-	rt-$CsCB_{11}H_{12}$	$R3$ ($R3/c$ above 323 K)	novel	Oct-bi, tri	15-fold	6-fold	*hcp*	[94]

Table 1. *Cont.*

Cation	Compound	Space Group	Structural Prototype	Cation Coordination by Anions	Cation Coordination by Hydrogen and Halogen	Anion Coordination	Aristotype of Anion Packing	Ref.
closo-	ht-CsCB$_{11}$H$_{12}$	$I\bar{4}3d$	Th$_3$P$_4$, cI28, 220	Oct-bi, tri	14-fold	8-fold	*hcp*	[94]
closo-	CsCB$_{11}$H$_6$Cl$_6$	$P2/c$	CsCl, cP2, 221	Cub-bi, tri	6cap-Cub, 16-fold	Cub	*aa*	[95]
closo-	CsCB$_{11}$H$_{11}$Br$_1$	$P2_1/n$	NaCl, cF8, 225	Oct-mon, bi, tri	12-fold	Oct	*ccp*	[96]
closo-	CsCB$_{11}$H$_6$Br$_6$	$P2/c$	CsCl, cP2, 221	Cub-bi, tri	6cap-Cub, 16-fold	Cub	*aa*	[95]
closo-	CsCB$_9$H$_5$Br$_5$	$P4/nmm$	sphalerite, cF8, 216	Tet-tri	Cuboct	Tet	*ccp*	[95]
nido-	α-CsC$_2$B$_9$H$_{12}$	$I4$	AgI, cI38, 229	Tet		SqPris	*bcc*	[97]
nido-	β-CsC$_2$B$_9$H$_{12}$	$P2_1/c$	NiAs, hP4, 194—CsCl, cP2, 221	8-fold		8-fold	*hcp*	[97]
nido-	γ-CsC$_2$B$_9$H$_{12}$	$P2_1/c$	NiAs, hP4, 194	Oct		TriPris	*hcp*	[97]
closo, *closo-*	Cs(B$_{21}$H$_{18-x}$F$_x$) (x = 2.55–2.85)	$P2_1/n$ ($P2_1/c$)	NiAs, hP4, 194	TriAntPris-bi, tri, tetra	18-fold	TriPris	*hcp*	[93]
closo, *closo-*	Cs(B$_{21}$H$_{18-x}$F$_x$) (x = 2.85–3)	$P2_12_12_1$	NiAs, hP4, 194	TriAntPris-bi, tri, tetra	18-fold	TriPris	*hcp*	[93]
nido, *closo-*	Cs$_3$BH$_4$B$_{12}$H$_{12}$	$P23$	Ag$_3$SI, (anti-perovskite) cP5, 221	Oct-bi	Cuboct	BH$_4$-Oct, B$_{12}$H$_{12}$-Cuboct	*bcc*	[98]
Be^{2+} *arachno-*	Be(B$_3$H$_8$)$_2$	$P2_1/c$		Lin-bi	Tet		*ccp* of Be(B$_3$H$_8$)$_2$ molecules	[99]
arachno- *nido-*	Be(B$_5$H$_{10}$)BH$_4$	$P2_1/c$		Lin-bi	Tet		*ccp* of Be(B$_5$H$_{10}$)BH$_4$ molecules	[100]
Ca^{2+} *closo-*	CaB$_{12}$H$_{12}$	$C2/c$	BN-b, hP4, 194	TriBiPyr-mon, bi	8-fold	TriBiPyr	*hcp*	[101]
closo-	α-CaB$_{10}$H$_{10}$	Cc	wurtzite, hP4, 186	Tet-mon, tri	10-fold	Tet	*hcp*	[102]
closo-	β-CaB$_{10}$H$_{10}$	$Pbca$	FePO$_4$, oP48, 61	Tet-bi, tri	10-fold	Tet	*hcp*	[102]
Sr^{2+} *closo-*	SrB$_{12}$H$_{12}$	$P31c$	wurtzite, hP4, 186	Tet-tri	Cuboct	Tet	*hcp*	[103]

Table 1. *Cont.*

Cation	Compound	Space Group	Structural Prototype	Cation Coordination by Anions	Cation Coordination by Hydrogen and Halogen	Anion Coordination	Aristotype of Anion Packing	Ref.
Ba^{2+} *closo-*	$BaB_{12}H_{12}$	$P31c$	wurtzite, hP4, 186	Tet-tri	Cuboct	Tet	*hcp*	[103]
Cr^{2+} *arachno-*	$Cr(B_3H_8)_2$	$P\bar{1}$		Lin-bi	Sq		*aa* of $Cr(B_3H_8)_2$ molecules	[104]
Mn^{2+} *closo-*	$MnB_{12}H_{12}$	$P31c$	BN-b, hP4, 194	Lin-tri	Oct	Lin	*hcp*	[105]
Fe^{2+} *closo-*	$FeB_{12}H_{12}$	$R\bar{3}$		Lin-tri	Oct	Tri	*bcc*	[105]
Co^{2+} *closo-*	$CoB_{12}H_{12}$	$R\bar{3}$		Lin-tri	Oct	Tri	*bcc*	[106]
Ni^{2+} *closo-*	$NiB_{12}H_{12}$	$R\bar{3}$		Lin-tri	Oct	Tri	*bcc*	[106, 107]
Cu^+ *closo-*	$Cu_2B_{12}H_{12}$	$Pn\bar{3}$	Cu_2O, cP6, 224	Lin-bi	4-fold	Tet	*bcc*	[106]
Ag^+ *closo-*	α-$Ag_2B_{12}H_{12}$	$Pa\bar{3}$	anti-CaF_2, cF12, 225	Tri-bi	TriAntPris	TriAntPris	*ccp*	[108]
closo-	β-$Ag_2B_{12}H_{12}$	$Fm\bar{3}m$	anti-CaF_2, cF12, 225	Tet		Cub	*ccp*	[108]
closo-	α-$Ag_2B_{10}H_{10}$	$P4/nnc$	anti-CaF_2, cF12, 225	Tet-tri	Cuboct	Cub	*ccp*	[108]
closo-	β-$Ag_2B_{10}H_{10}$	$Fm\bar{3}m$	anti-CaF_2, cF12, 225	Tet		Cub	*ccp*	[108]
closo-	$AgCB_9H_{10}$	$P2_1/m$	NaCl, cF8, 225	TriAntPris-mon, tri	10-fold	TriPris	*ccp*	[109]
closo-	$AgCB_{11}H_6Br_6$	$Pnma$	NaCl, cF8, 225	Lin-tri	Oct	NLin	*ccp*	[110]
Hg^{2+} *closo-*	$HgB_{12}H_{12}$	$P\bar{1}$	CsCl, cP2, 221	Cub		Cub	*ccp*	[111]
Tl^+ *closo-*	$Tl_2B_{10}H_{10}$	$Fm\bar{3}$	anti-CaF_2, cF12, 225	Tet-tri	Cuboct	Cub	*ccp*	[112]
Pb^{2+} *closo-*	$PbB_{12}H_{12}$	$P31c$	wurtzite, hP4, 186	Tet-tri	Cuboct	Tet	*hcp*	[113]

Abbreviations: Term = terminal, Lin = collinear, NLin = non-collinear, Tri = coplanar triangular, NTri = non-coplanar triangular, Sq = square, Sad = saddle-like, Tet = tetrahedron or tetragonal, Pent = pentagonal, Oct = octahedron, Hex = hexagonal, Rho = rhombic, Cuboct = cuboctahedron, Dodeca = dodecahedron, Pris = prism, Pyr = pyramid, Ant = anti, cap = capped. The degree of deformation of the coordination polyhedron from ideal shape is not specified. mon = monodentate, bi = bidentate, tri = tridentate. *lt* = low temperature, *rt* = room temperature, *ht* = high temperature, *hp* = high pressure, *m* = monoclinic, *o* = orthorhombic, *h* = hexagonal (refers to different polymorphs). *ccp* = cubic close packed, *hcp* = hexagonal close packed, *bcc* = body-centered cubic, *aa* = simple sequence of hexagonal layers. 1: gyroelongated square bipyramid; 2: triangular orthobicupola; 3: snub disphenoid (dodecadeltahedron). Color coding: alkali metals: brown; alkali earth metals: orange; rare earths: magenta; transition elements: blue; actinides: dark green; other metals: light green.

Table 2. Structural classification of double cation metal hydroborates. See the caption of Table 1 for details.

Cation	Compound	Space Group	Structural Prototype	Cation Coordination by Anions	Cation Coordination by Hydrogen	Anion Coordination	Anion Packing	Ref.
Li^+, Na^+ *closo-*	rt-$(Li_x,Na_{1-x})_2B_{12}H_{12}$ ($x = 0.33$–0.67)	$Pa\bar{3}$	anti-CaF_2, cF12, 225	Tri-bi	Oct	Oct	*ccp*	[114]
closo-	ht-$(Li_x,Na_{1-x})_2B_{12}H_{12}$ ($x = 0.377$)	$Fm\bar{3}m$	anti-CaF_2, cF12, 225	Tri			*ccp*	[115]
nido-, *closo-*	$(Li_{0.7}Na_{0.3})_3BH_4B_{12}H_{12}$	$Pna2_1$	novel	Tet			*FeB*	[23]
Li^+, K^+ *closo-*	rt-$LiKB_{12}H_{12}$	$P6_3mc$	Ni_2In ($Co_{1.75}Ge$), hP6, 194	Li-Tri-bi, K-Oct-bi, tri	Li-TriPris K-15-fold	3cap-TriPris	*hcp*	[116]
closo-	ht-$(Li_x,K_{1-x})_2B_{12}H_{12}$ ($x = 0.8$)	$P6_3mc$	Ni_2In ($Co_{1.75}Ge$), hP6, 194	Li-Tri, (K,Li)-Oct		3cap-TriPris	*hcp*	[21]
Li^+, Cs^+ *closo-*	rt-$LiCsB_{12}H_{12}$	$Pn2_1a$	Ni_2In ($Co_{1.75}Ge$), hP6, 194	Li-Tri, Cs-Oct		3cap-TriPris	*hcp*	[21]
closo-	ht-$LiCsB_{12}H_{12}$	$P6_3mc$	Ni_2In ($Co_{1.75}Ge$), hP6, 194	Li-Tri, Cs-Oct		3cap-TriPris	*hcp*	[21]
closo-	$meta$-$(Li_x,Cs_{1-x})_2B_{12}H_{12}$ ($x = 0.11$)	$P6_3mc$	Ni_2In ($Co_{1.75}Ge$), hP6, 194	(Cs,Li)-Tri, Tet Cs-Oct		10-fold	*hcp*	[21]
Na^+, K^+ *closo-*	rt-$(Na_x,K_{1-x})_2B_{12}H_{12}$ ($x = 0.5$)	$Fm\bar{3}$	anti-CaF_2, cF12, 225	Tet-tri	Cuboct	Cub	*ccp*	[21]
closo-	ht-$(Na_x,K_{1-x})_2B_{12}H_{12}$ ($x = 0.5$)	$Pm\bar{3}n$	Ag_2S, cI20, 229	Tet			*bcc*	[21]
closo-	rt-$NaKB_{12}H_{12}$	$P1$	Ni_2In ($Co_{1.75}Ge$), hP6, 194	Na-Tet, K-Oct		4cap-TriPris	*hcp*	[21]
closo-	ht-$NaKB_{12}H_{12}$	$P6_3mc$	Ni_2In ($Co_{1.75}Ge$), hP6, 194	Na-Tet, K-Oct		4cap-TriPris	*hcp*	[21]
Na^+, Cs^+ *closo-*	rt-$NaCsB_{12}H_{12}$	$P\bar{1}$	Ni_2In ($Co_{1.75}Ge$), hP6, 194	Na-Tet, Cs-Oct		4cap-TriPris	*hcp*	[21]
closo-	$ht1$-$NaCsB_{12}H_{12}$	$P2_1/c$	Ni_2In ($Co_{1.75}Ge$), hP6, 194	Na-Tet, Cs-Oct		4cap-TriPris	*hcp*	[21]
closo-	$ht2$-$NaCsB_{12}H_{12}$	$Pn2_1a$	Ni_2In ($Co_{1.75}Ge$), hP6, 194	Na-Tet, Cs-Oct		4cap-TriPris	*hcp*	[21]
closo-	$ht3$-$NaCsB_{12}H_{12}$	$P6_3mc$	Ni_2In ($Co_{1.75}Ge$), hP6, 194	Na-Tet, Cs-Oct		4cap-TriPris	*hcp*	[21]
Cs^+, Cu^+ *closo-*	$Cs(CuB_{10}H_{10})$	$Pbcn$	novel	Cs- Oct-mon, bi, tri Cu-Lin-bi	Cs-15-fold Cu-Tet	8-fold	*ccp*	[117]
Cs^+, Ag^+ *closo-*	$Cs(AgB_{10}H_{10})$	$Pbcm$	anti-BiF_3, cF16, 225	Cs-Oct-mon, bi, tri Ag-Tet-bi	Cs-15-fold Ag-8-fold	2cap-SqAntPris	*ccp*	[118]

Color coding: alkali metals: brown; transition elements: blue.

Table 3. Structural classification of hydroborates with complex inorganic cations. See the caption of Table 1 for details.

Cation	Compound	Space Group	Structural Prototype	Cation Coordination by Anions	Cation Coordination by Hydrogen	Anion Coordination	Anion Packing	Ref.
NH_4^+ *closo-*	$(NH_4)_2B_{12}H_{12}$	$Fm\bar{3}$	anti-CaF_2, cF12, 225	Tet-tri	Cuboct	Cub	*ccp*	[83]
closo-	$(NH_4)_2B_{10}H_{10}$	$P2_1/c$	Ni_2In ($Co_{1.75}Ge$), hP6, 194	Tet, Oct		8-fold	*hcp*	[119]
arachno-	$NH_4B_3H_8$	$Cmcm$	NaCl, cF8, 225	Oct-bi, tri, tetra	18-fold	Oct	*ccp*	[120]
$N_2H_5^+$ *closo-*	$(N_2H_5)_2B_{12}H_{12}$	$C2/c$	anti-AlB_2, hP3, 191	TriPris		HexPris	*aa*	[121]
$N(n\text{-}Bu_4)^+$ *closo-*	$N(n\text{-}Bu_4)B_8H_9$	$P4_1$	LiSe subnet in NH_4LiSeO_4, oP28, 33	Tet		Sq	*aa*	[91]
closo-	$N(n\text{-}Bu_4)_2B_7H_7$	$P2_1/n$ ($P2_1/c$)	anti-NiP subnet in $Ba_2Ni(PO_4)_2$, mP52, 14	Tri, Lin		TriBiPyr	*hcp*	[122]
closo-	$N(n\text{-}Bu_4)B_7H_8$	$P4_1$	LiSe subnet in NH_4LiSeO_4, oP28, 33	Sq		Sq	*aa*	[122]
Ph_4P^+ *closo-*	lt-$(Ph_4P)B_8H_9$	$P2/c$	CsCl, cP2, 221	SqAntPris		Cub	*aa*	[91]
closo-	rt-$(Ph_4P)B_8H_9$	$P4/n$	CsCl, cP2, 221	SqAntPris		Cub	*aa*	[91]
Ph_3P^+ *nido-*	$(Ph_3P)B_{11}H_{14}$	$P2_1/m$	BN-b, hP4, 194	TriBiPyr		TriBiPyr	*aa*	[123]
PNP^+ *closo-*	$(PNP)B_7H_8$	$P2_1/c$	novel	9-fold		9-fold	*aa*	[122]
$HAgu^+$ *closo-*	$(HAgu)_2B_{12}H_{12}$	$P\bar{1}$	anti-AlB_2, hP3, 191	TriPris		HexPris	*aa*	[124]
$HAgu^+$ Cu^{2+} *closo-*	$(Cu_{0.61}H_{0.78}Agu_2)B_{12}H_{12}$	$P\bar{1}$	CsCl, cP2, 221	RhoPris		RhoPris	*aa*	[124]

Color coding: transition elements: blue.

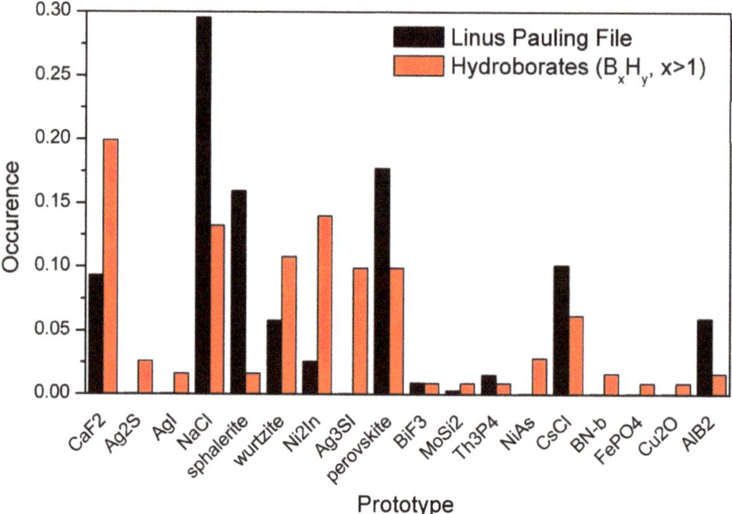

Figure 4. Occurrence of selected structural prototypes among binary inorganic compounds extracted from the Linus Pauling File [51], and among the hydroborates (excluding borohydrides with BH_4^- anion). Both graphs include ternary solid solutions. The fractions are renormalized, to sum up to one for each graph.

8. Conclusions

Contrary to borohydrides, where all types of inorganic solid compounds can be found ranging from salts based on anions packing up to 2D and 3D coordination polymers, the higher hydroborates usually form salts where the large hydroborate anions pack, in most cases, according to one of the three common packings. The exception are hydroborates of Be that form molecular compounds, and transition metals hydroborates, which have still a packed anion sublattice, but the resulting metal–anion interaction is covalent, classifying these compounds as coordination polymers. The same is valid also for double cation hydroborates, while the situation is different for poly-cations, where the compact anion sublattice is broken as a consequence of the directional anion–cation bonds. The exception is the nearly spherical (if orientationally disordered) NH_4^+ cation.

Contrary to other inorganic compounds, the hydroborate salts crystallize very often with the *bcc* anion packing, favoring 3D pathways based on tetrahedral sites for high cation mobility. The anion packing can be best controlled by anion mixing and to a certain extend by anion modification. The mixing of cations and inclusion of neutral molecules in the crystal has a negative effect on cations mobility. The knowledge of anions packing and the ability to control it play an important role in designing the materials with desired properties such as high cation mobility at *rt* in solid-state electrolytes. We hope that this review clarifies the existing relation between different hydroborates poly-anions and the resulting crystal structures, sustaining the material science community in designing novel hydroborate-based compounds.

Supplementary Materials: The following are available online at http://www.mdpi.com/2624-8549/2/4/53/s1, Figure S1: Cation coordination in single metal hydroborates as a function of cation/anion radius ratio C/A. Figure S2: Analogy between temperature polymorphs of $Li_2B_{12}H_{12}$ and C60, and between $Na_2B_{12}H_{12}$ and Ag2S.

Author Contributions: Conceptualization, R.Č.; writing—original draft preparation, R.Č., M.B. and F.M.; writing—review and editing, R.Č., M.B. and F.M.; All authors have read and agreed to the published version of the manuscript.

Funding: This research received no external funding and it was funded by the University of Geneva.

Acknowledgments: We acknowledge all colleagues who shared with us their unpublished results, i.e., the references marked "in preparation".

Conflicts of Interest: The authors declare no conflict of interest.

References and Note

1. Adams, R.M. Nomenclature of inorganic boron compounds. *Pure Appl. Chem.* **1972**, *30*, 681–710. [CrossRef]
2. Lipscomb, W.N. *Boron Hydrides*; W.A. Benjamin Inc.: New York, NY, USA, 1963.
3. Wade, K. The structural significance of the number of skeletal bonding electron-pairs in carboranes, the higher boranes and borane anions, and various transition-metal carbonyl cluster compounds. *J. Chem. Soc. D* **1971**, 792–793. [CrossRef]
4. Mingos, D.M.P. A General Theory for Cluster and Ring Compounds of the Main Group and Transition Elements. *Nat. Phys. Sci.* **1972**, *236*, 99–102. [CrossRef]
5. Clark, J.D. *Ignition!: An Informal History of Liquid Rocket Propellants*; Rutgers University Press: New Brunswick, NJ, USA, 1972.
6. Brown, H.C. *Hydroboration*; W.A. Benjamin Inc.: New York, NY, USA, 1962.
7. Douvris, C.; Ozerov, O.V. Hydrodefluorination of Perfluoroalkyl Groups Using Silylium-Carborane Catalysts. *Science* **2008**, *321*, 1188–1190. [CrossRef] [PubMed]
8. Shao, B.; Bagdasarian, A.L.; Popov, S.; Nelson, H.M. Arylation of Hydrocarbons Enabled by Organosilicon Reagents and Weakly Coordinating Anions. *Science* **2017**, *355*, 1403–1407. [CrossRef]
9. Barfh, R.F.; Soloway, A.H.; Fairchild, R.G.; Brugger, R.M. Boron Neutron Capture Therapy for Cancer. Realities and Prospects. *Cancer* **1992**, *70*, 2995–3007. [CrossRef]
10. Bondarev, O.; Khan, A.A.; Tu, X.; Sevryugina, Y.V.; Jalisatgi, S.S.; Hawthorne, M.F. Synthesis of [closo-$B_{12}(OH)_{11}NH_3$]$^-$: A New Heterobifunctional Dodecaborane Scaffold for Drug Delivery Applications. *J. Am. Chem. Soc.* **2013**, *135*, 13204–13211. [CrossRef]
11. Goswami, L.N.; Ma, L.; Chakravarty, S.; Cai, Q.; Jalisatgi, S.S.; Hawthorne, M.F. Discrete Nanomolecular Polyhedral Borane Scaffold Supporting Multiple Gadolinium(III) Complexes as a High Performance MRI Contrast Agent. *Inorg. Chem.* **2013**, *52*, 1694–1700. [CrossRef]
12. Johnson, J.W.; Brody, J.F. Lithium Closoborane Electrolytes: III. Preparation and Characterization. *J. Electrochem. Soc.* **1982**, *129*, 2213–2219. [CrossRef]
13. Udovic, T.J.; Matsuo, M.; Unemoto, A.; Verdal, N.; Stavila, V.; Skripov, A.V.; Rush, J.J.; Takamura, H.; Orimo, S. Sodium superionic conduction in $Na_2B_{12}H_{12}$. *Chem. Commun.* **2014**, *50*, 3750–3752. [CrossRef]
14. Udovic, T.J.; Matsuo, M.; Tang, W.S.; Wu, H.; Stavila, V.; Soloninin, A.V.; Skoryunov, R.V.; Babanova, O.A.; Skripov, A.V.; Rush, J.J.; et al. Exceptional Superionic Conductivity in Disordered Sodium Decahydro-closo-decaborate. *Adv. Mater.* **2014**, *26*, 7622–7626. [CrossRef] [PubMed]
15. Černý, R.; Schouwink, P. The crystal chemistry of inorganic metal borohydrides and their relation to metal oxides. *Acta Crystallogr. Sect. B Struct. Sci. Cryst. Eng. Mater.* **2015**, *71*, 619–640. [CrossRef] [PubMed]
16. Pauling, L. The Principles Determining the Structure of Complex Ionic Crystals. *J. Am. Chem. Soc.* **1929**, *51*, 1010–1026. [CrossRef]
17. Larsen, P.M.; Schmidt, S.; Schiøtz, J. Robust structural identification via polyhedral template matching. *Model. Simul. Mater. Sci. Eng.* **2016**, *24*, 055007. [CrossRef]
18. Brighi, M.; Murgia, F.; Černý, R. *Closo*-hydroborate sodium salts: An emerging class of room-temperature solid electrolytes. *Cell Rep. Phys. Sci.* **2020**, *1*, 100217d.
19. Kweon, K.E.; Varley, J.B.; Shea, P.; Adelstein, N.; Mehta, P.; Heo, T.W.; Udovic, T.J.; Stavila, V.; Wood, B.C. Structural, Chemical, and Dynamical Frustration: Origins of Superionic Conductivity in closo-Borate Solid Electrolytes. *Chem. Mater.* **2017**, *29*, 9142–9153. [CrossRef]
20. Bradley, J.N.; Green, P.D. Relationship of structure and ionic mobility in solid MAg_4I_5. *Trans. Faraday Soc.* **1967**, *63*, 2516–2521. [CrossRef]
21. Sadikin, Y.; Schouwink, P.; Brighi, M.; Łodziana, Z.; Černý, R. Modified Anion Packing of $Na_2B_{12}H_{12}$ in Close to Room Temperature Superionic Conductors. *Inorg. Chem.* **2017**, *56*, 5006–5016. [CrossRef]
22. Brighi, M.; Murgia, F.; Łodziana, Z.; Schouwink, P.; Wolczyk, A.; Černý, R. A mixed anion borane/carborane as a room temperature Na-ion solid electrolyte. *J. Power Sour.* **2018**, *404*, 7–12. [CrossRef]

23. Sadikin, Y.; Brighi, M.; Schouwink, P.; Černý, R. Superionic Conduction of Sodium and Lithium in Anion-Mixed Hydroborates $Na_3BH_4B_{12}H_{12}$ and $(Li_{0.7}Na_{0.3})_3BH_4B_{12}H_{12}$. *Adv. Energy Mater.* **2015**, *5*, 1501016. [CrossRef]

24. Hansen, B.R.S.; Tumanov, N.; Santoru, A.; Pistidda, C.; Bednarcik, J.; Klassen, T.; Dornheim, M.; Filinchuk, Y.; Jensen, T.R. Synthesis, structures and thermal decomposition of ammine $M_xB_{12}H_{12}$ complexes (M = Li, Na, Ca). *Dalton Trans.* **2017**, *46*, 7770–7781. [CrossRef] [PubMed]

25. Jørgensen, M.; Hansen, B.R.S.; Lee, Y.-S.; Cho, Y.-W.; Jensen, T.R. Crystal Structures and Energy Storage Properties of Ammine Sodium Decahydro-closo-decaboranes ($Na_2B_{10}H_{10}\cdot nNH_3$, *n* = 1,2). *J. Phys. Chem. C* **2019**, *123*, 20160–20166. [CrossRef]

26. Shannon, R.D. Revised effective ionic radii and systematic studies of interatomic distances in halides and chalcogenides. *Acta Crystallogr. Sect. A Cryst. Phys. Diffr. Theor. Gen. Crystallogr.* **1976**, *32*, 751–767. [CrossRef]

27. Hansen, B.R.S.; Paskevicius, M.; Li, H.W.; Akiba, E.; Jensen, T.R. Metal boranes: Progress and applications. *Coord. Chem. Rev.* **2016**, *323*, 60–70. [CrossRef]

28. George, J.; Waroquiers, D.; Di Stefano, D.; Petretto, G.; Rignanese, G.-M.; Hautier, G. The Limited Predictive Power of the Pauling Rules. *Angew. Chem. Int. Ed.* **2020**, *59*, 7569–7575. [CrossRef]

29. Wu, H.; Tang, W.S.; Stavila, V.; Zhou, W.; Rush, J.J.; Udovic, T.J. Structural Behavior of $Li_2B_{10}H_{10}$. *J. Phys. Chem. C* **2015**, *119*, 6481–6487. [CrossRef]

30. Verdal, N.; Her, J.-H.; Stavila, V.; Soloninin, A.V.; Babanova, O.A.; Skripov, A.V.; Udovic, T.J.; Rush, J.J. Complex high-temperature phase transitions in $Li_2B_{12}H_{12}$ and $Na_2B_{12}H_{12}$. *J. Solid State Chem.* **2014**, *212*, 81–91. [CrossRef]

31. Hansen, B.R.S.; Paskevicius, M.; Jørgensen, M.; Jensen, T.R. Halogenated Sodium-closo-Dodecaboranes as Solid-State Ion Conductors. *Chem. Mater.* **2017**, *29*, 3423–3430. [CrossRef]

32. Skripov, A.V.; Soloninin, A.V.; Babanova, O.A.; Skoryunov, R.V. Nuclear Magnetic Resonance Studies of Atomic Motion in Borohydride-Based Materials: Fast Anion Reorientations and Cation Diffusion. *J. Alloys Compd.* **2015**, *645*, S428–S433. [CrossRef]

33. Varley, J.B.; Kweon, K.; Mehta, P.; Shea, P.; Heo, T.W.; Udovic, T.J.; Stavila, V.; Wood, B.C. Understanding Ionic Conductivity Trends in Polyborane Solid Electrolytes from Ab Initio Molecular Dynamics. *ACS Energy Lett.* **2017**, *2*, 250–255. [CrossRef]

34. Skripov, A.V.; Soloninin, A.V.; Babanova, O.A.; Skoryunov, R.V. Anion and Cation Dynamics in Polyhydroborate Salts: NMR Studies. *Molecules* **2020**, *25*, 2940. [CrossRef]

35. Verdal, N.; Udovic, T.J.; Stavila, V.; Tang, W.S.; Rush, J.J.; Skripov, A.V. Anion Reorientations in the Superionic Conducting Phase of $Na_2B_{12}H_{12}$. *J. Phys. Chem. C* **2014**, *118*, 17483–17489. [CrossRef]

36. Callear, S.K.; Nickels, E.A.; Jones, M.O.; Matauo, M.; Orimo, S.-I.; Edwards, P.; David, W.I.F. Order and disorder in lithium tetrahydroborate. *J. Mater. Sci.* **2011**, *46*, 566–569. [CrossRef]

37. Heere, M.; Hansen, A.L.; Payandeh, S.H.; Aslan, N.; Gizer, G.; Sørby, M.H.; Hauback, B.C.; Ristidda, C.; Dornheim, M.; Lohstroh, W. Dynamics of porous and amorphous magnesium borohydride to understand solid state Mg-ionconductors. *Sci. Rep.* **2020**, *10*, 9080. [CrossRef] [PubMed]

38. Tiritiris, I.; Schleid, T.; Müller, K. Solid-State NMR Studies on Ionic *closo*-Dodecaborates. *Appl. Magn. Reson.* **2007**, *32*, 459–481. [CrossRef]

39. Verdal, N.; Udovic, T.J.; Rush, J.J.; Cappelletti, R.L.; Zhou, W. Reorientational Dynamics of the Dodecahydro-closo-dodecaborate Anion in $Cs_2B_{12}H_{12}$. *J. Phys. Chem. A* **2011**, *115*, 2933–2938. [CrossRef]

40. Skripov, A.V.; Babanova, O.A.; Soloninin, A.V.; Stavila, V.; Verdal, N.; Udovic, T.J.; Rush, J.J. Nuclear Magnetic Resonance Study of Atomic Motion in $A_2B_{12}H_{12}$ (A = Na, K, Rb, Cs): Anion Reorientations and Na^+ Mobility. *J. Phys. Chem. C* **2013**, *117*, 25961–25968. [CrossRef]

41. Bürgi, H.B.; Restori, R.; Schwarzenbach, D. Structure of C60: Partial orientational order in the room-temperature modification of C60. *Acta Crystallogr. Sect. B Struct. Sci. Cryst. Eng. Mater.* **1993**, *49*, 832–838. [CrossRef]

42. Williams, R.E. The polyborane, carborane, carbocation continuum: Architectural patterns. *Chem. Rev.* **1992**, *92*, 177–207. [CrossRef]

43. Aihara, J. Three-dimensional aromaticity of polyhedral boranes. *J. Am. Chem. Soc.* **1978**, *100*, 3339–3342. [CrossRef]

44. Von Ragué Schleyer, P.; Najafian, K. Stability and Three-Dimensional Aromaticity of *closo*-Monocarbaborane Anions, $CB_{n-1}H_n^-$, and *closo*-Dicarboranes, $C_2B_{n-2}H_n$. *Inorg. Chem.* **1998**, *37*, 3454–3470. [CrossRef] [PubMed]

45. Douvris, C.; Michl, J. Update 1 of: Chemistry of the Carba-*closo*-dodecaborate(–) Anion, $CB_{11}H_{12}^-$. *Chem. Rev.* **2013**, *113*, PR179–PR233. [CrossRef] [PubMed]

46. Hofmann, M.; Fox, M.A.; Greatrex, R.; von Ragué Schleyer, P.; Williams, R.E. Empirical and ab Initio Energy/Architectural Patterns for 73 nido-6⟨V⟩-Carborane Isomers, from $B_6H_9^-$ to $C_4B_2H_6$. *Inorg. Chem.* **2001**, *40*, 1790–1801. [CrossRef] [PubMed]

47. Payandeh, S.H.; Asakura, R.; Avramidou, P.; Rentsch, D.; Łodziana, Z.; Černý, R.; Remhof, A.; Battaglia, C. Nido-Borate/Closo-Borate Mixed-Anion Electrolytes for All-Solid-State Batteries. *Chem. Mater.* **2020**, *32*, 1101–1110.

48. Tang, W.S.; Matsuo, M.; Wu, H.; Stavila, V.; Zhou, W.; Talin, A.A.; Soloninin, A.V.; Skoryunov, R.V.; Babanova, O.A.; Skripov, A.V.; et al. Liquid-Like Ionic Conduction in Solid Lithium and Sodium Monocarba—closo—Decaborates Near or at Room Temperature. *Adv. Energy Mater.* **2016**, *6*, 1502237. [CrossRef]

49. Bergerhoff, G.; Brown, I.D.; Allen, F. *Crystallographic Databases*; (Hrsg.) International Union of Crystallography: Chester, UK, 1987. Available online: http://www.fiz-karlsruhe.de/icsd_home.html.

50. Villars, P.; Cenzual, K. *Pearson's Crystal Data-Crystal Structure Database for Inorganic Compounds*; ASM International Materials Park: Russel Township, OH, USA, 2015.

51. Villars, P.; Cenzual, K.; Gladyshevskii, R.; Iwata, S. Pauling File—Towards a holistic view. *Chem. Met. Alloys* **2018**, *11*, 43–76. [CrossRef]

52. Paskevicius, M.; Jepsen, L.H.; Schouwink, P.; Černý, R.; Ravnsbaeck, D.; Filinchuk, Y.; Dornheim, M.; Besenbacher, F.; Jensen, T.R. Metal borohydrides and derivatives—Synthesis, structure and properties. *Chem. Soc. Rev.* **2017**, *46*, 1565–1634. [CrossRef]

53. Axtell, J.C.; Saleh, L.M.A.; Qian, E.A.; Wixtrom, A.I.; Spokoyny, A.M. Synthesis and Applications of Perfunctionalized Boron Clusters. *Inorg. Chem.* **2018**, *57*, 2333–2350. [CrossRef]

54. Muetterties, E.L.; Balthis, J.H.; Chia, Y.T.; Knoth, W.H.; Miller, H.C. Chemistry of Boranes. VIII. Salts and Acids of $B_{10}H_{10}^{-2}$ and $B_{12}H_{12}^{-2}$. *Inorg. Chem.* **1964**, *3*, 444–451. [CrossRef]

55. Sivaev, I.B.; Prikaznov, A.V.; Naoufal, D. Fifty years of the closo-decaborate anion chemistry. *Collect. Czechoslov. Chem. Commun.* **2010**, *75*, 1149–1199. [CrossRef]

56. Sivaev, I.B.; Bregadze, V.I.; Sjöberg, S. Chemistry of closo-Dodecaborate Anion $[B_{12}H_{12}]^{2-}$: A Review. *Collect. Czechoslov. Chem. Commun.* **2002**, *67*, 679–727. [CrossRef]

57. Housecroft, C.E. Carboranes as guests, counterions and linkers in coordination polymers and networks. *J. Organomet. Chem.* **2015**, *798*, 218–228. [CrossRef]

58. Malinina, E.A.; Avdeeva, V.V.; Goeva, L.V.; Kuznetsov, N.T. Coordination Compounds of Electron-Deficient Boron Cluster Anions $B_nH_n^{2-}$ (n = 6, 10, 12). *Russ. J. Inorg. Chem.* **2010**, *55*, 2148–2202. [CrossRef]

59. Müller, U. *Symmetry Relationships between Crystal Structures*; Oxford University Press: Oxford, UK, 2013.

60. Blatov, V.A. Multipurpose christalochemical analysis with the program package TOPOS. *IUCr Comp. Comm. Newsl.* **2006**, *7*, 4–38.

61. Topospro. Available online: https://topospro.com/databases/ttd/ (accessed on 29 September 2020).

62. Her, J.-H.; Yousufuddin, M.; Zhou, W.; Jalisatgi, S.S.; Kulleck, J.G.; Zan, J.A.; Hwang, S.-J.; Bowman, R.C.; Udovic, T.J. Crystal structure of $Li_2B_{12}H_{12}$: A possible intermediate species in the decomposition of $LiBH_4$. *Inorg. Chem.* **2008**, *47*, 9757–9759. [CrossRef]

63. Paskevicius, M.; Pitt, M.P.; Brown, D.H.; Sheppard, D.A.; Chumphongphan, S.; Buckley, C.E. First-order phase transition in the $Li_2B_{12}H_{12}$ system. *Phys. Chem. Chem. Phys.* **2013**, *15*, 15825–15828. [CrossRef]

64. This work. The symmetry of *Pa*-3 from ref. [63] was corrected to *Fm*-3.

65. Tang, W.S.; Unemoto, A.; Zhou, W.; Stavila, V.; Matsuo, M.; Wu, H.; Orimo, S.-I.; Udovic, T.J. Unparalleled lithium and sodium superionic conduction in solid electrolytes with large monovalent cage-like anions. *Energy Environ. Sci.* **2015**, *8*, 3637–3645. [CrossRef]

66. Jørgensen, M.; Shea, P.T.; Tomich, A.; Varley, J.B.; Bercx, M.; Lovera, S.; Černý, R.; Zhou, W.; Udovic, T.J.; Lavallo, V.; et al. Understanding Superionic Conductivity in Lithium and Sodium Salts of Weakly Coordinating *Closo*-Hexahalocarbaborate Anions. *Chem. Mater.* **2020**, *32*, 1475–1487. [CrossRef]

67. Payandeh, S.H.; Rentsch, D.; Asakura, R.; Łodziana, Z.; Bigler, L.; Černý, R.; Remhof, A.; Battaglia, C. Nido-Hydroborate based Electrolytes for All-Solid-State Lithium Batteries. 2020; in preparation.

68. Tang, W.S.; Yoshida, K.; Soloninin, A.V.; Skoryunov, R.V.; Babanova, O.A.; Skripov, A.V.; Dimitrievska, M.; Stavila, V.; Orimo, S.-I.; Udovic, T.J. Stabilizing Superionic-Conducting Structures via Mixed-Anion Solid Solutions of Monocarba-closo-borate Salts. *ACS Energy Lett.* **2016**, *1*, 659–664. [CrossRef]

69. Moury, R.; Łodziana, Z.; Remhof, A.; Duchêne, L.; Roedern, E.; Gigante, A.; Hagemann, H. Pressure-induced phase transitions in $Na_2B_{12}H_{12}$, structural investigation on a candidate for solid-state electrolyte. *Acta Crystallogr. Sect. B Struct. Sci. Cryst. Eng. Mater.* **2019**, *75*, 406–413. [CrossRef]

70. Wu, H.; Tang, W.S.; Zhou, W.; Stavila, V.; Rush, J.J.; Udovic, T.J. The structure of monoclinic $Na_2B_{10}H_{10}$: A combined diffraction, spectroscopy, and theoretical approach. *CrystEngComm* **2015**, *17*, 3533–3540. [CrossRef]

71. Dunbar, A.C.; Macor, J.A.; Girolami, G.S. Synthesis and single crystal structure of sodium octahydrotriborate, NaB_3H_8. *Inorg. Chem.* **2014**, *53*, 822–826. [CrossRef] [PubMed]

72. Wu, H.; Tang, W.S.; Zhou, W.; Tarver, J.D.; Stavila, V.; Brown, C.M.; Udovic, T.J. The low-temperature structural behavior of sodium 1-carba-closo-decaborate: $NaCB_9H_{10}$. *J. Solid State Chem.* **2016**, *243*, 162–167. [CrossRef]

73. Tang, W.S.; Dimitrievska, M.; Stavila, V.; Zhou, W.; Wu, H.; Talin, A.A.; Udovic, T.J. Order−Disorder Transitions and Superionic Conductivity in the Sodium nido—Undeca(carba)borates. *Chem. Mater.* **2017**, *29*, 10496–10509. [CrossRef]

74. Yoshida, K.; Sato, T.; Unemoto, A.; Matsuo, M.; Ikeshoji, T.; Udovic, T.J.; Orimo, S.-I. Fast sodium ionic conduction in $Na_2B_{10}H_{10}$-$Na_2B_{12}H_{12}$ pseudo-binary complex hydride and application to a bulk-type all-solid-state battery. *Appl. Phys. Lett.* **2017**, *110*, 103901. [CrossRef]

75. Duchêne, L.; Kuhnel, R.S.; Rentsch, D.; Remhof, A.; Hagemann, H.; Battaglia, C. A highly stable sodium solid-state electrolyte based on a dodeca/deca-borate equimolar mixture. *Chem. Commun.* **2017**, *53*, 4195–4198. [CrossRef] [PubMed]

76. Wunderlich, J.A.; Lipscomb, W.N. Structure of $B_{12}H_{12}^{-2}$ ion. *J. Am. Chem. Soc.* **1960**, *82*, 4427–4428. [CrossRef]

77. Hofmann, K.; Albert, B. Crystal structures of $M_2[B_{10}H_{10}]$ (M = Na, K, Rb) via real space simulated annealing powder techniques. *Z. Kristall.* **2005**, *220*, 142–146. [CrossRef]

78. Grinderslev, J.B.; Møller, K.T.; Yan, Y.; Chen, X.-M.; Li, Y.; Li, H.-W.; Zhou, W.; Skibsted, J.; Chen, X.; Jensen, T.R. Potassium octahydridotriborate: Diverse polymorphism in a potential hydrogen storage material and potassium ion conductor. *Dalton Trans.* **2019**, *48*, 8872–8881. [CrossRef]

79. Kuznetsov, I.Y.; Vinitskii, D.M.; Solntsev, K.A.; Kuznetsov, N.T. The Crystal Structure of $K_2B_6H_6$ and $Cs_2B_6H_6$. *Russ. J. Inorg. Chem.* **1987**, *32*, 1803–1804.

80. Dimitrievska, M.; Wu, H.; Stavila, V.; Babanova, O.A.; Skoryunov, R.V.; Soloninin, A.V.; Zhou, W.; Trump, B.A.; Andersson, M.S.; Skripov, A.V.; et al. Structural and Dynamical Properties of Potassium Dodecahydro-monocarba-closo-dodecaborate: $KCB_{11}H_{12}$. *J. Phys. Chem. C* **2020**, *124*, 17992–18002. [CrossRef]

81. Bernhardt, E.; Brauer, D.; Finze, M.; Willner, H. Closo-$[B_{21}H_{18}]^-$: A Face-Fused Diicosahedral Borate Ion. *Angew. Chem. Int. Ed.* **2007**, *46*, 2927–2930. [CrossRef] [PubMed]

82. Sadikin, Y.; Skoryunov, R.V.; Babanova, O.A.; Soloninin, A.V.; Lodziana, Z.; Brighi, M.; Skripov, A.V.; Černý, R. Anion disorder in $K_3BH_4B_{12}H_{12}$ and its effect on cation mobility. *J. Phys. Chem. C* **2017**, *121*, 5503–5514. [CrossRef]

83. Tiritiris, I.; Schleid, T. The Dodecahydro-*closo*-Dodecaborates $M_2[B_{12}H_{12}]$ of the Heavy Alkali Metals (M = K$^+$, Rb+, NH$_4^+$, Cs$^+$) and their Formal Iodide Adducts $M_3I[B_{12}H_{12}]$ (= MI·$M_2[B_{12}H_{12}]$). *Z. Anorg. Allg. Chem.* **2003**, *629*, 1390–1402. [CrossRef]

84. Dimitrievska, M.; Stavila, V.; Soloninin, A.V.; Skoryunov, R.V.; Babanova, O.A.; Wu, H.; Zhou, W.; Tang, W.S.; Faraone, A.; Tarver, J.D.; et al. Nature of Decahydro-Closo-Decaborate Anion Reorientations in an Ordered Alkali-Metal Salt: $Rb_2B_{10}H_{10}$. *J. Phys. Chem. C* **2018**, *122*, 15198–15207. [CrossRef]

85. Guggenberger, L.J. Chemistry of boranes. XXXIII. The crystal structure of $Rb_2B_9H_9$. *Inorg. Chem.* **1968**, *7*, 2260–2264. [CrossRef]

86. DeBoer, B.G.; Zalkin, A.; Templeton, D.H. The crystal structure of the rubidium salt of an octa-decahydroeicosaborate(2-) photoisomer. *Inorg. Chem.* **1968**, *7*, 1085–1090. [CrossRef]

87. Schouwink, P.; Sadikin, Y.; van Beek, W.; Černý, R. Experimental observation of polymerization from BH_4 to $B_{12}H_{12}$ in mixed-anion $A_3BH_4B_{12}H_{12}$ (A = Rb, Cs). *Int. J. Hydrogen Energy* **2015**, *40*, 10902–10907. [CrossRef]

88. Tiritiris, I.; Schleid, T.; Müuller, K.; Preetz, W. Structural Investigations on $Cs_2[B_{12}H_{12}]$. *Z. Anorg. Allg. Chem.* **2000**, *626*, 323–325. [CrossRef]

89. Verdal, N.; Wu, H.; Udovic, T.J.; Stavila, V.; Zhou, W.; Rush, J.J. Evidence of a transition to reorientational disorder in the cubic alkali-metal dodecahydro-closo-dodecaborates. *J. Solid State Chem.* **2011**, *184*, 3110–3116. [CrossRef]

90. Deiseroth, H.J.; Sommer, O.; Binder, H.; Wolfer, K.; Frei, B. CsB_3H_8: Crystal Structure and Upgrading of the Synthesis. *Z. Anorg. Allg. Chem.* **1989**, *571*, 21–28. [CrossRef]

91. Schlüter, F.; Bernhardt, E. Syntheses and Crystal Structures of the *closo*-Borate $M[B_8H_9]$ (M = $[PPh_4]^+$ and $[N(n\text{-}Bu_4)]^+$). *Inorg. Chem.* **2012**, *51*, 511–517. [CrossRef] [PubMed]

92. Greenwood, N.N.; McGinnety, J.A.; Owen, J.D. Crystal structure of caesium tetradecahydrononaborate(1–), CsB_9H_{14}. *J. Chem. Soc. Dalton Trans.* **1972**, 986–989. [CrossRef]

93. Schlüter, F.; Bernhardt, E. Fluorination of *closo*-$[B_{21}H_{18}]^-$ with *a*HF and F_2 to yield *closo*-$[B_{21}H_{18-x}F_x]^-$ (x = 1–3) and *closo*-$[B_{21}F_{18}]^-$. *Z. Anorg. Allg. Chem.* **2012**, *638*, 594–601. [CrossRef]

94. Černý, R.; Brighi, M.; Dimitrievska, M.; Udovic, T.J. Dodeca-carbahydroborates of Caesium and Rubidium, crystal structure and anions dynamic. 2020; in preparation.

95. Franken, A.; Bullen, N.J.; Jelínek, T.; Thornton-Pett, M.; Teat, S.J.; Clegg, W.; Kennedy, J.D.; Hardie, M.J. Structural chemistry of halogenated monocarbaboranes: The extended structures of $Cs[1\text{-}HCB_9H_4Br_5]$, $Cs[1\text{-}HCB_{11}H_5Cl_6]$ and $Cs[1\text{-}HCB_{11}H_5Br_6]$. *New J. Chem.* **2004**, *28*, 1499–1505. [CrossRef]

96. Jelínek, T.; Baldwin, P.; Scheldt, W.R.; Reed, C.A. New Weakly Coordinating Anions. 2. Derivatization of the Carborane Anion $CB_{11}H_{12}^-$. *Inorg. Chem.* **1993**, *32*, 1982–1990. [CrossRef]

97. Rius, J.; Romerosa, A.; Teixidor, F.; Casabo, J.; Miravitlles, C. Phase transitions in cesium 7,8-dicarbaundecaborate(12): A new one-dimensional cesium solid electrolyte at 210 °C. *Inorg. Chem.* **1991**, *30*, 1376–1379. [CrossRef]

98. Tiritiris, I. Untersuchungen zu Reaktivität, Aufbau und Struktureller Dynamik von Salzartigen Closo Dodekaboraten. Ph.D. Thesis, Universität Stuttgart Fakultät Chemie, Stuttgart, Germany, 2003.

99. Calabrese, J.C.; Gaines, D.F.; Hildebrandt, S.J.; Morris, J.H. The Low Temperature Crystal and Molecular Structure of Beryllium Bis(octahydrotriborate), $Be(B_3H_8)_2$. *J. Am. Chem. Soc.* **1976**, *98*, 5489–5492. [CrossRef]

100. Gaines, D.F.; Walsh, J.L.; Calabrese, J.C. Low-Temperature Crystal and Molecular Structures of 2-Tetrahydroborato-2- berylla-nido-hexaborane(11) and 2,2'-*commo*-Bis[2-berylla-nido-hexaborane1(1)]. *Inorg. Chem.* **1978**, *17*, 1242–1248. [CrossRef]

101. Stavila, V.; Her, J.H.; Zhou, W.; Hwang, S.J.; Kim, C.; Ottley, L.A.M.; Udovic, T.J. Probing the structure, stability and hydrogen storage properties of calcium dodecahydro-*closo*-dodecaborate. *J. Solid State Chem.* **2010**, *183*, 1133–1140. [CrossRef]

102. Jørgensen, M.; Zhou, W.; Wu, H.; Udovic, T.J.; Jensen, T.R.; Paskevicius, M.; Černý, R. Polymorphism of CaB10H10 and Its Hydrates. 2020; in preparation.

103. Her, J.H.; Wu, H.; Verdal, N.; Zhou, W.; Stavila, V.; Udovic, T.J. Structures of the strontium and barium dodecahydro-*closo*-dodecaborates. *J. Alloys Compd.* **2012**, *514*, 71–75. [CrossRef]

104. Goedde, D.M.; Windler, G.K.; Girolami, G.S. Synthesis and Characterization of the Homoleptic Octahydrotriborate Complex $Cr(B_3H_8)_2$ and Its Lewis Base Adducts. *Inorg. Chem.* **2007**, *46*, 2814–2823. [CrossRef] [PubMed]

105. Didelot, E.; Łodziana, Z.; Murgia, F.; Černý, R. Ethanol—and methanol—Coordinated and solvent—free dodecahydro closo-dodecaborates of 3d transition metals and of magnesium. *Crystals* **2019**, *9*, 372. [CrossRef]

106. Didelot, E.; Sadikin, Y.; Łodziana, Z.; Černý, R. Hydrated and anhydrous dodecahydro closo-dodecaborates of 3d transition metals and of magnesium. *Solid State Sci.* **2019**, *90*, 86–94. [CrossRef]

107. Sadikin, Y.; Didelot, E.; Łodziana, Z.; Černý, R. Synthesis and crystal structure of solvent-free dodecahydro *closo*-dodecaborate of nickel, $NiB_{12}H_{12}$. *Dalton Trans.* **2018**, *47*, 5843–5849. [CrossRef]

108. Paskevicius, M.; Hansen, B.R.S.; Jørgensen, M.; Richter, B.; Jensen, T.R. Multifunctionality of Silver *closo*-Boranes. *Nat. Commun.* **2017**, *8*, 15136. [CrossRef]

109. Xie, Z.; Jelinek, T.; Bau, R.; Reed, C.A. New weakly coordinating anions. 3. Useful silver and trityl salt reagents of carborane anions. *J. Am. Chem. Soc.* **1994**, *116*, 1907–1913. [CrossRef]

110. Xie, Z.; Wu, B.; Mak, T.C.W.; Reed, C.A. Structural Diversity in Silver Salts of Hexahalogenocarborane Anions, $Ag(CB_{11}H_6X_6)$ (X = Cl, Br or I). *J. Chem. Soc. Dalton Trans.* **1997**, *640*, 1213–1217. [CrossRef]

111. Nguyen-Duc, V. New Salt-Like Dodecahydro-Closo-Dodecaborates and Efforts for the Partial Hydroxylation of $[B_{12}H_{12}]^{2-}$ Anions. Ph.D. Thesis, Universität Stuttgart Fakultät Chemie, Stuttgart, Germany, 2009.

112. Tiritiris, I.; Van, N.D.; Schleid, T. Two Dodecahydro-*closo*-Dodecaborates with *Lone-Pair* Cations of the 6th Period in Comparison: $Tl_2[B_{12}H_{12}]$ and $Pb(H_2O)_3[B_{12}H_{12}]\cdot3H_2O$. *Z. Anorg. Allg. Chem.* **2011**, *637*, 682–688. [CrossRef]

113. Kleeberg, F.M.; Zimmermann, L.W.; Schleid, T. Synthesis and Crystal Structure of anhydrous $Pb_2[B_{12}H_{12}]$. *Z. Anorg. Allg. Chem.* **2014**, *640*, 2352.

114. Tang, W.S.; Udovic, T.J.; Stavila, V. Altering the structural properties of $A_2B_{12}H_{12}$ compounds via cation and anion modifications. *J. Alloys Compd.* **2015**, *645*, S200–S204. [CrossRef]

115. He, L.Q.; Li, H.W.; Nakajima, H.; Tumanov, N.; Filinchuk, Y.; Hwang, S.J.; Sharma, M.; Hagemann, H.; Akiba, E. Synthesis of a Bimetallic Dodecaborate $LiNaB_{12}H_{12}$ with Outstanding Superionic Conductivity. *Chem. Mater.* **2015**, *27*, 5483–5486. [CrossRef]

116. Tumanov, N.; He, L.; Sadikin, Y.; Brighi, M.; Li, H.W.; Akiba, E.; Černý, R.; Filinchuk, Y. Structural study of a bimetallic dodecaborate LiKB12H12 synthesized from decaborane B10H14. in preparation.

117. Polyakova, I.N.; Malinina, E.A.; Kuznetsov, N.T. Crystal Structures of Cesium and Dimethylammonium Cupradecaborates, $Cs[CuB_{10}H_{10}]$ and $(CH_3)_2NH_2[CuB_{10}H_{10}]$. *Crystallogr. Rep.* **2003**, *48*, 84–91. [CrossRef]

118. Malinina, E.A.; Zhizhin, K.Y.; Polyakova, I.N.; Lisovskii, M.V.; Kuznetsov, N.T. Silver(I) and Copper(I) Complexes with the closo-Decaborate Anion $B_{10}H_{10}^-$ as a Ligand. *Russ. J. Inorg. Chem.* **2002**, *47*, 1158–1167.

119. Yisgedu, T.B.; Huang, Z.; Chen, X.; Lingam, H.K.; King, G.; Highley, A.; Maharrey, S.; Woodward, P.M.; Behrens, R.; Shore, S.G.; et al. The structural characterization of $(NH_4)_2B_{10}H_{10}$ and thermal decomposition studies of $(NH_4)_2B_{10}H_{10}$ and $(NH_4)_2B_{12}H_{12}$. *Int. J. Hydrogen Energy* **2012**, *37*, 4267–4273. [CrossRef]

120. Huang, Z.; Chen, X.; Yisgedu, T.; Meyers, E.A.; Shore, S.G.; Zhao, J.C. Ammonium Octahydrotriborate $(NH_4B_3H_8)$: New Synthesis, Structure, and Hydrolytic Hydrogen Release. *Inorg. Chem.* **2011**, *50*, 3738–3742. [CrossRef]

121. Derdziuk, J.; Malinowski, P.J.; Jaron, T. The synthesis, structural characterization and thermal decomposition studies of $(N_2H_5)_2B_{12}H_{12}$ and its solvates. *Int. J. Hydrogen Energy* **2019**, *44*, 27030–27038. [CrossRef]

122. Schlüter, F.; Bernhardt, E. Syntheses and Crystal Structures of the closo-Borates $M_2[B_7H_7]$ and $M[B_7H_8]$ (M = PPh$_4$, PNP, and N(n-Bu$_4$)): The Missing Crystal Structure in the Series $[B_nH_n]_2$- (*n* = 6–12). *Inorg. Chem.* **2011**, *50*, 2580–2589. [CrossRef]

123. McGrath, T.D.; Welch, A.J. $[Ph_3PH][nido-B_{11}H_{14}]$. *Acta Crystallogr. Sect. C Struct. Chem.* **1997**, *53*, 229–231. [CrossRef]

124. Polyakova, I.N.; Malinina, E.A.; Drozdova, V.V.; Kuznetsov, N.T. The Isomorphous Substitution of $2H^+$ for the Cu^{2+} Cation in the Complex of Bis(aminoguanidine) copper (II): Crystal Structures of $(Cu_{0.61}H_{0.78}Agu_2)B_{12}H_{12}$ and $(HAgu)_2B_{12}H_{12}$. *Crystallogr. Rep.* **2009**, *54*, 831–836. [CrossRef]

Publisher's Note: MDPI stays neutral with regard to jurisdictional claims in published maps and institutional affiliations.

Article

Stereosopecificity in [Co(sep)][Co(edta)]Cl$_2$·2H$_2$O

Peter Osvath, Allen Oliver and A. Graham Lappin *

Department of Chemistry and Biochemistry, University of Notre Dame, Notre Dame, IN 46556-5670, USA;
peter.osvath@gmail.com (P.O.); aoliver2@nd.edu (A.O.)
* Correspondence: alappin@nd.edu

Abstract: The X-ray structure of racemic [Co(sep)][Co(edta)]Cl$_2$·2H$_2$O is reported and reveals hetero-chiral stereospecificity in the interactions of [Co(sep)]$^{3+}$ with [Co(edta)]$^-$. Hydrogen-bonding along the molecular C_2-axes of both complexes accounts for the stereospecificity. The structure of Λ-[Co(en)$_3$]Δ-[Co(edta)]$_2$Cl·10H$_2$O has been re-determined. Previous structural data for this compound were collected at room temperature and the model did not sufficiently describe the disorder in the structure. The cryogenic temperature used in the present study allows the disorder to be conforma-tionally locked and modeled more reliably. A clearer inspection of other, structurally interesting, interactions is possible. Again, hydrogen-bonding along the molecular C_2-axis of [Co(en)$_3$]$^{3+}$ and the equatorial carboxylates of [Co(edta)]$^-$ is the important interaction. The unique nature of the equa-torial carboxylates and molecular C_2-axis in [Co(edta)]$^-$, straddled by two pseudo-C_3-faces where the arrangement of the carboxylate groups conveys the same helicity, is highlighted. Implications of these structures in understanding stereoselectivity in ion-pairing and electron transfer reactions are discussed.

Keywords: stereoselectivity; hydrogen-bonding; complex ion

check for
updates

Citation: Osvath, P.; Oliver, A.; Lappin, A.G. Stereosopecificity in [Co(sep)][Co(edta)]Cl$_2$·2H$_2$O. *Chemistry* **2021**, *3*, 228–237. https://doi.org/10.3390/chemistry3010017

Received: 11 December 2020
Accepted: 2 February 2021
Published: 6 February 2021

Publisher's Note: MDPI stays neutral with regard to jurisdictional claims in published maps and institutional affil-iations.

1. Introduction

Structural studies have played a critical role in the determination of absolute con-figuration [1], an essential component of investigations of the chiral discriminations in-volving transition metal complexes. For over forty years, there have been attempts to understand the underlying principles that govern the diastereoselectivity that results from hydrogen-bonded interactions between complex cations such as [Co(en)$_3$]$^{3+}$ (en = ethane-1,2-diamine) and complex anions such as [Co(edta)]$^-$ (edta^{4-} = 2,2′,2″3,2‴-(ethane-1,2-diyldinitrilo)tetraacetate(−4)) [2–5]. Indeed, the first reported resolution of [Co(edta)]$^-$ involved formation of a diastereoselective salt with Λ-[Co(en)$_3$]$^{3+}$ as resolving agent [6].

Yoneda and co-workers employed the structurally related cations Δ-[Co(en)$_3$]$^{3+}$, Δ-[Co(sep)]$^{3+}$, and $\Delta(\lambda,\lambda,\lambda)$-[Co((RR)-chxn)$_3$]$^{3+}$ (sep = (1,3,6,8,10,13,16,19-octaazabicyclo [6.6.6]icosane), chxn = *trans*-1,2-cyclohexanediamine) to infer orientation effects in a series of careful chromatographic experiments, and suggested the dominant interactions in solution align major symmetry axes in the participating complexes that are homochiral [2,3]. Thus, Δ-[Co(sep)]$^{3+}$, where the P(C_3)-axis is sterically inhibited, interacts predominantly through the M(C_2)-axis and with an anion interacting through its C_3-axis, will have a M(C_2)M(C_3) or ΔΛ-preference, while with an anion interacting through its C_2-axis, the preference will be M(C_2)M(C_2) or ΔΔ, Figure 1.

The complex Δ-[Co(en)$_3$]$^{3+}$, like Δ-[Co(sep)]$^{3+}$, is found to interact through the M(C_2)-axis while, conversely, the sterically more restricted $\Delta(\lambda,\lambda,\lambda)$-[Co((RR)-chxn)$_3$]$^{3+}$ has a preference for interactions through the P(C_3)-axis. It is through such arguments that [Co(ox)$_3$]$^{3-}$, [Co(gly)(ox)$_2$]$^{2-}$, and [Co(edta)]$^-$ (ox^{2-} = oxalate(−2), gly$^-$ = glycinate(−1)) have been identified as using their C_3 or pseudo-C_3 carboxylate faces in hydrogen bonding with the cations, see Figure 2.

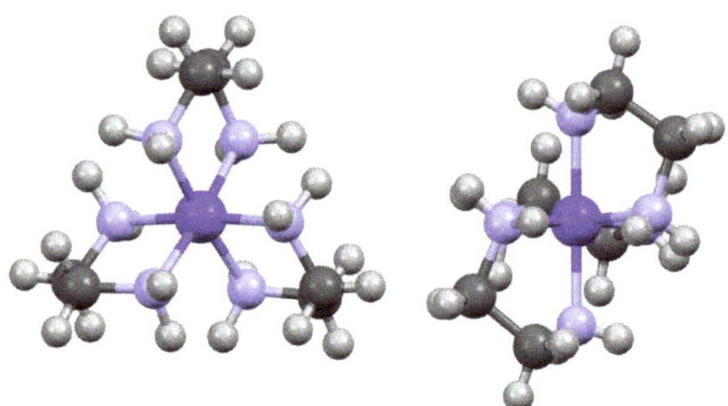

Figure 1. View down the molecular C_3- and C_2-axes of $\Delta(\lambda,\lambda,\lambda)$-[Co(en)$_3$]$^{3+}$ showing the different helicities, P and M respectively, established by the arrangement of the chelating ligands. In $\Delta(\lambda,\lambda,\lambda)$-[Co(sep)]$^{3+}$, the molecular C_3-axis is capped.

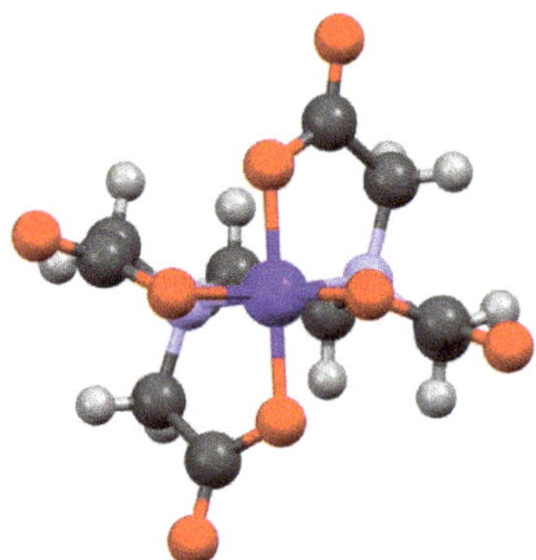

Figure 2. View down the molecular C_2-axis of $\Delta\Lambda\Delta$-[Co(edta)]$^-$ showing how it is flanked by the carboxylate pseudo-C_3-faces.

Quantitative studies of ion pairing by conductivity, notably by Tatehata and co-workers, are consistent with this simple explanation [4], although the discrimination is at the limit of detection. Thus, the ion pairing constant in solutions containing Λ-[Co(en)$_3$]$^{3+}$, Δ-[Co(edta)]$^-$, is 125(5) M^{-1}, whereas for the pair Λ-[Co(en)$_3$]$^{3+}$, Λ-[Co(edta)]$^-$, it is 119(5) M^{-1} (25 °C, 0.01 M ionic strength (KI)) consistent with a $\Lambda\Delta$-preference. Limited solution structural data from NMR relaxation measurements in the presence of [Cr(en)$_3$]$^{3+}$, are also consistent and reveal [Co(edta)]$^-$ to use the pseudo-C_3 carboxylate face in discriminations. However, in [Co(edta)]$^-$, it must be noted that the chirality resulting from the arrangement of the carboxylate groups of the equivalent pseudo-C_3 faces straddle the molecular C_2 axis, see Figure 2. Consequently, unlike in the case of the tris-bidentate chelates, the helicity conveyed by the ligands along the principal C_2-axis and the pseudo-C_3 faces is the same [5].

Our interest derives from the useful correlation between the chiral recognition in these ion pairs and the chiral induction in the electron transfer reactions of [Co(edta)]$^-$ and other complexes with the pseudo-C_3 face with [Co(en)$_3$]$^{2+}$ and derivatives [7–13]. A conclusion might be that the ion pairs serve as reasonable analogues for the precursor complex for the electron transfer process, despite the difference in the charge on the cation. The [Co(edta)]$^-$ system is particularly apt, as the complex has two very distinct sides; the carbon CH$_2$-backbone of the ligand that is not capable of forming hydrogen-bonds, and the molecular C_2-carboxylate axis surrounded by the two-pseudo-C_3 faces, all, in the parlance of Yoneda, with the same handedness. Dipole considerations dictate that it is this latter side that should interact with hydrogen bonding cations.

Information derived from crystal structures on the interactions of $[Co(edta)]^-$ with $[Co(en)_3]^{3+}$ and its derivatives or their analogues is very limited. In an earlier study, the structure of Λ-$[Co(en)_3]\Delta$-$[Co(edta)]_2Cl\cdot10H_2O$ was reported [13]. In the present paper, this structure has been re-determined at cryogenic temperature to provide a better model for a chelate ring conformational disorder and improve an understanding of the hydrogen bonding between the complexes. Indeed, a characteristic of the chiral induction in electron transfer with $[Co(edta)]^-$ and diastereomeric derivatives of $[Co(en)_3]^{3+}$ is a dependence on chelate ring conformation. The structure of racemic $[Co(sep)][Co(edta)]Cl_2\cdot2H_2O$ has been determined for the first time. The relevance of these static structures in understanding electron transfer is discussed.

2. Materials and Methods

The compounds $[\Lambda$-$(+)_D$-$[Co(en)_3]Cl_3$, $Na[\Delta$-$(+)_{546}Co(edta)]\cdot4H_2O$, $K[Co(edta)]\cdot2H_2O$, and $[Co(sep)]Cl_3$ [13,14] were prepared by literature methods. Crystals of Λ-$[Co(en)_3]\Delta$-$[Co(edta)]_2Cl\cdot10H_2O$ were obtained by diffusion of propan-2-ol into a solution prepared by the addition of 0.051 g (0.12 mmol) of $Na[\Delta$-$(+)_{546}Co(edta)]\cdot4H_2O$, to a solution of 0.038 g (0.10 mmol) of $[\Lambda$-$(+)_D$-$[Co(en)_3]Cl_3$, in 3 mL water as previously described [13]. An arbitrary sphere of data was collected on a violet rod-like crystal, having approximate dimensions of $0.267 \times 0.062 \times 0.040$ mm, on a Bruker APEX-II diffractometer using a combination of ω- and φ-scans of $0.5°$ [15].

A sample of red-pink block microcrystals of $[Co(sep)][Co(edta)]Cl_2\cdot2H_2O$ (calc. (found): C 33.1 (33.5), H 5.81 (6.03); N 17.5 (17.2); Cl 8.87 (8.72) was obtained by slow evaporation of an aqueous solution containing 0.281 g (0.635 mM) of $Na[Co(edta)]\cdot4H_2O$ and 0.10 g (0.212 mM) of $[Co(sep)]Cl_3\cdot H_2O$. An arbitrary sphere of data was collected on a red-pink block-like crystal, having approximate dimensions of $0.031 \times 0.018 \times 0.010$ mm, on a Bruker PHOTON-2 CMOS diffractometer using a combination of ω- and φ-scans of $0.5°$ [15].

For both structures, data were corrected for absorption and polarization effects, and analyzed for space group determination [16]. The structures were solved by dual-space methods and expanded routinely [17]. Models were refined by full-matrix least-squares analysis of F^2 against all reflections [18]. All non-hydrogen atoms were refined with anisotropic atomic displacement parameters.

Atomic displacement parameters for hydrogen atoms in Λ-$[Co(en)_3]\Delta$-$[Co(edta)]_2Cl\cdot10H_2O$ were modeled as a mixture of refined and constrained geometries. Hydrogen atoms on the Co complexes were modeled with atoms riding on the coordinates of the atom to which they are bonded with atomic displacement parameters tied to that of the atom to which they are bonded ($U_{iso}(H) = 1.2\ U_{eq}$ (C/N)). Water hydrogen atoms were included in positions located from a difference Fourier map. Most water hydrogen atoms were refined freely; several that did not model well were modeled with atomic displacement parameters tied to that of the oxygen to which they are bonded ($U_{iso}(H) = 1.5\ U_{eq}(O)$).

In $[Co(sep)][Co(edta)]Cl_2\cdot2H_2O$, one water was found to be disordered over two positions during refinement. In the final structure, this atom was modeled with two, half-occupancy oxygen atoms, and concomitant hydrogen atoms, at sites suggested as the loci of the original extended displacement parameters. Residual electron density ($2.24\ e^-/Å^3$) is located near (0.87 Å) Co4 of one of the $[Co(edta)]^-$ anions. It is unclear what this residual density might be, and is likely due to Fourier ripple or a very small amount of otherwise unresolvable molecular disorder. Hydrogen atoms for $[Co(sep)][Co(edta)]Cl_2\cdot2H_2O$ were treated as a mixture of freely refined and geometrically constrained atoms. Hydrogen atoms bonded to carbon and nitrogen were treated as riding models with $U_{iso}(H) = 1.2\ U_{eq}(C)$. Water hydrogen atoms were modeled at locations initially located from a difference Fourier map and subsequently tied to the coordinates of the oxygen to which they are bonded. Atomic displacement parameters for water hydrogen atoms in this model were restrained to $U_{iso}(H) = 1.5\ U_{eq}(O)$.

The disorder in Λ-[Co(en)$_3$]Δ-[Co(edta)]$_2$Cl·10H$_2$O was resolved more thoroughly with a cryogenic measurement of the data for the complex. The ethane-1,2-diamine was observed to be a twist disorder of the ethylene backbone, across the crystallographic two-fold axis that bisects the ethylene chain. Only carbon atom C3 is the unique atom in the model. The two sites were modeled from density observed in a Fourier difference map and the site occupancy ratios summed to unity yielding an approximately 0.77:0.23 ratio. The major component was refined with anisotropic displacement parameters and the minor component with an isotropic atom. Hydrogen atoms about the disorder (on nitrogen and the disordered carbon) were modeled using routine methodology (occupancies tied to the disorder component, riding atom positions, and displacement parameters).

3. Results

3.1. [Co(sep)][Co(edta)]Cl$_2$·2H$_2$O

The structure of [Co(sep)][Co(edta)]Cl$_2$·2H$_2$O (Supplementary Materials) consists of two crystallographically independent, yet chemically identical, [Co(sep)]$^{3+}$ cations, two [Co(edta)]$^-$ anions, four chlorine anions, and four water molecules of crystallization in the asymmetric unit of the primitive, centrosymmetric, triclinic space group P-1. Crystal data are summarized for [Co(sep)][Co(edta)]Cl$_2$·2H$_2$O [19].

The cations have a cobalt atom encapsulated in an octahedral fashion by a sepulchrate ligand. The cobalt is coordinated by the amine nitrogen atoms, that retain their hydrogen atoms. The "apical" nitrogen atoms of the sepulchrate are non-coordinating. The conformation of the complex is $\Delta(\lambda,\lambda,\lambda)$ or $\Lambda(\delta,\delta,\delta)$ (*lel*$_3$). In the anion, the cobalt is chelated by edta^{4-} in a six-coordinate coordination geometry with geometry $\Delta\Lambda\Delta$ or $\Lambda\Delta\Lambda$, abbreviated Δ and Λ, respectively.

The complex ions are arranged in a hetero-chiral pairwise fashion Δ-[Co(sep)]Λ-[Co(edta)] or Λ-[Co(sep)]Δ-[Co(edta)]. The Co-Co distances are 5.162(1) Å and 5.170(1) Å for the two independent pairs. The chlorine atoms form hydrogen bonds with two neighboring sepulchrate amide nitrogen atoms, Table 1, occupying two of the molecular C_2-axes of the complex cation, but do not participate in further H-bonding. Each chloride atom is hydrogen bonded by two N-H atoms from a single sepulchrate. The third molecular C_2-axis of the [Co(sep)]$^{3+}$ is occupied in stereospecific fashion by [Co(edta)]$^-$ with a pair of N-H hydrogen bonds to the equatorially coordinated G-ring oxygens of its neighboring [Co(edta)]$^-$ anion (graph set notation $R_2^2(8)$), see Figure 3. Thus, they also do not propagate the hydrogen bonded network.

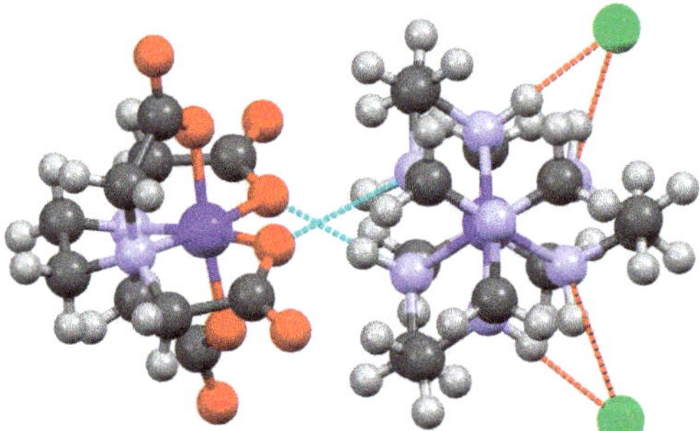

Figure 3. Hydrogen bonding between Δ-[Co(sep)]$^{3+}$ and Λ-[Co(edta)]$^-$. The view is down the molecular C_3-axis of [Co(sep)]$^{3+}$ and shows the interaction with Cl$^-$ on the molecular C_2-axes. Bond distances for complexes (Å): Co-N (Δ-[Co(sep)]$^{3+}$) 1.973(3), 1.974(3) 1.979(3) 1.987(3) 1.987(3) 1.993(3); Co-N (Λ-[Co(edta)]$^-$) 1.921(3), 1.930(3); Co-O (G-ring) 1.898(3), 1.911(3); Co-O (R-ring) 1.879(3), 1.898(3).

Table 1. Hydrogen bonds between metal-ion complexes [Å and °].

D-H$\cdots$A	d(D-H)	d(H$\cdots$A)	d(D$\cdots$A)	<(DHA)
Λ-[Co(sep)]Δ-[Co(edta)]Cl$_2$				
N(7)-H(7)$\cdots$Cl(1)	1.00	2.21	3.204(3)	175.2
N(3)-H(3)$\cdots$Cl(1)	1.00	2.27	3.215(3)	157.5
N(2)-H(2)$\cdots$Cl(2)	1.00	2.19	3.138(3)	158.0
N(6)-H(6)$\cdots$Cl(2)	1.00	2.17	3.171(3)	174.9
N(5)-H(5)$\cdots$O(1)	1.00	2.03	2.919(4)	146.8
N(8)-H(8)$\cdots$O(5)	1.00	2.03	2.947(4)	151.7
Δ-[Co(sep)]Λ-[Co(edta)]Cl$_2$				
N(10)-H(10)$\cdots$Cl(3)	1.00	2.11	3.075(3)	162.9
N(16)-H(16)$\cdots$Cl(3)	1.00	2.36	3.346(3)	168.1
N(11)-H(11)$\cdots$Cl(4)	1.00	2.17	3.139(3)	163.6
N(13)-H(13)$\cdots$Cl(4)	1.00	2.25	3.238(3)	169.4
N(14)-H(14)$\cdots$O(15)	1.00	2.02	2.908(4)	147.4
N(15)-H(15)$\cdots$O(9)	1.00	1.97	2.883(4)	151.0
Λ-[Co(en)$_3$]Δ-[Co(edta)]$_2$Cl				
G-ring interactions				
N(1)-H(1NA)$\cdots$O(5)#2 [a]	0.87(4)	3.07(4)	3.757(4)	137(3)
N(1)-H(1NA)$\cdots$O(6)#2 [b]	0.87(4)	2.23(4)	3.009(3)	149(3)
N(3)-H(3NB)$\cdots$O(5)#4(*ob*) [a]	0.91	2.10	2.985(3)	162.8
N(3)-H(3NB)$\cdots$O(6)#4(*ob*) [b]	0.91	3.13	3.783(3)	129.9
N(1)-H(1NA)$\cdots$O(5)#2 [a]	0.87(4)	3.07(4)	3.757(4)	137(3)
N(1)-H(1NA)$\cdots$O(6)#2 [b]	0.87(4)	2.23(4)	3.009(3)	149(3)
R-ring interactions				
N(1)-H(1NA)$\cdots$O(2)#1 [b]	0.87(4)	3.02(4)	3.623(4)	129(3)
N(1)-H(1NB)$\cdots$O(8)#3 [b]	0.77(5)	3.08(5)	3.558(4)	123(4)
N(3)-H(3NC)$\cdots$O(2)(*lel*) [b]	0.91	1.96	2.850(3)	165.1
N(3)-H(3NA)$\cdots$O(8)#4(*ob*) [b]	0.91	3.39	3.820(3)	111.4
N(3)-H(3ND)$\cdots$O(7)#4(*lel*) [a]	0.91	3.02	3.489(3)	114.1

[a] Coordinated oxygen, [b] terminal oxygen.

The hydrogen bonded network is extended through the structure with water molecules linking non-coordinated acetate oxygen atoms of [Co(edta)]$^-$. The arrangement of molecules results in a 2D sheet of H-bonded molecules parallel to the b/c plane. Each [Co(edta)]$^-$ accepts four hydrogen bonds and is the "corner" of a 4-connected square. Located in the center of each square is a [Co(sep)]$^{3+}$ from an adjacent sheet, related by inversion symmetry. Due to the orientation of the ligands, these sheets are bi-layers, with hydrophobic regions between layers, see Figure 4.

3.2. (Λ-[Co(en)$_3$]Δ-[Co(edta)]$_2$Cl·10(H$_2$O))

The complex, Λ-[Co(en)$_3$]Δ-[Co(edta)]$_2$Cl·10(H$_2$O) crystallizes as violet rod-like crystals from an aqueous solution. There are, collectively, two molecules of the cation, four molecules of [Co(edta)]$^-$ anion, two chloride anions, and 20 water molecules of crystallization in the unit cell of the primitive, acentric, orthorhombic space group P2$_1$2$_1$2. The correct enantiomorph of the space group and absolute stereochemistry of the complex were determined both by comparison with the known configuration of the complex and by comparison of intensities of Friedel pairs of reflections. Friedel pair analysis (Flack x parameter = 0.006(6) [20] and Hooft y parameter = 0.001(4) [21] support the assignment. Crystal data are summarized for Λ-[Co(en)$_3$]Δ-[Co(edta)]$_2$Cl·10(H$_2$O) [22].

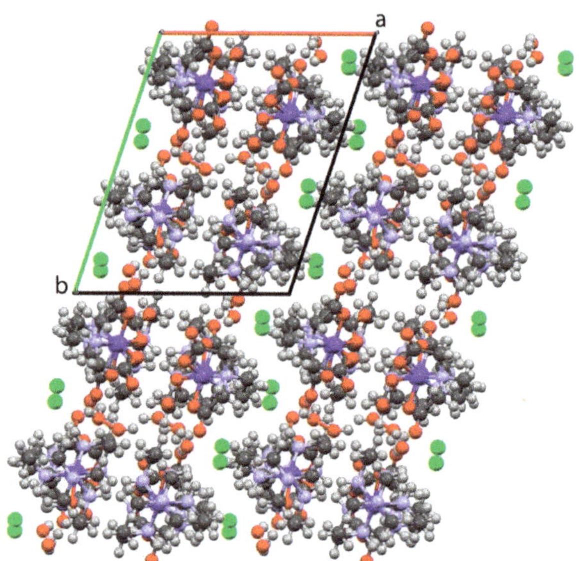

Figure 4. Packing diagram for [Co(sep)][Co(edta)]Cl$_2$·2H$_2$O looking down the crystallographic c-axis. Note the hydrophobic channels parallel to the b/c plane alternating with the hydrogen-bonded channels containing the Cl$^-$ ions.

The [Co(en)$_3$]$^{3+}$ cation is located on the crystallographic two-fold axis at [0, 0.5, z] and the chloride anion on the two-fold axis at [1, 0, z], while the [Co(edta)]$^-$ anion is located in a general position. Thus, the charge carrying species in the asymmetric unit consist of one half [Co(en)$_3$]$^{3+}$ cation, one half chloride anion, and one [Co(edta)]$^-$ anion. One ethylene diamine ligand is disordered over two sites (the ligand in question bisects the two-fold axis) and was routinely modeled with partial occupancy atoms (0.77:0.23).

The chloride ions are coordinated by eight water molecules, part of an extensive hydrogen bonding network forming channels, parallel to the c-axis. The amine nitrogen atoms of the [Co(en)$_3$]$^{3+}$ cations also participate, with alternating [Co(en)$_3$]$^{3+}$ and Cl$^-$ in the ac-plane. The [Co(edta)]$^-$ anions alternate in orientation in a parallel ac-plane, completing a layered structure along the b-axis. The layers are held together by hydrogen bonding between the [Co(en)$_3$]$^{3+}$ and [Co(edta)]$^-$. The two [Co(edta)]$^-$ anions interact with [Co(en)$_3$]$^{3+}$ in different fashion. One interaction, with a Co-Co distance of 5.844(1) Å lies roughly along a molecular C_2-axis of [Co(en)$_3$]$^{3+}$. There is hydrogen-bonding between one N-H proton along the C_2-axis, and a second nitrogen on the C_3-face of [Co(en)$_3$]$^{3+}$ with the coordinated and un-coordinated oxygen atoms of the carboxylate group of one of the G-rings on [Co(edta)]$^-$ (graph set notation $R_2^2(8)$). The interaction with the C_2-axis nitrogen involves the disordered ethane-1,2-diamine ring on [Co(en)$_3$]$^{3+}$, and examination of the linearity of the hydrogen bonds for the two conformations reveals that the λ conformation is preferred, Table 1, and that the 5.844 Å interaction favors $\Lambda(\lambda,\lambda,\lambda)$ (ob_3), see Figure 5.

The other interaction with a Co-Co distance of 7.509(1) Å shows a hydrogen bonding interaction between an uncoordinated carbonyl oxygen of an out-of-plane R-ring of [Co(edta)]$^-$ with an N-H proton from the disordered 1,2-diaminoethane ring on [Co(en)$_3$]$^{3+}$. Again, examination of the linearity of the hydrogen bonds for the two conformations reveals that the δ conformation is preferred, and that the 7.509 Å interaction favors $\Lambda(\delta,\lambda,\lambda)$ ($lelob_2$).

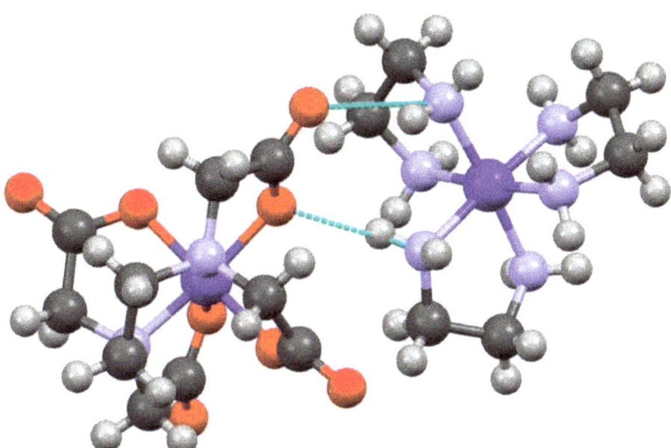

Figure 5. Hydrogen bonding between Λ-[Co(en)$_3$]$^{3+}$ and Δ-[Co(edta)]$^-$ with a Co-Co distance of 5.844 (1) Å. The view is roughly down the molecular C_3-axis of Λ-[Co(en)$_3$]$^{3+}$ and shows the $\Lambda(\lambda,\lambda,\lambda)$ (ob_3) conformation. Bond distances for complexes (Å): Co-N (Λ-[Co(en)$_3$]$^{3+}$) 1.955(2), 1.965(2), 1.960(3); Co-N (Δ-[Co(edta)]$^-$) 1.921(3), 1.927(2); Co-O (G-ring) 1.914(2), 1.921(3); Co-O (R-ring) 1.888(2), 1.877(2).

4. Discussion

As expected, bond distances and angles within the molecular ions in both structures are comparable with related studies. However, the focus of this communication is the interactions between the metal-ion complexes.

It has been a generally accepted concept that the chiral discriminations of tris-bidentate chelate complexes such as [Co(en)$_3$]$^{3+}$ are the result of different orientations of hydrogen-bonding interactions since the C_3 axis has the opposite helicity to the C_2 axis [2–4]. Thus, Δ-[Co(en)$_3$]$^{3+}$ is P(C_3)M(C_2) using the nomenclature for P or positive helicity referring to a right-handed screw and M to the left-handed screw. Ion pairing discrimination studies with metal-complex carboxylate liganded anions carried out by chromatography and conductivity measurements have highlighted the importance of the hydrogen-bonded match of the C_3 and C_2-axes of [Co(en)$_3$]$^{3+}$ with a pseudo-C_3 carboxylate face where three carboxylate groups form a facial motif that is not subtended by a chelate ring. The helicity of the pseudo-C_3 carboxylate face in the anion projects to the axis with the same helicity in the cation. Thus the pseudo-C_3 carboxylate face of Δ-[Co(edta)]$^-$ interacts preferentially with the N-H hydrogens on the C_3-axes of Δ-[Co(chxn)$_3$]$^{3+}$ where the C_2-axes are sterically encumbered, but with the N-H hydrogens on C_2-axes of Λ-[Co(sep)]$^{3+}$ where the C_3-axes are encumbered.

However, the hexa-coordinated complex ion, [Co(edta)]$^-$, differs in symmetry from a tris-bidentate chelate. The two pseudo-C_3 carboxylate faces flank the C_2 axis, and the ligand arrangement is such that all three present the same overall helicity. What is notable in both of the structures reported here is that the interactions involving the closest Co-Co distances involve the C_2 axis or the in-plane G-rings of the anion, strongly suggesting a more important role for the helicity conveyed along the C_2-axis of [Co(edta)]$^-$ in determining the discriminations. This observation is also consistent with the solution NMR structure of the ion pair, {[Cr(en)$_3$]$^{3+}$[Co(edta)]$^-$}, where by symmetry, the paramagnetic cation straddles the C_2-axis of [Co(edta)]$^-$ [5].

There is a structural comparison with Δ-[Ni(en)$_3$]Δ-[Ni(edta)]·4H$_2$O that is also relevant [23]. The cation and anion share two interactions. The closest Ni-Ni distance is 5.40 Å and reveals a direct hydrogen bond formed between the non-coordinated G-ring oxygen of [Ni(edta)]$^{2-}$ and an N-H on the C_3-axis of [Ni(en)$_3$]$^{2+}$ with a second interaction involving the coordinated O of the other G-ring, a bridging water molecule, and a second N-H on the C_3-axis of [Ni(en)$_3$]$^{2+}$ (Graph set notation $R_2^2(12)$). There is no direct pseudo-C_3 carboxylate interaction involving the three N-H groups of the C_3-axis of $[Ni(en)_3]^{2+}$. Instead, it is the in-plane G-ring carboxylates that again play a dominant role. A longer Ni-Ni distance at 6.14 Å involves a non-coordinated R-ring oxygen of $[Ni(edta)]^{2-}$ with two N-H protons on the C_3-axis of $[Ni(en)_3]^{2+}$ (Graph set notation $R_2^1(6)$).

While the distinction between the ligand arrangement conveying the same helicity through the C_2-axis and pseudo-C_3 faces in $[Co(edta)]^-$ with the same helicity may seem semantic, it should be noted that the dipole moment of $[Co(edta)]^-$ projects along the C_2-axis. The idea that discriminations by $[Co(edta)]^-$ can be projected through the C_2-axis and not only by the pseudo-C_3-faces has implications in the interpretation of the extensive data on related outer-sphere stereoselective electron transfer. The reductions of $[Co(edta)]^-$ with both $[Co(en)_3]^{2+}$ and $[Co(sep)]^{2+}$ have been shown to occur by an outer-sphere mechanism [7,9].

Computational work on the effects of distance and orientation in outer-sphere electron-transfer reactions between metal ion complexes has focused predominantly on the $[Fe(OH_2)_6]^{3+/2+}$ and $[Ru(OH_2)_6]^{3+/2+}$ self-exchange reactions [24–26]. Increasingly, sophisticated calculations [27–31] have not markedly changed the conclusions first reached that face-to-face interactions along the C_3-axes represent the closest approach of the two metal centers, at distances roughly 5–6 Å, and the most favorable configuration for overlap of the donor and acceptor orbitals resulting in electron transfer. Other orientations over a range of distances provide less favorable pathways with super-exchange mechanisms involving the ligands more likely at distances in excess of 6 Å.

Unlike the self-exchange reactions, the outer-sphere oxidations of $[Co(en)_3]^{2+}$ and $[Co(sep)]^{2+}$ by $[Co(edta)]^-$ involve complexes with opposite charges, and the additional electrostatic attraction can provide a more intimate interaction, generally at hydrogen-bonded distances. An attractive model for a C_3-C_3 interaction is provided by the structure of $[Cr(en)_3]^{3+}[Cr(ox)_3]^{3+}$ where the metal-metal distance is 4.98 Å [32], shorter than the closest approach distances found in the present structural studies with $[Co(en)_3]^{3+}$ (5.84 Å) and $[Co(sep)]^{3+}$ (5.17 Å). Although the electron transfer precursor is dynamic and comparisons with static structures are fraught with problems, were distance the only factor, then for $[Co(edta)]^-$ projecting discrimination through the pseudo-C_3-face, one might well expect the reaction stereoselectivity to reflect a dominant C_3-C_3 interaction and hence a homochiral, $\Delta\Delta$ or $\Lambda\Lambda$, preference. That is not what is observed, providing a further piece of evidence that the discrimination by $[Co(edta)]^-$ is more complex.

Further, a detailed analysis of structural, charge, and dipolar effects on the stereoselective electron transfer data also revealed [33] that there is a distinction between $[Co(edta)]^-$ and the oxalate containing reagents that possess the pseudo-C_3 motif such as C_1-*cis*(N)-$[Co(gly)_2(ox)]^-$ (gly$^-$ = glycinate(-1)). Stereoselectivity and chiral discriminations involving $[Co(edta)]^-$ should be considered in a class of their own.

It is noted that the chelate ring conformations of $[Co(en)_3]^{3+}$ (*ob*$_3$ or *lelob*$_2$) and $[Co(sep)]^{3+}$ (*lel*$_3$) differ in the two structures presented. Both cations employ the C_2-axis in the closest interaction, and it is not unreasonable to expect that $[Co(en)_3]^{3+}$ (*lel*$_3$) would interact in similar fashion to $[Co(sep)_3]^{3+}$ (*lel*$_3$) since they differ mainly along the C_3-axis. In stereoselective electron-transfer studies where $[Co(edta)]^-$ is used as an oxidant for $[Co(R,S-pn)_3]^{2+}$, (R,S-pn = R,S-propane-1,2-diamine), $[Co(RR,SS-bn)_3]^{2+}$ (RR,SS-bn = RR,SS-butane-2,3-diamine), and $[Co(RR,SS-chxn)_3]^{2+}$, the individual stereoselectivities for the conformation isomers can be determined and show a trend from homo-chiral to hetero-chiral with increasing *ob*-character, previously ascribed to differences in hydrogen-bonding [9,11]. The weighted average stereoselectivities are $[Co(RR,SS-chxn)_3]^{2+}$ (8% homo-), $[Co(RR,SS-bn)_3]^{2+}$ (3% homo-), $[Co(R,S-pn)_3]^{2+}$ (4% hetero-), and can be compared with those of the conformationally labile $[Co(en)_3]^{2+}$ (10% hetero-), and $[Co(sep)]^{2+}$ (17% hetero-), reflecting the clear trend in discrimination in the tris-bidentate derivatives of $[Co(en)_3]^{2+}$. However, this does not provide insight into the discriminating nature of the oxidant $[Co(edta)]^-$. The most apt comparison is with the oxidant Λ-$[Co((R)-tacntp)]$ ((R)-tacntp^{3-} = 1,4,7-tri-aza-cyclo-nonane-1,4,7-tris [2'(R)-2'-propionate](-3)), which has

3-fold symmetry with a strong dipole, but no C_2-axis [34]. While the selectivity in the oxidation of $[Co(en)_3]^{2+}$ (11% hetero-) is comparable in sense and magnitude with the value for reaction with $[Co(edta)]^-$, the weighted average selectivity with $[Co(RR,SS\text{-chxn})_3]^{2+}$ (31% homo-) reflects a much stronger C_3-C_3 preference.

5. Conclusions

In conclusion, the structures presented highlight an important role for hydrogen bonding involving the unique C_2-axis of $[Co(edta)]^-$ in chiral discriminations with $[Co(en)_3]^{3+}$ and derivatives. Stereoselectivity and chiral discriminations involving $[Co(edta)]^-$ should be considered in a class of their own. This has implications in the interpretation of data for related stereoselective electron transfer reactions and suggest that generalizations should be avoided.

Supplementary Materials: The following are available online at https://www.mdpi.com/2624-854 9/3/1/17/s1, Figure S1: Labelling diagram for $[Co(sep)][Co(edta)]Cl_2 \cdot 2H_2O$; Figure S2: Packing diagram for $[Co(sep)][Co(edta)]Cl_2 \cdot 2H_2O$; Figure S3: View along c-axis for $[Co(sep)][Co(edta)]Cl_2 \cdot 2H_2O$; Figure S4: Labelling diagram for Λ-$[Co(en)_3]\Delta$-$[Co(edta)]_2Cl \cdot 10H_2O$. X-Ray crystallographic files of $[Co(sep)][Co(edta)]Cl_2 \cdot 2H_2O$ and Λ-$[Co(en)_3]\Delta$-$[Co(edta)]_2Cl \cdot 10H_2O$, CCDC 2049761, 2049762.

Author Contributions: Conceptualization, P.O. and A.G.L.; methodology, P.O., A.O., and A.G.L.; validation, P.O., A.O., and A.G.L.; formal analysis, A.O. and A.G.L.; investigation, P.O., A.O., and A.G.L.; resources, P.O., A.O., and A.G.L.; data curation, A.O., and A.G.L.; writing—original draft preparation, A.G.L.; writing—review and editing, A.G.L.; visualization, P.O., A.O., and A.G.L.; supervision, A.G.L.; project administration, A.G.L.; funding acquisition, A.O., and A.G.L. All authors have read and agreed to the published version of the manuscript.

Funding: This research was funded under National Science Foundation (Grant No. CHE84-06113). The ALS is supported by the U.S. Dept. of Energy, Office of Energy Sciences, under contract DE-AC02-05CH11231.

Acknowledgments: Samples for synchrotron crystallographic analysis were submitted through the SCrALS (Service Crystallography at Advanced Light Source) program. Crystallographic data were collected at Beamline 12.2.1 at the Advanced Light Source (ALS), Lawrence Berkeley National Laboratory.

Conflicts of Interest: The authors declare no conflict of interest.

References

1. Flack, H.D. On Enantiomorph-Polarity Estimation. *Acta Crystallogr.* **1983**, *A39*, 876–881. [CrossRef]
2. Sakaguchi, Y.; Yamamoto, I.; Izumoto, S.; Yoneda, H. The Mode of Stereoselective Association between Complex Cation and Complex Anion. *Bull. Chem. Soc. Jpn.* **1983**, *56*, 153–156. [CrossRef]
3. Miyoshi, K.; Sakamoto, Y.; Ohguni, A.; Yoneda, H. Stereoselective Interaction of Chiral Metal Complexes in Solution as Studied by Chromatography. i. Modes of Chiral Discrimination and Optical Resolution of anion Complexes. *Bull. Chem. Soc. Jpn.* **1985**, *58*, 2239–2246. [CrossRef]
4. Tatehata, A.; Fujita, M.; Ando, K.; Asaba, Y. Chiral Recognition in Ion Pairing of an Optically Active Tris(Ethylenediamine)Cobalt(Iii) Cation and Applications to Chromatographic Resolution of Metal Complex Anions. *J. Chem Soc. Dalton Trans.* **1987**, *8*, 1977–1982. [CrossRef]
5. Marusak, R.A.; Lappin, A.G. Observations on the Structure of the Ion Pair Formed between $[Co(en)_3]^{3+}$ and $[Co(edta)]^-$. *J. Phys. Chem.* **1989**, *93*, 6856–6859. [CrossRef]
6. Dwyer, F.P.; Gyarfas, E.C.; Mellor, D.P. The Resolution and Racemization of Potassium Ethylenediaminetetraacetatocobaltate (III). *J. Phys Chem.* **1955**, *59*, 296–297. [CrossRef]
7. Geselowitz, D.A.; Taube, H. Stereoselectivity in Electron-Transfer Reactions. *J. Am. Chem Soc.* **1980**, *102*, 4525–4526. [CrossRef]
8. Osvath, P.; Lappin, A.G. Conformational Effects in Stereoselective Electron Transfer between Metal Ion Complexes. *J. Chem. Soc. Chem. Commun.* **1986**, *14*, 1056–1057. [CrossRef]
9. Osvath, P.; Lappin, A.G. Stereoselectivity as a Probe of Reaction Mechanism in the Oxidation of $[Co(en)_3]^{2+}$ and Its Derivatives by $[Co(edta)]^-$. *Inorg. Chem.* **1987**, *26*, 195–202. [CrossRef]
10. Geselowitz, D.A.; Hammershøi, A.; Taube, H. Stereoselective Electron-Transfer Reactions of (Ethylenediaminetetraacetato)cobaltate(III), (Propylenediaminetetraacetato)cobaltate(III), and (1,2-Cyclohexanediaminetetraacetato)cobaltate(III) with Tris(ethylenediamine)cobalt(II). *Inorg. Chem.* **1987**, *26*, 1842–1845. [CrossRef]

11. Tatehata, A.; Mitani, T. Stereoselectivity in Electron-Transfer Reactions of Tris(ethylenediamine)cobalt(II) with Several Anionic Cobalt(III) Complexes. *Chem. Lett.* **1989**, *18*, 1167–1170. [CrossRef]

12. Tatehata, A.; Muraida, A. Temperature Dependence of Chiral Recognition in the Oxidation of Tris(ethylenediamine)cobalt(II) Complex by Optically Active Anionic Cobalt(III) Complexes. *Chem. Lett.* **1996**, *25*, 461–462. [CrossRef]

13. Warren, R.M.L.; Haller, K.J.; Tatehata, A.; Lappin, A.G. Chiral Discrimination in the Reduction of $[Co(edta)]^-$ by $[Co(en)_3]^{2+}$ and $[Ru(en)_3]^{2+}$. X-ray Structure of $[\Lambda\text{-}Co(en)_3][\Delta\text{-}Co(edta)]_2Cl\cdot10H_2O$. *Inorg. Chem.* **1994**, *33*, 227–232. [CrossRef]

14. Harrowfield, J.M.; Herlt, A.J.; Sargeson, A.M. Caged Metal Ions: Cobalt Sepulchrates. *Inorg. Synth.* **1980**, *20*, 85–86.

15. *APEX-3*; Bruker AXS: Madison, WI, USA, 2016.

16. Krause, L.; Herbst-Irmer, R.; Sheldrick, G.M.; Stalke, D. Comparison of silver and molybdenum microfocus X-ray sources for single-crystal structure determination. *J. Appl. Cryst.* **2015**, *48*, 3–10. [CrossRef] [PubMed]

17. Sheldrick, G.M. SHELXT—Integrated space-group and crystal-structure determination. *Acta Cryst.* **2015**, *A71*, 3–8. [CrossRef]

18. Sheldrick, G.M. Crystal structure refinement with SHELXL. *Acta Cryst.* **2015**, *C71*, 3–8.

19. Crystal data for $C_{22}H_{46}C_{12}Co_2N_{10}O_{10}$; Mr = 799.45; Triclinic; space group P-1; a = 13.0187(8) Å; b = 15.4433(10) Å; c = 17.2326(11) Å; α = 90.310(2)°; β = 108.030(2)°; γ = 107.509(2)°; V = 3123.3(3) Å^3; Z = 4; T = 150(2) K; λ(synchrotron) = 0.72880 Å; μ(synchrotron) = 1.387 mm^{-1}; d_{calc} = 1.700 g.cm^{-3}; 162884 reflections collected; 13890 unique (R_{int} = 0.1250); giving R_1 = 0.0561, wR_2 = 0.1480 for 10514 data with [I > 2σ(I)] and R_1 = 0.0768, wR_2 = 0.1643 for all 13890 data. Residual electron density (e–·Å^{-3}) max/min: 2.239/-0.632. Available online: https://rruff.geo.arizona.edu/AMS/amcsd.php (accessed on 6 February 2021).

20. Parsons, S.; Flack, H.D.; Wagner, T. Use of intensity quotients and differences in absolute structure refinement. *Acta Cryst.* **2013**, *B69*, 249–259. [CrossRef]

21. Hooft, R.W.W.; Straver, L.H.; Spek, A.L. Determination of absolute structure using Bayesian statistics on Bijvoet differences. *J. Appl. Cryst.* **2008**, *41*, 96–103. [CrossRef]

22. Crystal data for $C_{26}H_{68}C_1Co_3N_{10}O_{26}$; M_r = 1149.14; Orthorhombic; space group P2$_1$2$_1$2; a = 12.8265(17) Å; b = 21.055(3) Å; c = 8.1536(11) Å; α = 90°; β = 90°; γ = 90°; V = 2202.0(5) Å^3; Z = 2; T = 120(2) K; λ(Mo-Kα) = 0.71073 Å; μ(Mo-Kα) = 1.280 mm^{-1}; d_{calc} = 1.733 g.cm^{-3}; 42751 reflections collected; 5478 unique (R_{int} = 0.0550); giving R_1 = 0.0271, wR_2 = 0.0642 for 5032 data with [I > 2σ(I)] and R_1 = 0.0323, wR_2 = 0.0662 for all 5478 data. Residual electron density (e–·Å^{-3}) max/min: 0.500/-0.533. Available online: https://rruff.geo.arizona.edu/AMS/amcsd.php (accessed on 6 February 2021).

23. Sysoeva, T.F.; Agre, V.M.; Trunov, V.K.; Dyatlova, N.M.; Fridman, A.Y. Crystal structure of complex compound of nickel(II) with ethylenediamine and ethylenediaminetetraacetic acid, $Ni_2en_3A.4H_2O$. *J. Struct. Chem. (Engl. Transl.)* **1986**, *27*, 97–103. [CrossRef]

24. Tembe, B.L.; Friedman, H.L.; Newton, M.D. The Theory of Fe^{2+}–Fe^{3+} Electron Exchange in Water. *J. Chem. Phys.* **1982**, *76*, 1490–4367. [CrossRef]

25. Logan, J.; Newton, M.D. Ab initio study of electronic coupling in the aqueous Fe^{2+}-Fe^{3+} electron exchange process. *J. Chem. Phys.* **1983**, *78*, 4086–4091. [CrossRef]

26. Newton, M.D. Electronic Structure Analysis of Electron-Transfer Matrix Elements for Transition-Metal Redox Pairs. *J. Phys. Chem.* **1988**, *92*, 3049–3056. [CrossRef]

27. Kumar, P.V.; Tembe, B.L. Solvation structure and dynamics of the Fe^{2+}–Fe^{3+} ion pair in water. *J. Chem. Phys.* **1992**, *97*, 4356–4367. [CrossRef]

28. Babu, C.S.; Madhusoodanan, M.; Sridhar, G.; Tembe, B.L. Orientations of $[Fe(H_2O)_6]^{2+}$ and $[Fe(H_2O)_6]^{3+}$ Complexes at a Reactive Separation in Water. *J. Am. Chem. Soc.* **1997**, *119*, 5679–5681. [CrossRef]

29. Migliore, A.; Sit, P.H.-L.; Klein, M.L. Evaluation of Electronic Coupling in Transition-Metal Systems Using DFT: Application to the Hexa-Aquo Ferric-Ferrous Redox Couple. *J. Chem. Theory Comput.* **2009**, *5*, 307–323. [CrossRef]

30. Oberhofer, H.; Blumberger, J. Insight into the Mechanism of the Ru^{2+}–Ru^{3+} Electron Self-Exchange Reaction from Quantitative Rate Calculations. *Angew. Chem. Int. Ed.* **2010**, *49*, 3631–3634. [CrossRef]

31. Miliordos, E.; Xantheas, S.S. Ground and Excited States of the $[Fe(H_2O)_6]^{2+}$ and $[Fe(H_2O)_6]^{3+}$ Clusters: Insight into the Electronic Structure of the $[Fe(H_2O)_6]^{2+}$-$[Fe(H_2O)_6]^{3+}$ Complex. *J. Chem. Theory Comput.* **2015**, *11*, 1549–1563. [CrossRef] [PubMed]

32. Hua, X.; Larsson, K.; Neal, T.J.; Wyllie, G.R.A.; Shang, M.; Lappin, A.G. Structure and magnetic properties of $[Cr(en)_2(ox)][Cr(en)(ox)_2]\cdot2H_2O$, $\Delta\text{-}[Cr(en)_2(ox)]\Delta\text{-}[Cr(en)(ox)_2]$ and $\Delta\text{-}[Cr(en)_2(ox)]\Delta\text{-}[Cr(en)(ox)_2]$. *Inorg. Chem. Commun.* **2001**, *4*, 635–639. [CrossRef]

33. Mitani, T.; Honma, N.; Tatehata, A.; Lappin, A.G. Shape selectivity in outer-sphere electron transfer reactions. *Inorg. Chim. Acta* **2002**, *331*, 39–71. [CrossRef]

34. Saenz, G.L.; Warren, R.M.L.; Shang, M.; Lappin, A.G. Stereoselectivity in the Reduction of Λ-[1,4,7-Triazacyclononane-1,4,7-tris[2'(R)-2'-propionate]cobalt(III). *J. Coord Chem.* **1995**, *34*, 129–137. [CrossRef]

 *chemistry*

Article

The Tyranny of Arm-Wrestling Methyls on Iron(II) Spin State in Pseudo-Octahedral [Fe(didentate)$_3$] Complexes [†]

Neel Deorukhkar [1], Timothée Lathion [1], Laure Guénée [2], Céline Besnard [2] and Claude Piguet [1],*

[1] Department of Inorganic and Analytical Chemistry, University of Geneva, 30 quai E. Ansermet,
    CH-1211 Geneva 4, Switzerland; Neel.deorukhkar@unige.ch (N.D.); timothee.lathion@gmail.com (T.L.)
[2] Laboratory of Crystallography, University of Geneva, 24 quai E. Ansermet, CH-1211 Geneva, 4 Switzerland;
    Laure.Guenee@unige.ch (L.G.); Celine.Besnard@unige.ch (C.B.)
* Correspondence: Claude.Piguet@unige.ch
[†] Dedicated to Dr. Howard Flack (1943–2017).

Received: 11 March 2020; Accepted: 31 March 2020; Published: 2 April 2020

Abstract: The connection of a sterically constrained 3-methyl-pyrazine ring to a *N*-methyl-benzimidazole unit to give the unsymmetrical α,α'-diimine ligand **L5** has been programmed for the design of pseudo-octahedral spin-crossover [Fe(**L5**)$_3$]$^{2+}$ units, the transition temperature ($T_{1/2}$) of which occurs in between those reported for related facial tris-didentate iron chromophores fitted with 3-methyl-pyridine-benzimidazole in a LaFe helicate ($T_{1/2} \sim$ 50 K) and with 5-methyl-pyrazine-benzimidazole **L2** ligands ($T_{1/2}$ ~350 K). A thorough crystallographic analysis of [Fe(**L5**)$_3$](ClO$_4$)$_2$ (**I**), [Ni(**L5**)$_3$](ClO$_4$)$_2$ (**II**), [Ni(**L5**)$_3$](BF$_4$)$_2$·H$_2$O (**III**), [Zn(**L5**)$_3$](ClO$_4$)$_2$ (**IV**), [Ni(**L5**)$_3$](BF$_4$)$_2$·1.75CH$_3$CN (**V**), and [Zn(**L5**)$_3$](BF$_4$)$_2$·1.5CH$_3$CN (**VI**) shows the selective formation of pure facial [M(**L5**)$_3$]$^{2+}$ cations in the solvated crystals of the tetrafluoroborate salts and alternative meridional isomers in the perchlorate salts. Except for a slightly larger intra-strand interannular twist between the aromatic heterocycles in **L5**, the metric parameters measured in [Zn(**L5**)$_3$]$^{2+}$ are comparable to those reported for [Zn(**L2**)$_3$]$^{2+}$, where **L2** is the related unconstrained ligand. This similitude is reinforced by comparable ligand-field strengths (Δ_{oct}) and nephelauxetic effects (as measured by the Racah parameters B and C) extracted from the electronic absorption spectra recorded for [Ni(**L5**)$_3$]$^{2+}$ and [Ni(**L2**)$_3$]$^{2+}$. In this context, the strictly high-spin behavior observed for [Fe(**L5**)$_3$]$^{2+}$ within the 5–300 K range contrasts with the close to room-temperature spin-crossover behavior of [Fe(**L2**)$_3$]$^{2+}$ ($T_{1/2}$ = 349(5) K in acetonitrile). This can be unambiguously assigned to an intraligand arm wrestling match operating in bound **L5**, which prevents the contraction of the coordination sphere required for accommodating low-spin Fe^{II}. Since the analogous 3-methyl-pyridine ring in [Fe(**L3**)$_3$]$^{2+}$ derivatives are sometimes compatible with spin-crossover properties, the consequences of repulsive intra-strand methyl–methyl interactions are found to be amplified in [Fe(**L5**)$_3$]$^{2+}$ because of the much lower basicity of the 3-methyl-pyrazine ring and the resulting weaker thermodynamic compensation. The decrease of the stability constants by five orders of magnitude observed in going from [M(**L2**)$_3$]$^{2+}$ to [M(**L5**)$_3$]$^{2+}$ (M = Ni^{II} and Zn^{II}) is diagnostic for the operation of this effect, which had been not foreseen by the authors.

Keywords: pyrazine-benzimidazole; spin crossover; iron(II); ligand field; nephelauxetic

---

1. Introduction

In line with the formulation of the ligand field theory [1,2], or as it was originally called by Bethe, crystal-field theory [3], it was realized that an open-shell metal with at least two valence electrons in a specific chemical environment could exist with either high-spin or low-spin configuration [4].

Following van Vleck's approach to magnetism [5], Pauling perceptively recognized that it would be feasible to obtain systems in which two spin states could be present simultaneously, while their ratio should depend on the energy difference between them [6,7]. The discovery of thermal spin-state equilibria operating in Fe^{III} dithiocarbamate by Cambi et al. [8–10] at the same period, indeed confirmed these predictions. Since then, a myriad of metal coordination complexes and polymeric materials have been shown to display spin transitions, often referred to as spin-crossover (SCO) materials. These have been studied in detail and extensively reviewed over the last two decades [11–20]. Due to the 'on–off' switching of the magnetic properties accompanying the spin transition from the low-spin diamagnetic configuration ($^1A_1$ label in octahedral symmetry) to the high-spin paramagnetic form ($^5T_2$ label in octahedral symmetry) for d^6 transition metals in pseudo-octahedral geometry (Scheme 1a), the 'magic' $[Fe^{II}N_6]$ chromophores, where N is a heterocyclic nitrogen donor atom, have been intensively investigated [11–20]. Various external stimulations such as changes in temperature [21,22], pressure [23,24], magnetic field [25] or light-irradiation [26,27] can be used for inducing the SCO processes, which makes these microscopic magneto-optical switches very attractive for their introduction into responsive macroscopic materials [12,13,16,28–31]. The most common and accepted approach for rationalizing the design of spin-crossover pseudo-octahedral Fe^{II} complexes relies on the energetic balance $\Delta E_{HL} = E^0_{hs} - E^0_{ls} = 2(\Delta_{oct} - P)$ between the ligand-field stabilization energy as measured by $\Delta_{oct} = 10\,Dq$ and the spin pairing energy modeled with the Racah parameters B and C with $P = 2B + 4C \approx 19B$ (Scheme 1a) [31]. When $\Delta_{oct} \gg P$, $\Delta E_{HL} = E^0_{hs} - E^0_{ls} = 2(\Delta_{oct} - P) \gg 0$ and the pseudo-octahedral Fe^{II} complex adopts a low spin configuration with a diamagnetic $^1A_1$ electronic ground state, as shown in the right part of the Tanabe–Sugano diagram built for the electronic d^6 configuration ($\Delta_{oct}/B > 20$ in Scheme 1b). The reverse situation occurs when $P \gg \Delta_{oct}$, which leads to $\Delta E_{HL} = E^0_{hs} - E^0_{ls} = 2(\Delta_{oct} - P) \ll 0$ and the existence of the paramagnetic high-spin $^5T_2$ ground state (left part of Tanabe–Sugano diagram with $\Delta_{oct}/B < 10$ in Scheme 1b). Finally, for intermediate values $0 \leq |\Delta E_{HL} = 2(\Delta_{oct} - P)| \leq mRT$ ($m \leq 10$), the two spin states coexist and are thermally populated at accessible temperatures. However, the latter statement is misleading and physically unsound since both Δ_{oct} and P change during the spin transition as a result of the population of the antibonding orbitals in the high-spin form. For pseudo-octahedral spin-crossover $[Fe^{II}N_6]$ complexes, the Fe–N bond lengths extend by approximately 10% upon the low-spin to high-spin transition and Δ_{oct} consequently decreases according to a $1/r^n$ dependence with $n = 5–6$ (Equation (1)) [31].

$$\frac{\Delta^{HS}_{oct}}{\Delta^{LS}_{oct}} = \left(\frac{r_{LS}}{r_{HS}}\right)^n \tag{1}$$

Taking typical Fe–N bond distances of $r_{LS} = 2.0$ Å and $r_{HS} = 2.2$ Å [11–20] leads to $\Delta^{LS}_{oct}/\Delta^{HS}_{oct} \approx 1.75$ accompanying the spin transition, whereas P changes very little ($P^{HS} : P^{LS} \cong 19B$), except for a faint reduction of the nephelauxetic effect with larger bond lengths [32].

Taking into account the 10% bond length expansion accompanying the spin transition, the simplistic zero-point energy differences between the two states summarized in Scheme 1a (i.e., $\Delta E_{HL} = E^0_{hs} - E^0_{ls} = 2(\Delta_{oct} - P)$) should be replaced with Equation (2), which is transformed into Equations (3) and (4) upon introducing $\Delta^{LS}_{oct}/\Delta^{HS}_{oct} \approx 1.75$.

$$\Delta E_{HL} = E^0_{hs} - E^0_{ls} = \frac{12}{5}\Delta^{LS}_{oct} - \frac{2}{5}\Delta^{HS}_{oct} - 38B \tag{2}$$

$$\Delta E_{HL} = E^0_{hs} - E^0_{ls} = 3.8\Delta^{HS}_{oct} - 38B \tag{3}$$

$$\Delta E_{HL} = E^0_{hs} - E^0_{ls} = 2.17\Delta^{LS}_{oct} - 38B \tag{4}$$

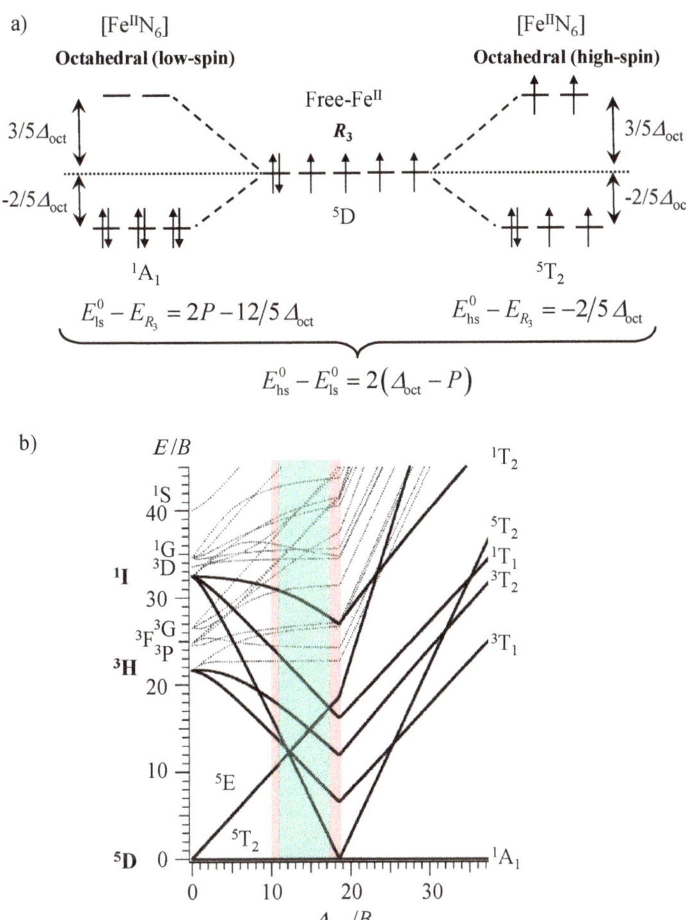

Scheme 1. (a) The thermodynamic origin of Fe^{II} spin state equilibrium in octahedral symmetry (P is the electron spin pairing energy). (b) Tanabe–Sugano diagram for a d^6 metal ion calculated using the electrostatic matrices in the strong-field basis and using the free ion Racah parameters $B = 917$ cm^{-1} and $C = 4.41B = 4040$ cm^{-1} [31]. The red areas illustrate the limiting domain of co-existence of low-spin and high-spin complexes for which $0 \le \left\{E^0_{hs} - E^0_{ls}\right\} \le 2000$ cm^{-1} with a Racah parameter fixed at 75% of the free ion values, whereas the dashed green domain corresponds to non-accessible ligand-field strengths (see text) [31].

Solving Equations (3) and (4) for $\Delta E_{HL} = 0$ provides $\Delta^{HS}_{oct}/B = 38/3.8 = 10$ and $\Delta^{LS}_{oct}/B = 38/2.17 = 17.5$ as the lower limit of the existence of thermally accessible spin state equilibria, while the higher limits can be estimated for $\Delta E_{HL} = 2000$ cm^{-1}, which gives $\Delta^{HS}_{oct}/B = 10 + [2000/(3.8B)]$ and $\Delta^{LS}_{oct}/B = 17.5 + [2000/(2.17B)]$ (red areas in Scheme 1b). More sophisticated calculations using Racah parameters B and C, reduced by 70–80% of their free ion values, predict narrow ranges of ligand field strengths $11{,}000 \le \Delta^{HS}_{oct} \le 12{,}500$ cm^{-1} and $19{,}000 \le \Delta^{LS}_{oct} \le 22{,}000$ cm^{-1}, for which the phenomenon of a thermal spin transition can be expected in Fe^{II} coordination complexes [31]. Following this theoretical approach, the toolkit of coordination chemists for programming and tuning the thermodynamic spin transition parameters in molecular [Fe^{II}N$_6$] complexes logically relied on the manipulation of Δ_{oct} and B via (i) some controlled distortions of the coordination geometry from a perfect octahedron by using chelating ligands with fixed bite angles [33] and (ii) specific programming of metal–ligand bonding interactions via ligand design [15,19,34]. Benefiting from the huge amount of experimental data collected during the last decades for $[Fe(N^{\cap}N)_3]^{2+}$ complexes, where $N^{\cap}N$ is an α,α'-diimine chelate ligand possessing two N-heterocyclic donor atoms, it was shown that the connection of a six-membered heterocycle to a five-membered heterocycle in $N^{\cap}N$ provides favorable ligand-field strengths around Fe^{II} for promoting spin-state equilibria (Equation (5)) with transition temperatures $T_{1/2} = \Delta H_{SCO}/\Delta S_{SCO}$ (i.e., the temperature at which $\Delta G_{SCO} = 0$ and $x_{hs} = x_{ls} = 0.5$) within the 30–500 K range [14,15,33].

$$Fe^{II}_{low-spin} \overset{K_{SCO}}{\rightleftharpoons} Fe^{II}_{high-spin} \quad K_{SCO} = x_{hs}/x_{ls} = e^{-(\Delta G_{SCO}/RT)} = e^{(\Delta S_{SCO}/R - \Delta H_{SCO}/RT)} \tag{5}$$

The didentate ligands **L1** and **L2** match the latter criteria and the associated pseudo-octahedral complexes $[Fe(Lk)_3]^{2+}$ indeed exhibit spin-crossover behaviors in acetonitrile solutions, the transition temperatures of which reveal the stronger Fe–N bonds induced by the strong-accepting pyrazine units in $[Fe(L2)_3]^{2+}$ ($T_{1/2}$ ~350 K) compared with pyridine units in $[Fe(L1)_3]^{2+}$ ($T_{1/2}$ ~310 K, Scheme 2) in the absence of steric constraints [35].

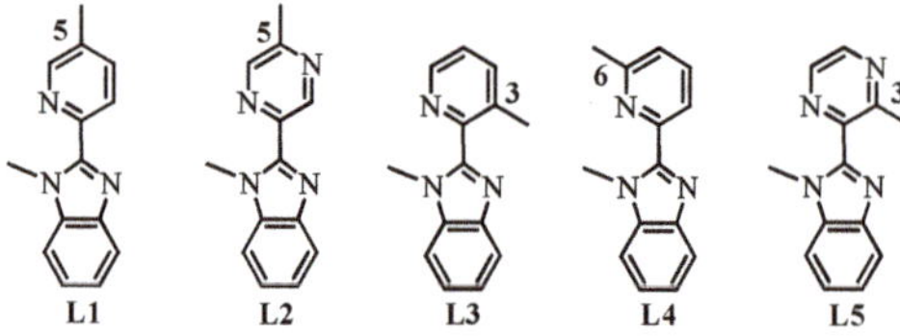

Complex	$[Ni(L1)_3]^{2+}$	$[Ni(L2)_3]^{2+}$	$[Ni(L3)_3]^{2+}$	$[Ni(L4)_3]^{2+}$	$[Ni(L5)_3]^{2+}$
Δ_{oct} /cm^{-1}	11423	11476	10681	9644	11555
B /cm^{-1}	889	866	918	994	866
Δ_{oct}/B	12.9	13.3	11.6	9.7	13.3
Complex	$[Fe(L1)_3]^{2+}$	$[Fe(L2)_3]^{2+}$	$[Fe(L3)_3]^{2+}$	$[Fe(L4)_3]^{2+}$	$[Fe(L5)_3]^{2+}$
ΔH_{SCO} / kJ·mol^{-1}	28.3(4)	35.0(3)	High-Spin[a]	High-Spin	High-Spin
ΔS_{SCO} / J·mol^{-1}·K^{-1}	91(1)	100(1)	High-Spin[a]	High-Spin	High-Spin
$T_{1/2}$ /K	309(6)	349(5)	High-Spin[a]	High-Spin	High-Spin

[a] The *fac*-$[Fe(L3)_3]^{5+}$ chromophore, when embedded into a non-constrained LaFe helicate, displayed $T_{1/2}$ ~ 50 K.

Scheme 2. Chemical structures of the didentate ligands **L1**–**L5**, together with the electronic properties of the associated pseudo-octahedral $[Ni(Lk)_3]^{2+}$ complexes and thermodynamic spin-crossover (SCO) properties of the pseudo-octahedral $[Fe(Lk)_3]^{2+}$ complexes in CD_3CN solutions [35,36].

Moving the methyl group bound to the pyridine ring from the 5-position in **L1** to the 3-position in **L3** and to the 6-position in **L4** (Scheme 2) is well-known to stepwise decrease the ligand-field strengths in the resulting $[Fe(Lk)_3]^{2+}$ complexes because the operation of additional sterical constraints, produced by intra-strand interactions in $[Fe(L3)_3]^{2+}$ [37] and by inter-strand interactions in $[Fe(L4)_3]^{2+}$, extends the Fe–N bond lengths (see Equation (1)) [38–40]. The associated trend $\Delta_{oct}(L1) \approx \Delta_{oct}(L2) > \Delta_{oct}(L3) > \Delta_{oct}(L4)$ observed for the isostructural $[Ni(Lk)_3]^{2+}$ complexes (Scheme 2), for which the determination of ligand field Δ_{oct} and Racah B parameters are not complicated by any SCO behavior, are in line with the observation of pure high-spin configurations for the $[Fe(L3)_3]^{2+}$ and $[Fe(L4)_3]^{2+}$ complexes in solution (Scheme 2) [35]. Whereas the connection of methyl groups adjacent to the donor nitrogen atom in the bound 6-methyl-pyridine groups in $[Fe(L4)_3]^{2+}$ produces such large inter-strand interactions that the contraction accompanying the high-spin to low-spin transition cannot be envisioned [41], the situation with the remote 3-methyl substituted pyridine units in [Fe(L3)$_3$]$^{2+}$ is less clear and a sophisticated triple-stranded heterometallic LaFe helicate containing the facial [Fe(L3)$_3$]$^{2+}$ chromophore has been shown to display partial SCO behavior at low temperature ($T_{1/2}$ ~50 K) [36]. Taking into account that (i) the replacement of a pyridine with a pyrazine ring in going from [Fe(L1)$_3$]$^{2+}$ and [Fe(L2)$_3$]$^{2+}$ stabilizes the low-spin state by $T_{1/2}$ = 40 K (Scheme 2) and (ii) moving the methyl group from the 5-position in [Fe(L1)$_3$]$^{2+}$ to the 3-position in *fac*-[Fe(L3)$_3$]$^{2+}$ (as found in the related LaFe helicate) produces an opposite trend with the stabilization of the high-spin form by $T_{1/2} \approx$ 50–310 = −260 K [36], we thus ingenuously explored the possibility of combining both aspects in the didentate ligand **L5** where the methyl group is now connected to the 3-position of a pyrazine ring (Scheme 3a) with the hope of pushing the transition temperature toward cryoscopic temperatures for [Fe(L5)$_3$]$^{2+}$ around $T_{1/2} \sim T_{1/2}$ ([Fe(L2)$_3$]$^{2+}$) − 260 = 350 − 260 = 90 K (3-methyl-pyrazine). This effort is justified by our long-term quest for designing a pseudo-octahedral spin-crossover [Fe(L*k*)$_3$]$^{2+}$ unit that can modulate the luminescence of adjacent emissive lanthanides in (supra)molecular assemblies via energy transfers within a temperature domain (77–150 K) accessible to optical reading and addressing [36]. Finally, since minor structural variations may induce large changes in ligand-field strength, the systematic exploration of unpredictable intermolecular packing interactions [42,43] operating in crystalline samples of [Fe(L5)$_3$]X$_2$ complexes (X$^-$ = monoanionic counter-ions) may contribute to the lucky search for some 'ideal' Fe^{II} complexes, which additionally exhibit hysteretic behavior and bistability [17,44,45].

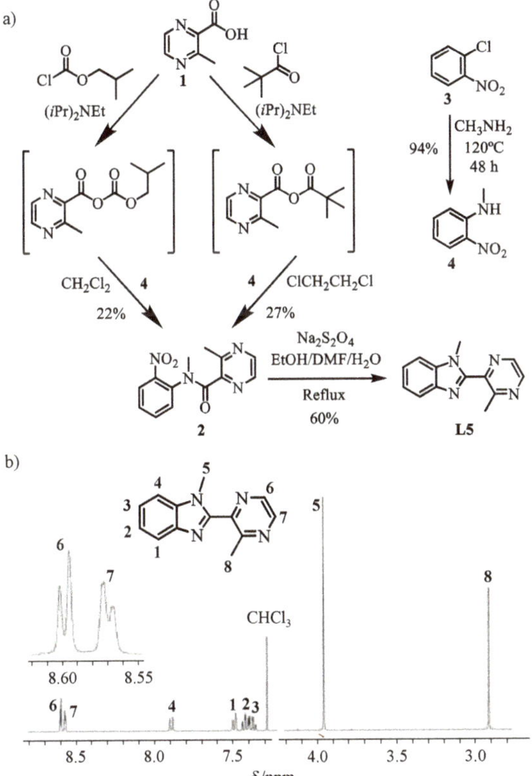

Scheme 3. (a) Chemical structure and synthesis of the didentate ligand **L5** shown in its anti-conformation and (b) associated ^{1}H NMR spectrum with numbering scheme (CDCl$_3$, 298 K).

2. Experimental

Chemicals were purchased from Sigma-Aldrich (Gmbh, Buchs) and Acros and used without further purification unless otherwise stated. Dichloromethane, 1,2-dichloroethane, tert-butylmethyl ether, and N,N-dimethylformamide were dried through an alumina cartridge. Silica-gel plates (Merck, 60 F_{254}) were used for thin-layer chromatography, SiliaFlash® silica gel P60 (0.04–0.063 mm,) and Acros silica gel 60 (0.035–0.07 mm) were used for preparative column chromatography.

Preparation of N-methyl-2-nitroaniline (4). 1-chloro-2-nitrobenzene (**3**, 31.75 g, 201.5 mmol, 1.0 eq) and methylamine (198 mL, 40% weight in H_2O, 2295.6 mmol, 11.4 eq) were introduced into a Carius tube equipped with a magnetic stirrer and heated at 120 °C for 48 h. Excess of methylamine was rotatory evaporated and the residual brown oil was partitioned between CH_2Cl_2 (300 mL) and half sat. aq. NH4Cl (300 mL). The organic layer was separated and the aq. phase was further extracted with CH_2Cl_2 (3 × 150 mL). The combined organic extracts were dried over anhydrous Na_2SO_4, filtered, and the solvent evaporated to dryness. The resulting red oil was purified by column chromatography (Silica, CH_2Cl_2) to give 28.99 g of N-methyl-2-nitroaniline (**4**, 190.5 mmol, yield 94%) as a deep red orange oil, which slowly crystallized within hours. ^{1}H NMR ($CDCl_3$, 400 MHz, 298 K) δ/ppm: 8.13 (1H, dd, ^3J = 8.8 Hz, ^4J = 1.6 Hz), 8.00 (1H, bs), 7.43 (1H, ddd, ^3J = 8.8 Hz, ^3J = 7.2 Hz, ^4J = 1.6 Hz), 6.81 (1H, dd, ^3J = 8.6 Hz, ^4J = 1.0 Hz), 6.62 (1H, ddd, ^3J = 8.4 Hz, ^3J = 7.2 Hz, ^4J = 1.2 Hz), 2.99 (3H, s).

Preparation of $N,3$-dimethyl-N-(2-nitrophenyl) pyrazine-2-carboxamide (2, left pathway in Scheme 3a). A suspension of 3-methyl-pyrazine-2-carboxylic acid (**1**, 5 g, 35.5 mmol, 1 eq) and di-isopropyl-ethylamine (7.2 mL, 5.5 g, 45 mmol, 1.26 eq) in CH_2Cl_2 (4 mL) was added dropwise into a two-necked flask containing isobutyl chloroformate (5.4 mL, 5.408 g, 45 mmol, 1.26 eq) dissolved in CH_2Cl_2 (1 mL). The mixture was stirred at −15 °C for 120 min, after which a solution of N-methyl-2-nitroaniline (**4**, 5.5 g, 0.0352 mol, 1 eq) in CH_2Cl_2 was added. After stirring for 15 h at room temperature, the solution was partitioned between CH_2Cl_2 (100 mL) and half-sat. aq. NH4Cl (250 mL). The organic layer was separated and the aqueous phase was further extracted with CH_2Cl_2 (3 × 100 mL). The organic fractions were dried over anhydrous Na_2SO_4, concentrated under vacuum, and purified by column chromatography (Silica, CH_2Cl_2/MeOH 99.2:0.8) to yield $N,3$-dimethyl-N-(2-nitrophenyl) pyrazine-2-carboxamide (**2**, 8.22 mmol, yield 22%). ^{1}H NMR ($CDCl_3$, 400 MHz, 298 K) δ/ppm: mixture of two rotamers A (72.5%) and B (27.5 %): 2.67 (3H, s, A), 2.72 (3H, s, B), 3.58 (3H, s, A), 3.33 (3H, s, B), 7.37–7.80 (3H, m, A and B), 8.29 (1H, d, ^3J = 2.5 Hz, A), 8.60 (1H, d, ^3J = 2.5 Hz, B), 8.0 (1H, dd, ^3J = 2.5 Hz, ^5J = 0.6 Hz, A), 8.50 (1H, dd, ^3J = 2.5 Hz, ^5J = 0.6 Hz, B), 7.86 (1H, dd, ^3J = 8.5 Hz, ^4J = 1.5 Hz, A), 8.13 (1H, dd, ^3J = 8.5 Hz, ^4J = 1.5 Hz, B). ESI-MS (soft-positive mode; MeOH+$CHCl_3$+HCOOH): m/z =273.0 ([2 + H]$^+$), 295.1 ([2 + Na]$^+$).

Preparation of $N,3$-dimethyl-N-(2-nitrophenyl) pyrazine-2-carboxamide (2, right pathway in Scheme 3a). A suspension of 3-methyl-pyrazine-2-carboxylic acid (**1**, 0.2 g, 1.44 mmol, 1 eq) and di-isopropyl-ethylamine (0.37 mL, 0.28 g, 2.16 mmol, 1.5 eq) in 1,2-dichloroethane (4 mL) was added dropwise into a two-necked flask containing trimethylacetyl chloride (0.194 mL, 0.19 g, 1.58 mmol, 1.1 eq) in 1,2-dichloroethane (1 mL). The mixture was stirred at −20 °C for 60 min, after which a solution of N-methyl-2-nitroaniline (**4**, 0.329 g, 2.16 mmol, 1.5 eq) in 1,2-dichloroethane (5 mL) was added. After refluxing for 16 h, the solution was concentrated under vacuum and partitioned between CH_2Cl_2 (80 mL) and half-saturated aqueous solution of NH4Cl (200 mL). The organic layer was separated and the aqueous phase was further extracted using CH_2Cl_2 (3 × 80 mL). The combined organic fractions were concentrated under vacuum after drying with anhydrous Na_2SO_4. Ultimate purification using column chromatography (Silica, CH_2Cl_2/MeOH 99.2:0.8) yielded 0.106 g of N, 3-dimethyl-N-(2-nitrophenyl) pyrazine-2-carboxamide (**2**, 0.387 mmol, yield 27%).

Preparation of 1-methyl-2-(3-methylpyrazin-2-yl)-1H-benzo[d]imidazole (L5). $N,3$-dimethyl-N-(2-nitrophenyl)pyrazine-2-carboxamide (**2**, 2.56 g, 7.58 mmol, 1 eq) was dissolved in EtOH:DMF (20 mL:25 mL). Sodium dithionite (8.0 g, 40 mmol, 5.2 eq) was added to the mixture and the temperature of the system was raised to 80 °C when 20 mL of water was added. After refluxing for 36 h, the mixture was neutralized using aqueous ammonia and the solvents were removed in vacuo.

The concentrate was dissolved in CH_2Cl_2 (50 mL) and washed with water (3×200 mL). The aqueous layers were collectively further extracted with CH_2Cl_2 (3×50 mL). The combined organic fraction was then concentrated under vacuum and purified by column chromatography (silica, CH_2Cl_2/MeOH 98:2) to yield **L5** (7.67 mmol, yield 60%). The compound was crystallized as needles by slow diffusion of n-hexane into a concentrated CH_2Cl_2 solution of **L5**. ^{1}H NMR (CDCl$_3$, 400 MHz, 298 K) δ/ppm: 2.91 (3H, s), 3.96 (3H, s), 7.35–7.44 (2H, m), 7.49 (1H, d, ^3J = 7 Hz), 7.88 (1H, d, ^3J = 7 Hz), 8.57 (1H-pz, dd, ^3J = 2.4 Hz, ^5J = 0.5 Hz), 8.60 (1H-pz, dd, ^3J = 2.4 Hz). ^{13}C NMR (CDCl$_3$, 101 MHz, 298K) δ/ppm: 155.68 (C_q), 149.12 (C_q), 144.54 (C_q), 143.84 (CH_{pz}), 142.45 (C_q), 140.86 (CH_{pz}), 136.15 (C_q), 123.85 (CH), 122.90 (CH), 120.45 (CH), 110.01 (CH), 31.83 (CH_3), 23.26 (CH_3). ESI-MS (soft-positive mode; MeOH+CHCl$_3$+HCOOH): m/z 225.1([**L5** + H]$^+$). Elemental analysis calculated for $C_{13}H_{12}N_4$ (%): C 69.62, H 5.39, N 24.98; Found (%): C 69.46, H 5.09, N 25.20.

Preparation of mononuclear Fe^{II}, Zn^{II}, and Ni^{II} complexes with 1-methyl-2-(3-methylpyrazin-2-yl)-1*H*-benzo[*d*]imidazole (L5). In a typical synthesis, 0.3 mmol (3 eq) of the ligand **L5** dissolved in acetonitrile (2 mL) was added to 0.1 mmol (1 eq) of Fe(ClO$_4$)$_2$·6H$_2$O or Fe(CF$_3$SO$_3$)$_2$ or Ni(BF$_4$)$_2$·6H$_2$O or Zn(CF$_3$SO$_3$)$_2$ in acetonitrile (2 mL). The resulting mixture was stirred under an inert atmosphere for 3 h, then evaporated to dryness under vacuum to yield microcrystalline powders of the respective complexes. These powders were dissolved in acetonitrile and allowed to crystallize by evaporation or by slow diffusion of tert-butyl methyl ether to give 64–78% of primary [Fe(**L5**)$_3$](CF$_3$SO$_3$)$_2$·1.5H$_2$O, [Ni(**L5**)$_3$](BF$_4$)$_2$·1.5H$_2$O·1.5CH$_3$CN and [Zn(**L5**)$_3$](BF$_4$)$_2$·4H$_2$O complexes (Table S1). Single crystals suitable for characterization by X-ray diffraction could be obtained by slow evaporation of acetonitrile solution containing 10 eq of (^{n}Bu)$_4$NClO$_4$ or (^{n}Bu)$_4$NBF$_4$ to give [Fe(**L5**)$_3$](ClO$_4$)$_2$ (I), [Ni(**L5**)$_3$](ClO$_4$)$_2$ (II), [Ni(**L5**)$_3$](BF$_4$)$_2$·H$_2$O (III), [Zn(**L5**)$_3$](ClO$_4$)$_2$ (IV), [Ni(**L5**)$_3$](BF$_4$)$_2$·1.75CH$_3$CN (V) and [Zn(**L5**)$_3$](BF$_4$)$_2$·1.5CH$_3$CN (VI).

Caution! Dry perchlorates may explode and should be handled in small quantities and with the necessary precautions [46,47].

2.1. Spectroscopic and Analytical Measurements

^{1}H and ^{13}C NMR spectra were recorded at 298 K on a Bruker Avance 400 MHz spectrometer. Chemical shifts are given in ppm with respect to tetramethylsilane. Spectrophotometric titrations were performed with a J&M diode array spectrometer (Tidas series) connected to an external computer. In a typical experiment, 25 cm^3 of ligand in acetonitrile (2×10^{-4} M) was titrated at 298 K with a solution of Fe(CF$_3$SO$_3$)$_2$ or Ni(BF$_4$)$_2$·6H$_2$O or Zn(CF$_3$SO$_3$)$_2$ (2×10^{-3} M) in acetonitrile under an inert atmosphere. After each addition of 33 μL, the absorbance was recorded using Hellma optrodes (optical path length 0.1 cm) immersed in the thermostated titration vessel and connected to the spectrometer. Mathematical treatment of the spectrophotometric titrations was performed with factor analysis [48–50] and with ReactLab^{TM} Equilibria (previously Specfit/32) [51–53]. Pneumatically-assisted electrospray (ESI-MS) mass spectra were recorded from 10^{-4} M (ligands) and 10^{-3} M (complexes) solutions on an Applied Biosystems API 150EX LC/MS System equipped with a Turbo Ionspray source. Elemental analyses were performed by K. L. Buchwalder from the Microchemical Laboratory of the University of Geneva. Elemental analysis was not conducted for perchlorate salts for security reasons, while crystals of the tetrafluoroborate salts lost their solvent upon separation from the mother liquor and were not further characterized. Electronic spectra in the UV–Vis region were recorded at 293 K from solutions in CH$_3$CN with a Perkin-Elmer Lambda 1050 using quartz cells of a 0.1 or 1.0 mm path length. Solid-state absorption spectra were recorded with a Perkin-Elmer Lambda 900 using capillaries. Solid-state magnetic data were recorded on a MPMS 3 or MPMS 5 QUANTUM DESIGN magnetometers using magnetic fields of 1000–5000 Oe at 1 K/min rates within the 5–300 K range. The magnetic susceptibilities were corrected for the magnetic response of the sample holder and for the diamagnetism of the compounds by using the approximation $\chi_D = -\frac{MW}{2} \cdot 10^{-6}$ cm^3·mol^{-1} [54].

2.2. X-Ray Crystallography

Summary of crystal data, intensity measurements, and structure refinements for compounds **L5**, [Fe(**L5**)$_3$](ClO$_4$)$_2$ (**I**), [Ni(**L5**)$_3$](ClO$_4$)$_2$ (**II**), [Ni(**L5**)$_3$](BF$_4$)$_2$·H$_2$O (**III**), and [Zn(**L5**)$_3$](ClO$_4$)$_2$ (**IV**) is presented in Tables S2–S4. Pertinent bond lengths, bond angles, and interplanar angles are collected in Tables S5–S14 together with ORTEP views and pertinent numbering schemes gathered in Figures S1–S5. The crystals were mounted on MiTeGen kapton cryoloops with protection oil. X-ray data collection was performed with an Agilent SuperNova Dual diffractometer equipped with a CCD Atlas detector (Cu[Kα] radiation). The structures were solved by using direct methods [55,56] or dual-space methods [57]. Full-matrix least-square refinements on F^2 were performed with SHELX2014 [58]. CCDC 1988655-1988659 contained the supplementary crystallographic data. These data can be obtained free of charge from the Cambridge Crystallographic Data Centre via www.ccdc.cam.ac.uk/. Single crystals of [Ni(**L5**)$_3$](BF$_4$)$_2$·1.75CH$_3$CN (**V**) and [Zn(**L5**)$_3$](BF$_4$)$_2$·1.5CH$_3$CN (**VI**) could also be obtained as inversion twins. The two complexes were isostructural and crystallized in the trigonal system (*P3c*1 space group) with five independent complexes in the asymmetric unit, all located on three-fold rotation axes (the metal content of the asymmetric unit is 5/3) and Z = 10 (Table S15). Although there is no doubt that the three ligands adopt facial arrangements around the metal to give exclusively *fac*-[M(**L5**)$_3$]$^{2+}$ cations (Figures S6–S7), we were only able to locate unambiguously three BF$_4^-$ counter-anions in the asymmetric unit. Despite numerous efforts, we were not able to obtain a satisfying model for the last third of a BF$_4^-$ counter-anion and gave up to further discuss these structures and to deposit the cif files.

3. Results and Discussion

Synthesis, characterization, and solid-state structures were obtained for the didentate ligand **L5** and its pseudo-octahedral complexes [M(**L5**)$_3$]X$_2$ (M = Fe, Ni, Zn and X = BF$_4$, ClO$_4$). Compared with pyridine-carboxylic acids, which are easily activated via their transformation into acyl chloride with the help of thionyl chloride or oxalyl chloride [59], the electron-rich pyrazine analogue **1** produced only negligible yield (<1%) of the target amide product **2** under these standard conditions [60]. The 3-methylpyrazine-2-carboxylic acid **1** was thus activated as its anhydride through reaction with either isobutyl chloroformate (left path in Scheme 3a) or pivaloyl chloride (right path in Scheme 3a). Subsequent nucleophilic attack with *N*-methyl-2-nitroaniline **4** yielded the ortho-nitroamide compound **2** in moderate yield. A subsequent reductive cyclisation reaction provided ligand **L5**, which was characterized by its ^{1}H-NMR spectrum (Scheme 3b). The lack of NOE effect observed between the methyl groups in positions 5 and 8 indicates an anti-conformation for the α,α'-diimine chelate unit, which was confirmed by (i) the crystal structure of **L5** (Figure 1a) and (ii) gas-phase calculations predicting a global energy minimum for the planar anti-conformation (interplanar angle between the two aromatic rings α = 180°, Figure 1b) [61]. Given that the same anti-conformations are (i) found in the solid state (Figure S8) and (ii) predicted in the gas phase for the ligands **L2** [60] and **L5**, their computed EHMO frontiers orbitals are comparable (Figure S9a) and lead to akin electronic absorption spectra dominated by intense $\pi^* \leftarrow \pi$ covering the near UV range (Figure S9b).

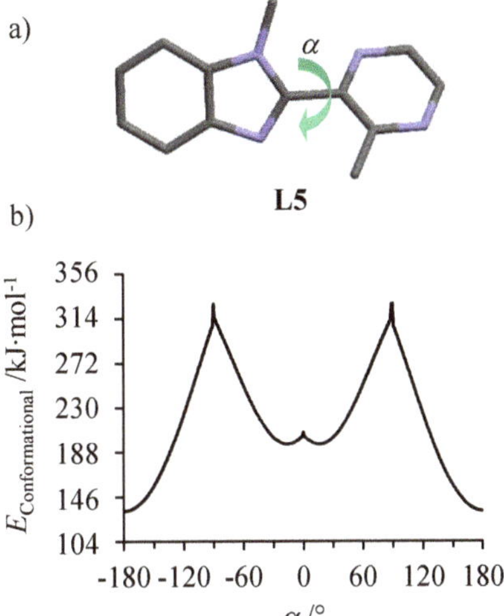

Figure 1. (**a**) Molecular structure of ligand **L5** in its crystal structure highlighting the dihedral angle $\alpha = 144.8°$ ($\alpha = 180°$ for a planar anti-conformation and $\alpha = 0°$ for a planar syn-conformation). (**b**) Gas-phase energy computed for **L5** at the MM2 level as a function of the interplanar angle [61].

Interestingly, the gas-phase energy of **L5** displayed two additional local energy minima for $\alpha = \pm16°$ (Figure 1b), which shifted from $\alpha = 0°$ previously reported for the second local minimum in **L2** (Figure S10) [60]. The larger interplanar angle of 16° in the optimized syn-conformation of **L5** was the result of the sterical crowding between the close methyl groups connected to the adjacent aromatic rings (positions 5 and 8 in the numbering of Scheme 3b). Taking the latter conformation as a limiting structural model when **L5** is bound to a metal cation provides $d_{N\cdots N} = 2.83$ Å between the two nitrogen donor atoms of the α,α'-diimine chelate. According to Phan et al. [33], the latter separation matches the $2.78 \leq d_{N\cdots N} \leq 2.93$ Å range for which a diimine ligand might be used to achieve spin-crossover behavior in tris-homoleptic Fe^{II} complexes.

Stoichiometric mixing of **L5** (3 eq.) with $Fe(CF_3SO_3)_2$, $Ni(BF_4)_2\cdot6H_2O$ or $Zn(BF_4)_2\cdot6H_2O$ (1 eq,) in acetonitrile gave fair yields of microcrystalline primary precipitates of $[Fe(\mathbf{L5})_3](CF_3SO_3)_2\cdot1.5H_2O$, $[Ni(\mathbf{L5})_3](BF_4)_2\cdot1.5H_2O\cdot1.5CH_3CN$, and $[Zn(\mathbf{L5})_3](BF_4)_2\cdot4H_2O$ complexes (Table S1). A series of isostructural complexes $[Fe(\mathbf{L5})_3](ClO_4)_2$ (**I**), $[Ni(\mathbf{L5})_3](ClO_4)_2$ (**II**), and $[Zn(\mathbf{L5})_3](ClO_4)_2$ (**IV**) could be obtained as single crystals by recrystallization in acetonitrile containing 10 eq of $(^nBu)_4NClO_4$.

Monocrystals suitable for x-ray diffractions were also obtained for $[Ni(\mathbf{L5})_3](BF_4)_2\cdot H_2O$ (**III**), $[Ni(\mathbf{L5})_3](BF_4)_2\cdot1.75CH_3CN$ (**V**), and $[Zn(\mathbf{L5})_3](BF_4)_2\cdot1.5CH_3CN$ (**VI**) using the same method except for the replacement of $(^nBu)_4NClO_4$ with $(^nBu)_4NBF_4$ (Tables S3 and S4). The crystal structures of the perchlorate salts systematically displayed the formation of *mer*-$[M(\mathbf{L5})_3]^{2+}$ cations (Figure 2), in which the $[MN_6]$ chromophores adopted a geometry close to the perfect octahedron as ascertained by SHAPE's scores close to zero (Table 1) [62–66].

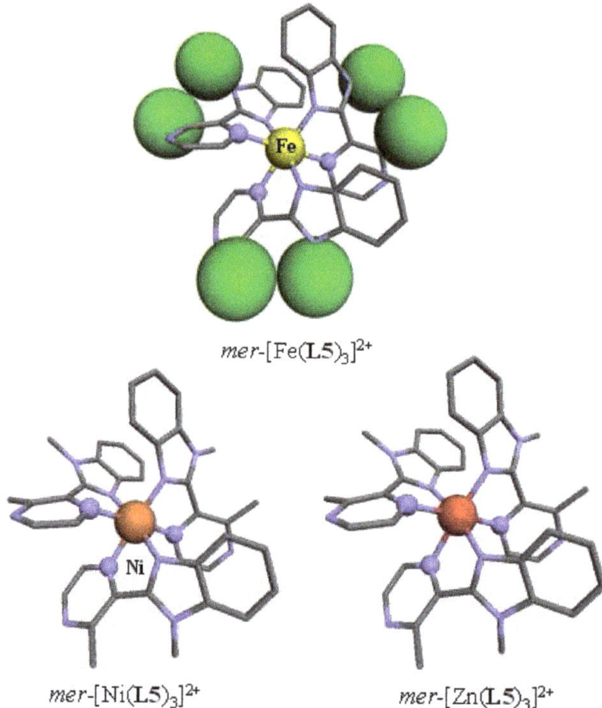

Figure 2. Perspective views of the molecular structures of the [M(**L5**)$_3$]$^{2+}$ cations in the crystal structures of [Fe(**L5**)$_3$](ClO$_4$)$_2$ (**I**), [Ni(**L5**)$_3$](ClO$_4$)$_2$ (**II**), and [Zn(**L5**)$_3$](ClO$_4$)$_2$ (**IV**). The nitrogen atoms of the pyrazine rings bound to the metals are displayed with blue spheres in order to highlight their meridional arrangements around the central cation. Color codes: C = grey, N = blue. Hydrogen atoms and ionic perchlorate counter-anions are omitted for clarity. For [Fe(**L5**)$_3$](ClO$_4$)$_2$ (**I**), the carbon atoms of the methyl groups are shown as green spheres using the Corey–Pauling–Koltun (CPK) model to highlight the intra-strand steric hindrance.

On the contrary, the crystal structures of the tetrafluoroborate salts showed the existence of *fac*-[M(**L5**)$_3$]$^{2+}$, in which the three didentate ligands adopted the same orientation along the pseudo-threefold axis passing through the metal (Figure 3). Having previously established that the energy gap between the facial (C_3-symmetry) and meridional (C_1-symmetry) geometries in [Zn(**L1**)$_3$]$^{2+}$ and [Zn(**L2**)$_3$]$^{2+}$ roughly followed a pure statistical (i.e., entropic) trend and does not overcome thermal energy at room temperature [60], we concluded that packing forces, specific to the use of perchlorate or tetrafluoroborate counter anions, are more than enough for the quantitative and selective crystallization of pure meridional, respectively facial isomers. For the [Ni(**L5**)$_3$]$^{2+}$ and [Zn(**L5**)$_3$]$^{2+}$ chromophores, the M–N bonds are systematically shorter for the more basic benzimidazole nitrogen donor (Table 1, entry 4; pK_a(bzim) = 5.68) than with its pyrazine counterpart (Table 1, entry 5; pK_a(pyrazine) = 0.65) [67], a trend in complete agreement with that reported for the analogous complex [Zn(**L2**)$_3$]$^{2+}$ (Table 1, column 8) [60]. Moreover, the shift of the methyl group bound to the pyrazine ring from the 5-position in **L2** to the 3-position in **L5** has globally no geometric influence on the [ZnN$_6$] coordination sphere, thus leading to Zn–N bond distances surrounding the standard value of Zn–N = 0.74 + 1.46 = 2.20 Å deduced from the effective ionic radii [68]. In other words, the close methyl groups found in the bound didentate ligand **L5** do not induce major intramolecular steric constraints in [Zn(**L5**)$_3$]$^{2+}$ and only a slight increase of the interannular intraligand angles can be detected in going from [Zn(**L2**)$_3$]$^{2+}$ (α =

21(13)°) to $[Zn(\mathbf{L5})_3]^{2+}$ ($\alpha = 38(4)°$, entry 5 in Table 1). The molecular structures of $[Ni(\mathbf{L5})_3]^{2+}$ were very similar to those observed for the Zn^{II} analogues (Figures 2 and 3), except for the slightly shorter Ni–N bond distances, a trend in line with the contraction of Shannon's effective ionic radii predicted to be 0.74 Å for six-coordinate Zn^{2+} and 0.69 Å for six-coordinate Ni^{2+} [68]. The detection of long Fe–N bond distances in $[Fe(\mathbf{L5})_3]^{2+}$ ($d(\text{Fe-N}_{bz}) = 2.14(1)$ Å and $d(\text{Fe-N}_{pz}) = 2.24(3)$ Å, entries 3–4 in Table 1) is more remarkable since it suggests that the Fe^{II} metal center adopts a pure high spin configuration at 180 K as previously found for the analogous 3-methylpyridine-benzimidazole ligand in $[Fe(\mathbf{L3})_3]^{2+}$ ($d(\text{Fe-N}_{bz}) = 2.14(2)$ Å and $d(\text{Fe-N}_{py}) = 2.26(5)$ Å) [35].

Spin-state, magnetic, and electronic properties of the pseudo-octahedral complexes $[M(\mathbf{L5})_3]X_2$ (M = Fe, Ni and X = BF_4, ClO_4) were obtained in the solid state. Molar magnetic susceptibilities (χ_M), corrected for diamagnetism of solid state samples of $[Fe(\mathbf{L5})_3](ClO_4)_2$ (I) were recorded at variable temperatures in a constant magnetic field of 5000 Oe. The $\chi_M T$ versus T plot shows a smooth and regular increase of the $\chi_M T$ product in the 50–300 K range (red trace in Figure 4), which can be fitted to Equation (6) using a high Curie constant $C = 3.5708(9)$ cm^3·K·mol^{-1} and a non-negligible temperature independent paramagnetism TIP = $848(5) \times 10^{-6}$ cm^3·mol^{-1}, a behavior diagnostic for a high-spin Fe(II) complex [69].

$$\chi_M T = C + T \cdot \text{TIP} \tag{6}$$

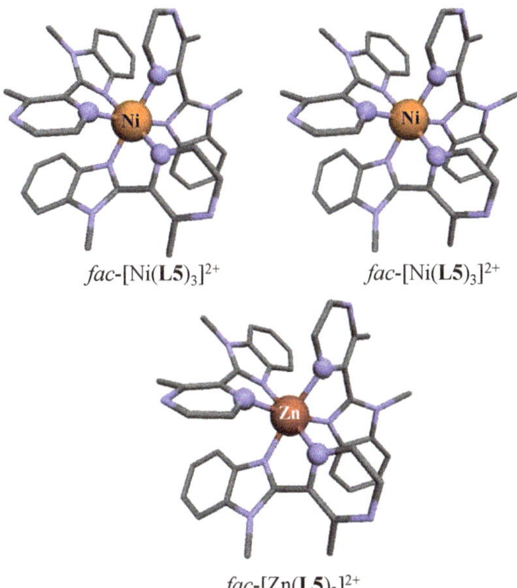

fac-$[Ni(\mathbf{L5})_3]^{2+}$ *fac*-$[Ni(\mathbf{L5})_3]^{2+}$

fac-$[Zn(\mathbf{L5})_3]^{2+}$

Figure 3. Perspective views of the molecular structures of the $[M(\mathbf{L5})_3]^{2+}$ cations in the crystal structures of $[Ni(\mathbf{L5})_3](BF_4)_2 \cdot H_2O$ (**III**, top left), $[Ni(\mathbf{L5})_3](BF_4)_2 \cdot 1.75CH_3CN$ (**V**, top right) and $[Zn(\mathbf{L5})_3](BF_4)_2 \cdot 1.5CH_3CN$ (**VI**). The nitrogen atoms of the pyrazine rings bound to the metals are displayed with blue spheres in order to highlight their facial arrangements around the central cation. Color codes: C = grey, N = blue. Hydrogen atoms and ionic tetrafluoroborate counter-anions are omitted for clarity.

Table 1. Structural data for hexa-coordinate metallic centers (Fe^{II}, Ni^{II} or Zn^{II}) found in the crystal structures of $[Fe(\mathbf{L5})_3](ClO_4)_2$ (**I**), $[Ni(\mathbf{L5})_3](ClO_4)_2$ (**II**), $[Ni(\mathbf{L5})_3](BF_4)_2 \cdot H_2O$ (**III**), $[Zn(\mathbf{L5})_3](ClO_4)_2$ (**IV**), $[Ni(\mathbf{L5})_3](BF_4)_2 \cdot 1.75CH_3CN$ (**V**), and $[Zn(\mathbf{L5})_3](BF_4)_2 \cdot 1.5CH_3CN$ (**VI**) at 180 K collecting average bond lengths, interannular intraligand angles (α), interchelate angles (β), chelate bite angles (γ), and SHAPE's scores.

	$[Fe(\mathbf{L5})_3](ClO_4)_2$	$[Ni(\mathbf{L5})_3](ClO_4)_2$	$[Ni(\mathbf{L5})_3](BF_4)_2$	$[Ni(\mathbf{L5})_3](BF_4)_2$	$[Zn(\mathbf{L5})_3](ClO_4)_2$	$[Zn(\mathbf{L5})_3](BF_4)_2$	$[Zn(\mathbf{L2})_3](CF_3SO_3)_2$
Crystal Structure	I	II	III	V	IV	VI	
Configuration	meridional	meridional	facial	facial	meridional	facial	meridional
$d(\text{M-N}_{bz})/\text{Å}$ [a]	2.14 (1)	2.08 (2)	2.08 (2)	2.07 (2)	2.11 (2)	2.09 (1)	2.09 (7)
$d(\text{M-N}_{pz})/\text{Å}$ [b]	2.24 (3)	2.14 (2)	2.12 (2)	2.12 (1)	2.25 (6)	2.26 (1)	2.27 (3)
$\alpha/°$	39 (3)	37 (3)	39 (3)	36 (4)	39 (3)	38 (4)	21 (13)
$\beta/°$	85 (9)	86 (7)	88 (8)	83.4 (9)	85 (8)	84 (1)	81 (8)
$\gamma/°$	76.6 (7)	78.8 (4)	79.1 (5)	79.2 (7)	77 (1)	77.8 (8)	76.6 (1)
Octahedron [c]	1.85	1.41	1.25 (3) [d]	1.21 (8) [d]	1.74	1.5 (1) [d]	1.95
Trigonal Prism [c]	15.65	15.42	15.5 (1) [d]	16.0 (2) [d]	15.39	15.9 (2) [d]	12.99
Reference	This work	This work	This work	This work	This work	This work	[60]

[a] bz = benzimidazole. [b] pz = pyrazine. [c] SHAPE's scores calculated with reference to ideal octahedral and trigonal prismatic geometries [62–66]. [d] Average for more than one complex in the asymmetric unit.

The additional abrupt decrease in the magnetic susceptibility occurring at low temperature ($T < 40$ K) can be assigned to zero-field splitting (ZFS) of high-spin Fe(II), which was modeled with Equation (7), where D and E are the axial and rhombic ZFS parameters, respectively [70–73].

$$E_n^0 = D \cdot \left(S_{\hat{z}}^2 - \frac{S(S+1)}{3}\right) + E \cdot \left(S_{\hat{x}}^2 - S_{\hat{y}}^2\right) \tag{7}$$

The pseudo-threefold axis characterizing the $[\text{FeN}_6]$ chromophore in $[\text{Fe}(\textbf{L5})_3](\text{ClO}_4)_2$ implies that E can be neglected ($E \sim 0$). Consequently, the electron–electron interaction splits the $S = 2$ manifold at zero magnetic field into three energy levels located at $E_1^0 = -2D$ ($m_s = 0$), $E_2^0 = -D$ ($m_s = \pm 1$), and $E_3^0 = 2D$ ($m_s = \pm 2$). Application of the van Vleck Equation (8), where $k_B = 0.695039$ cm$^{-1}\cdot$K^{-1} is the Boltzmann constant and N_A is Avogadro number, E_n^1 are first-order spin-only Zeeman effects given in Equation (9), where $\mu_B = -4.6686 \times 10^{-5}$ cm$^{-1}\cdot$G^{-1} is the Bohr magneton, and E_n^2 stands for the second-order Zeeman effects, leads to a satisfying fit (dotted black trace in Figure 4 with agreement factor AF $= 3.73 \times 10^{-3}$) with Landé factor $g = 2.20(1)$, $D = 0.52(1)$ cm^{-1} and TIP $= -N_A E_n^2 = 319(4) \times 10^{-6}$ cm$^3\cdot$mol^{-1}.

$$\chi_M T = T \cdot N_A \cdot \frac{\sum_n \left[\left(\frac{(E_n^1)^2}{k_B T} - 2E_n^2\right) \cdot \exp\left(-\frac{E_n^0}{k_B T}\right)\right]}{\sum_n \left[\exp\left(-\frac{E_n^0}{k_B T}\right)\right]} \tag{8}$$

$$E_n^1 = -g \cdot \mu_B \cdot m_S \tag{9}$$

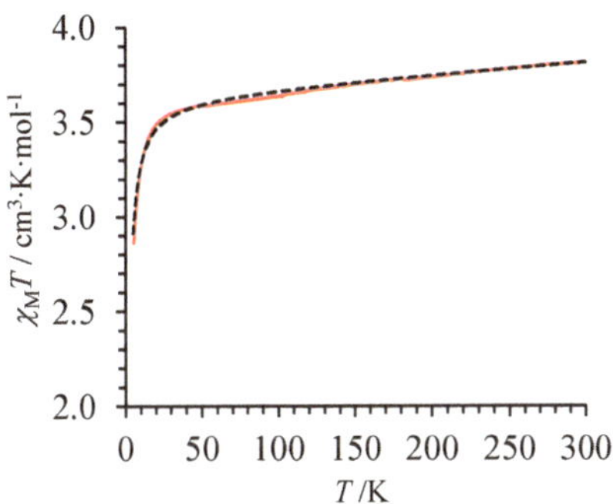

Figure 4. $\chi_M T$ versus T plot of the molar magnetic susceptibility (χ_M) between 5–300 K, corrected for diamagnetism recorded for the complex $[\text{Fe}(\textbf{L5})_3](\text{ClO}_4)_2$ (**I**, red trace) at 5000 Oe. The dotted black trace shows the best fit (agreement factor AF $= 3.73 \times 10^{-3}$) obtained using Equation (8) with $g = 2.20(1)$, $D = 0.52(1)$ cm^{-1} and TIP $= -N_A E_n^2 = 319(4) \times 10^{-6}$ cm$^3\cdot$mol^{-1}.

The latter magnetic data closely matched those reported for the analogous $[\text{Fe}(\textbf{L3})_3](\text{CF}_3\text{SO}_3)_2$ complex ($g = 2.20(2)$, $D = 0.85(1)$ cm^{-1} [36]), and demonstrate that our novel $[\text{Fe}(\textbf{L5})_3](\text{ClO}_4)_2$ complex, in which the 3-methyl-pyridine group of **L3** is replaced with a 3-methyl-pyrazine group in **L5**, is also purely high-spin within the 5–300 K range with no trace of SCO behavior. A careful inspection of the experimental curve around 80 K (Figure 4) showed a very minor deviation from the theoretical model, which could be tentatively assigned to traces of trapped low-spin form as previously reported for *fac*-$[\text{Fe}(\textbf{L3})_3]^{2+}$ when it is incorporated into a LaFe triple-stranded helicate [36].

The electronic absorption spectrum recorded for $[\text{Fe}(\textbf{L5})_3](\text{ClO}_4)_2$ (**I**) in the solid state shows the expected Jahn–Teller split $\text{Fe}^{II}(^5\text{E}\leftarrow{}^5\text{T}_2)$ ligand-field transition (Figure 5a) [31]. A deconvolution using two Gaussian functions gives $\tilde{v}_{max} = 8881$ cm^{-1} and 11,887 cm^{-1}, thus leading to a barycenter at 10,384 cm^{-1}, which provides a direct estimation of $\Delta_{oct}^{HS}(\text{Fe}^{II}) = 10$ Dq, a value only 700 cm^{-1} below the minimum of $\Delta_{oct}^{HS} \approx 11{,}000$ cm^{-1} suggested to be the lower limit for inducing spin state equilibria [31].

However, the latter criterion strongly depends on the choice of the Racah parameter B, which is not easily extracted from the single intrashell d–d transition observed in the electronic spectra of high-spin Fe^{II} complexes. For this reason, we have recorded the electronic absorption spectrum of the analogous Ni^{II} complex $[Ni(\mathbf{L5})_3](BF_4)_2 \cdot H_2O$ (**III**), for which the combination of two spin-allowed d–d transitions $Ni(^3T_2 \leftarrow {}^3A_2)$ and $Ni(^3T_1 \leftarrow {}^3A_2)$ and one spin-forbidden transition $Ni(^1E \leftarrow {}^3A_2)$ (Figure 5b) allows a complete characterization of the electronic parameters $\Delta_{oct}(Ni^{II})$, $B(Ni^{II})$, and $C(Ni^{II})$ with the help of Equations (10)–(13) [74,75].

$$E\left({}^3A_{2g}\right) = 0 \tag{10}$$

$$E\left({}^1E_g\right) = 8B + 2C - \frac{6B^2}{\Delta_{oct}} \tag{11}$$

$$E\left({}^3T_{2g}\right) = \Delta_{otc} \tag{12}$$

$$E\left({}^3T_{1g}\right) = 1.5\Delta_{oct} + 7.5B - 0.5\sqrt{225B^2 + \Delta_{oct}{}^2 - 18\Delta_{oct}B} \tag{13}$$

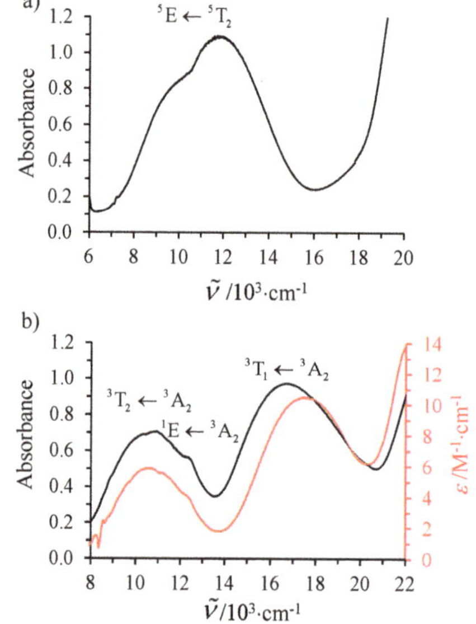

Figure 5. Electronic absorption spectra recorded at 298 K for (**a**) $[Fe(\mathbf{L5})_3](ClO_4)_2$ (**I**) in the solid state and (**b**) $[Ni(\mathbf{L5})_3](BF_4)_2 \cdot H_2O$ (**III**) in the solid state (black trace, left axis) and in acetonitrile solution (0.1 M, red trace, right axis).

A Gaussian deconvolution of the visible part of the absorption spectrum into three peaks yielded two broad bands, diagnostic for the spin-allowed, but parity-forbidden, transitions at 10,672 cm^{-1} $({}^3T_2 \leftarrow {}^3A_2)$ and 16,763 cm^{-1} $({}^3T_1 \leftarrow {}^3A_2$; Table S16), together with a third weaker band at 12,584 cm^{-1}, which can be ascribed to the spin-forbidden $^1E \leftarrow {}^3A_2$ component (Figure 5b). Subsequent non-linear least-squares fits of the energies of these transitions with Equations (10)–(13) provides a first rough set of ligand field strength $\Delta_{oct} = 10,672$ cm^{-1} and Racah parameters $B = 760$ cm^{-1} and $C = 3413$ cm^{-1} (Table S16). However, the mixing of the spin-allowed $^3T_2 \leftarrow {}^3A_2$ transition with the spin-forbidden $^1E \leftarrow {}^3A_2$ transition via spin-orbit coupling for apparent ligand field strengths around 11,000–12,000 cm^{-1}, as found for $[Ni(\mathbf{L5})_3](BF_4)_2 \cdot H_2O$, requires further refinements [76]. A detailed analysis of a series of Ni^{II} complexes led *Hancock* and coworkers to propose three empirical Equations (14)–(16) to obtain more reliable ligand field strengths Δ_{oct} and Racah parameters B and C in cm^{-1} units (ε_1 and

ε_2 are the extinction coefficients at the observed frequencies of the $^1E \leftarrow {}^3A_2$ transition and $^3T_2 \leftarrow {}^3A_2$ transition, respectively) [76]. The analysis of the experimental absorption spectra using this model gives the corrected parameters gathered in Table 2 for the $[Ni(L5)_3]^{2+}$ and $[Ni(L2)_3]^{2+}$ chromophores.

$$\Delta_{oct} = 10Dq = 10630 + 1370(\varepsilon_1/\varepsilon_2) \tag{14}$$

$$B = 1120 - 0.022 \cdot \Delta_{oct} \tag{15}$$

$$C = 15B - 9975 \tag{16}$$

Table 2. Ligand-field strengths ($_{oct}$) and Racah parameters (B, C) computed with Equations (14)–(16) for $[Ni(L5)_3](BF_4)_2 \cdot H_2O$ (**III**) in the solid state and in 0.1 M acetonitrile solution at 298 K.

	$[Ni(L5)_3](BF_4)_2$ (solid)	$[Ni(L5)_3]^{2+}$ (0.1 M CH_3CN)	$[Ni(L2)_3]^{2+}$ (0.1 M CH_3CN)
Δ_{oct}/cm^{-1}	11,631	11,555	11,605
B/cm^{-1}	864	866	865
C/cm^{-1}	2986	3012	2995
Δ_{oct}/B	13.5	13.3	13.4
C/B	3.45	3.48	3.46
$\beta^{\,a}$	0.83	0.83	0.83

^{a} Nephelauxetic parameter $\beta = B/B^\circ$ using $B^\circ = 1042\ cm^{-1}$ for free Ni^{2+} ion [77].

The refined Δ_{oct}, B and C parameters computed for $[Ni(L5)_3]^{2+}$ (Table 2) almost exactly matched those previously reported for the unconstrained $[Ni(L2)_3]^{2+}$ complex, for which the related $[Fe(L2)_3]^{2+}$ complex displayed spin-crossover behavior above room temperature ($T_{1/2}$ ~400 K in the solid state, $T_{1/2}$ ~ 350 K in acetonitrile solution [35]). Moreover Δ_{oct} ($[Ni(L5)_3]^{2+}$) = 11,630 cm^{-1} is compatible with the ligand field range $11,200 \leq \Delta_{oct}$ (Ni^{II}) $\leq 12,400$ cm^{-1} established by Busch and coworkers [78] as a reliable and useful benchmark for predicting and rationalizing the spin-crossover of the related Fe^{II} complexes [40]. The absence of SCO behavior depicted by $[Fe(L5)_3](ClO_4)_2$ is thus difficult to assign to some inadequate electronic properties of the $[FeN_6]$ chromophore, but more probably to the impossibility of the coordination sphere to shrink for adopting short-enough Fe–N bonds compatible with low-spin Fe^{II}. This pure sterical limitation can be tentatively assigned to the intraligand sterical constraints programmed to occur between the methyl groups bound to the pyrazine and benzimidazole rings in each coordinated syn-L5 ligand in $[Fe(L5)_3]^{2+}$. However, packing forces operating in the solid state may be as important, or even much larger than intramolecular constraints and a definitive assessment requires the extension of our analysis to isolated complexes in solution, where intermolecular interactions are significantly reduced.

Stabilities and electronic properties of the pseudo-octahedral complexes $[M(L5)_3]X_2$ (M = Fe, Ni, Zn and X = BF$_4$, CF$_3$SO$_3$) were obtained in acetonitrile solutions. Following the procedure previously detailed for analogous $[Zn(Lk)_3]^{2+}$ [60] and $[Fe(Lk)_3]^{2+}$ [35] with the didentate ligands L1 and L2, spectrophotometric titrations of submillimolar concentrations of L5 with $M(CF_3SO_3)_2$ (M = Ni, Zn) in dry acetonitrile (Figure 6a,b and Figure S11a,b) showed the successive formation of two absorbing complexes $[M(L5)_n]^{2+}$ (n = 2, 1; equilibria (17)–(18)) as ascertained by their independent eigenvectors found in the factor analyses (Figure 6c and Figures S11c–S12c) [48–50] and their satisfying re-constructed absorption spectra (Figure 6d and Figure S11d) [51–53].

$$M^{2+} + L5 \; [M(L5)]^{2+} \quad \beta_{1,1}^{M,L5} \tag{17}$$

$$M^{2+} + 2\,L5 \; [M(L5)_2]^{2+} \quad \beta_{1,2}^{M,L5} \tag{18}$$

$$M^{2+} + 3\,L5 \; [M(L5)_3]^{2+} \quad \beta_{1,3}^{M,L5} \tag{19}$$

The 4000 cm−1 red-shift of the ligand-centered $\pi^* \leftarrow \pi$ transition observed upon complexation to M^{II} (Figure 6a and Figure S11a) is diagnostic for the anti→syn conformational change of the α,α'diimine unit accompanying the coordination of **L5** to M^{II} [79,80]. Non-linear least-square fits [51–53] of the spectrophotometric data to equilibria (17)–(18) provide the macroscopic cumulative formation constants gathered in Table 3 (entries 3–4) together with speciation curves [81] showing a maximum formation of ca. 50% of the ligand speciation under the form of $[M(L5)_2]^{2+}$ at submillimolar concentrations (Figure 6e and Figure S11e). Attempts to consider the formation of an additional $[M(L5)_3]^{2+}$ complex according to equilibrium (19) only failed in our hands, which suggests that the $\beta_{1,3}^{M,L5}$ constant is too low for providing significant quantities of the latter complex at this concentration.

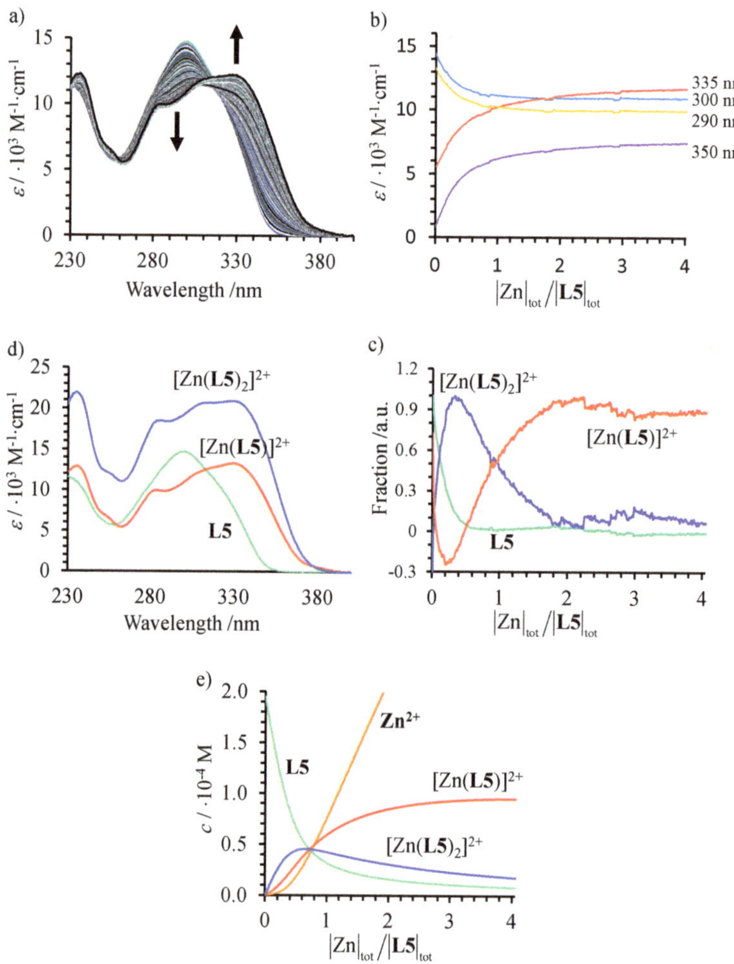

Figure 6. (**a**) Variation of absorption spectra and (**b**) corresponding variation of observed molar extinctions at different wavelengths recorded for the spectrophotometric titration of **L5** with $Zn(CF_3SO_3)_2$ (total ligand concentration: 2.0×10^{-4} mol.dm^{-3} in acetonitrile, 298 K). (**c**) Evolving factor analysis using four absorbing eigenvectors [48–50], (**d**) re-constructed individual electronic absorption spectra [51–53] and (**e**) associated computed speciation [81].

Table 3. Cumulative stability constants ($\log(\beta_{1,n}^{M,Lk})$, Equations (17)–(19)), intrinsic affinities ($\Delta G^{M,Lk} = -RT\ln(f^{M,Lk})$) and global interligand interactions ($\Delta E^{Lk,Lk} = -RT\ln(u^{Lk,Lk})$) estimated in acetonitrile at 298 K.

Ligand	L5		L2	
Metal	Ni(II)	Zn(II)	Zn(II)	Fe(II)
$\log(\beta_{1,1}^{M,Lk})$	4.27 (1)	4.38 (1)	6.89 (3)	6.045 (9)
$\log(\beta_{1,2}^{M,Lk})$	8.73 (1)	8.73 (1)	12.76 (5)	11.49 (2)
$\log(\beta_{1,3}^{M,Lk})$	12.0 [a]	12.7 [a]	17.64 (5)	16.88 (2)
$\Delta G^{M,Lk}$/kJ/mol	−16.5 (1)	−17.1 (7)	−31.1 (3)	−25.9 (8)
$\Delta E^{Lk,Lk}$/kJ/mol	−4.9 (1)	−3.6 (1)	1.1 (4)	−2.7 (9)
Reference	This work	This work	[60]	[35]

[a] Computed with Equation (22).

According to the site-binding model [82,83], the first stability constant $\beta_{1,1}^{M,Lk}$ reflects the simple intermolecular affinity $f^{M,Lk}$ between the didentate ligand **Lk** and the entering M^{II} cation (including the change in solvation, Equation (20)) modulated by a pure entropic contribution $\omega_{1,1} = 24$ [60] produced by the change in rotational entropies accompanying the transformation of the reactants into products, a parameter often referred to as the statistical factor [84,85].

$$\beta_{1,1}^{M,Lk} = 24\,f^{M,Lk} \tag{20}$$

Applying Equation (20) to the stability constants $\beta_{1,1}^{M,Lk}$ collected in Table 3 (entry 3) leads to intrinsic free energy affinities $\Delta G^{M,Lk} = -RT\ln(f^{M,Lk}) = -RT\ln(\beta_{1,1}^{M,Lk}/24)$ (entry 6), which are roughly reduced by a factor two in going from the 5-methyl-pyrazine ligand **L2** to the 3-methyl-pyrazine analogue **L5**, as illustrated for $[Zn(L2)]^{2+}$ ($\Delta G^{Zn,L2} = -31.1(3)$ kJ·mol^{-1}) and $[Zn(L5)]^{2+}$ ($\Delta G^{Zn,L5} = -17.1(7)$ kJ·mol^{-1}).

The fixation of two ligands to give $[M(Lk)_2]^{2+}$ obeying equilibrium (18) requires twice the intermolecular metal-ligand affinity, a statistical factor of $\omega_{1,2} = 120$, which takes into account all the possible geometric isomers [60] and the operation of allosteric cooperativity factors $u^{Lk,Lk}$ measuring the extra energy cost ($u^{Lk,Lk} < 1$), respectively, energy benefit ($u^{Lk,Lk} > 1$) produced by the binding of two ligands to the same metal (Equation (21)) [82,83,86].

$$\beta_{1,2}^{M,Lk} = 120\left(f^{M,Lk}\right)^2 u^{Lk,Lk} \tag{21}$$

Applying Equation (21) to the stability constants $\beta_{1,2}^{M,Lk}$ (entry 4 in Table 3) with the help of the intrinsic affinities $f^{M,Lk} = \exp\left(-\Delta G^{M,Lk}/RT\right)$ deduced from $\Delta G^{M,Lk}$ (entry 6 in Table 3) provides interligand free energies interactions $\Delta E^{Lk,Lk} = -RT\ln(u^{Lk,Lk}) = -RT\ln(\beta_{1,2}^{M,Lk}/120) - 2\Delta G^{M,Lk}$ (entry 7 in Table 3) close to zero (non-cooperativity) for $[M(L2)_2]^{2+}$ complexes, but negative (positive cooperativity) for $[M(L5)_2]^{2+}$. However, the neglect of the expected $[M(L5)_3]^{2+}$ for modeling the spectrophotometric titrations results in a slight overestimation of the second cumulative constant $\beta_{1,2}^{M,Lk}$, fully compatible with the apparent, but probably not pertinent, minor positive cooperativity observed for the successive binding of **L5** ligands to M^{2+} centers. Finally, the introduction of the estimated intrinsic affinities and interligand interactions into Equation (22) allows some predictions concerning the inaccessible third cumulative stability constants [35].

$$\beta_{1,3}^{M,Lk} = 64\left(f^{M,Lk}\right)^3 \left(u^{Lk,Lk}\right)^3 \tag{22}$$

The resulting values of $\log\left(\beta_{1,3}^{M,L5}\right) \approx 12.0$ (Table 3, entry 5) correspond to a reduction by five orders of magnitude with respect to $\log\left(\beta_{1,3}^{M,L2}\right) \approx 17.0$ experimentally found with the less constrained ligand **L2**. There is no doubt that the hindered planar arrangement of the two aromatic heterocycles in **L5** has a deep impact on the strength of the M–N bonds because of the misalignment of the nitrogen lone pairs with metal d-orbitals in pseudo-octahedral geometry. Nevertheless, at 1 M concentration of **L5** in acetonitrile, the ligand speciation curve computed [81] by using $\beta_{1,n}^{M,L5}$ (n = 1–3) gathered in Table 3 shows that $[M(L5)_3]^{2+}$ (M = Ni, Zn) corresponds to more than 90% of the distribution at the stoichiometric M:**L5** = 1/3 ratio (Figure S12). At a total ligand concentration of 0.1 M in acetonitrile, $[Ni(L5)_3]^{2+}$ stands for 84% of the ligand speciation and its absorption spectrum closely matches that recorded for related solid state samples (Figure 5). Repeating the detailed analysis described in the previous section (Equations (10)–(16)) provides ligand-field strengths (Δ_{oct}) and Racah parameters (B, C) similar to those found in the solid state (Table 2), especially for the crucial ratio $\Delta_{oct}/B = 13.3$, which is identical for $[Ni(L2)_3]^{2+}$ and $[Ni(L5)_3]^{2+}$, both in solution and in the solid state. In other words, the moving of the methyl group connected to the pyrazine ring from the 5-position in **L2** to the 3-position in **L5** indeed strongly reduces the affinity of the didentate ligand for Ni(II) in solution, but only has a weak effect on the ligand field strengths and on the Ni–N (or Zn–N) bond length. Interestingly, the variable-temperature ^{1}H NMR spectra recorded for diamagnetic $[Zn(L5)_3]^{2+}$ (>80% for a total concentration of 0.1 M in CD_3CN in the 233–333 K range, Figure S13) showed a single set of signals compatible with the exclusive formation of an averaged C_3-symmetrical species with no contribution from either blocked facial and meridional isomers or from partial decomplexation to give $[Zn(L5)_2]^{2+}$ + **L5** (equilibrium (19)). These observations are in contrast with the detection at low temperature in CD_3CN of two well-resolved spectra characteristics of a slow exchange operating between *fac*-$[Zn(L2)_3]^{2+}$ and *mer*-$[Zn(L2)_3]^{2+}$ [60], and suggests that the weaker stability constants are accompanied by faster ligand exchange processes around Zn^{2+} in $[Zn(L5)_3]^{2+}$. This decrease in affinity reaches its paroxysm for the coordination of Fe^{2+} since the spectrophotometric titration of **L5** with $Fe(CF_3SO_3)_2$ conducted at submillimolar concentration displays only a minor drift of the absorption spectra with no pronounced end point (Figure 7).

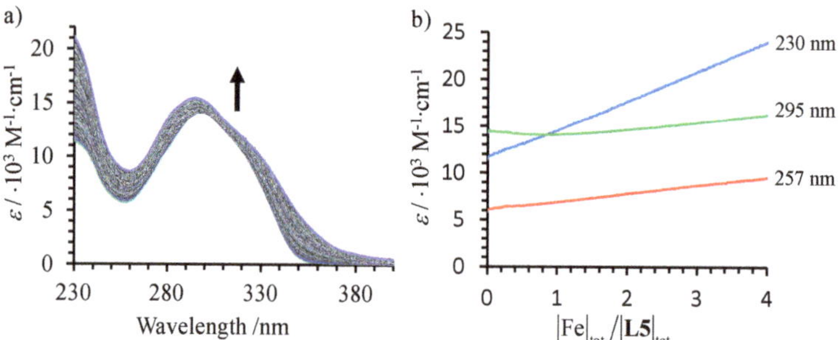

Figure 7. (**a**) Variation of absorption spectra and (**b**) corresponding variation of observed molar extinctions at different wavelengths recorded for the spectrophotometric titration of **L5** with $Fe(CF_3SO_3)_2$ (total ligand concentration: $2.0\cdot10^{-4}$ mol.dm^{-3} in acetonitrile, 298 K.

Attempts to model these limited variations within the frame of equilibria (17)–(18) only failed. The amount of $[Fe(L5)_n]^{2+}$ in solution is strongly limited by (very) low cumulative stability constants, a situation produced by the impossibility for Fe^{II} to adopt a compact low-spin configuration in the sterically constrained complex [35]. In the absence of a significant amount of $[Fe(L5)_3]^{2+}$ complexes in solution, no spin state equilibria could be investigated for this system.

4. Conclusions

Having established that the shift of the methyl group connected to the pyridine ring in going from the didentate ligand **L1** (5-position) to **L3** (3-position) was accompanied by a drift of the transition temperature in the associated spin-crossover complexes from $T_{1/2}$ ~310 K in [Fe(**L1**)$_3$]$^{2+}$ [34] to $T_{1/2}$ ~50 K in a pure facial version of the [Fe(**L3**)$_3$]2+ chromophore [36], we attempted in this work to transpose this trend for pyrazine analogues **L2** and **L5** with the preparation of the missing member of the series [Fe(**L5**)$_3$]$^{2+}$ (3-methyl-pyrazine), for which an ideal $T_{1/2}$ ~ 90 K could be naively predicted since $T_{1/2}$ ~ 350 K in [Fe(**L2**)$_3$]2+ (5-methyl-pyrazine). At first sight, this approach appeared to be promising since the molecular structures of pseudo-octahedral [M(**L2**)$_3$]$^{2+}$ and [M(**L5**)$_3$]$^{2+}$ (M = Ni^{II} and Zn^{II}) were comparable, except for the expected larger interannular twist between the connected aromatic rings produced by the close methyl groups in bound **L5**. The diagnostic ratio $\Delta/B = 13.3$ was identical for both Ni^{II} complexes, thus pointing to electronic properties also compatible with the induction of SCO behavior in the analogous Fe^{II} complexes. Surprisingly and disappointingly, [Fe(**L5**)$_3$]$^{2+}$ exists as a pure high spin complex within the 5–300 K range with no trace of spin state equilibrium. A thorough analysis of its thermodynamic formation in solution highlights a huge decrease in affinity of the ligand **L5**, compared with **L2**, for its binding to M^{2+} cations despite the presence of the same nitrogen donor atoms. Compared with the pyridine analogues [M(**L3**)$_3$]$^{2+}$, which possess similar intra-strand sterical constraints (3-methyl substituents), the much weaker σ-donating N-pyrazine donor atoms were unable to compensate for the additional interstrand constraints required for chelating **L5** around M^{2+}. This limiting factor, which can be compared to a sort of arm wrestling match, is amplified with small cations and low-spin Fe^{II} cannot be complexed to **L5**.

Supplementary Materials: The following are available online at http://www.mdpi.com/2624-8549/2/2/15/s1. Tables S1–S16 collecting elemental analyses, crystallographic, and photophysical data. Figures S1–S13 showing the crystal structures, ^{1}H NMR spectra, spectrophotometric titrations, and theoretical calculations.

Author Contributions: Conceptualization, T.L. and C.P.; Methodology and practical chemical and spectroscopic studies, N.D. and T.L.; Crystallography, L.G. and C.B.; Writing draft report, N.D.; Writing of manuscript and editing, C.P., N.D., T.L., and C.B.; Project administration and funding acquisition C.P. All authors have read and agreed to the published version of the manuscript.

Funding: Financial support from the Swiss National Science Foundation is gratefully acknowledged (grant number: 200020_178758).

Acknowledgments: The authors thank Kerry Buchwalder for performing the elemental analysis and for efficient technical support.

Conflicts of Interest: The authors declare no conflicts of interest.

References

1. Jorgensen, C.K. *Modern Aspects of Ligand Field Theory*; North Holland Publishing Company: Amsterdam, The Netherland, 1971.
2. Mingos, D.M.P. Structure and Bonding: The Early Days. *Struct. Bond.* **2016**, *172*, 1–18.
3. Bethe, H. Termaufspaltung in Kristallen. *Ann. Phys.* **1929**, *3*, 133–208. [CrossRef]
4. Gütlich, P.; Goodwin, H.A. Spin-crossover—An Overall Perspective. *Top. Curr. Chem.* **2004**, *233*, 1–47.
5. Van Vleck, J.H. *The Theory of Electric and Magnetic Susceptibilities*; Clarendon Press: Oxford, UK, 1932.
6. Pauling, L. The Nature of the Chemical Bond. III. The Transition from One Extreme Bond Type to Another. *J. Am. Chem. Soc.* **1932**, *54*, 988–1003. [CrossRef]
7. Corryell, C.D.; Stittand, F.; Pauling, L. The Magnetic Properties and Structure of Ferrihemoglobin (Methemoglobin) and Some of its Compounds. *J. Am. Chem. Soc.* **1937**, *59*, 633–642. [CrossRef]
8. Cambi, L.; Szego, L. The Magnetic Susceptibility of Complex Compounds. *Ber. Dtsch. Chem. Ges.* **1931**, *64*, 2591–2598. [CrossRef]
9. Cambi, L.; Szego, L.; Cagnasso, A. The Magnetic Susceptibility of Complexes. IV. Ferric N,N-Dipropyldithiocarbamates. *Atti Accad. Lincei* **1932**, *15*, 266–271.

10. Cambi, L.; Szego, L.; Cagnasso, A. The Magnetic Susceptibility of Complexes. V. Iron dibutyldithiocarbamates. *Atti Accad. Lincei* **1932**, *15*, 329–335.

11. *Spin-Crossover in Transition Metal Compounds I-III*; Gütlich, P.; Goodwin, H.A. (Eds.) Springer: Berlin/Heidelberg, Germany, 2004.

12. *Spin-Crossover Materials*; Halcrow, M.A. (Ed.) John Wiley & Sons, Ltd: Chichester, UK, 2013.

13. Olguin, J. Unusual Metal Centres/Coordination Spheres in Spin Crossover Compounds. *Coord. Chem. Rev.* **2020**, *407*, 213148. [CrossRef]

14. Gütlich, P.; Garcia, Y.; Goodwin, H.A. Spin-crossover Phenomena in Fe(II) Complexes. *Chem. Soc. Rev.* **2000**, *29*, 419–427. [CrossRef]

15. Halcrow, M.A. The Spin-States and Spin Transitions of Mononuclear Iron(II) Complexes of Nitrogen-Donor Ligands. *Polyhedron* **2007**, *26*, 3523–3576. [CrossRef]

16. Gaspar, A.B.; Seredyuk, M. Spin-crossover in Soft Matter. *Coord. Chem. Rev.* **2014**, *268*, 41–58. [CrossRef]

17. Brooker, S. Spin-crossover with Thermal Hysteresis: Practicalities and Lessons Learnt. *Chem. Soc. Rev.* **2015**, *44*, 2880–2892. [CrossRef] [PubMed]

18. Hogue, R.W.; Singh, S.; Brooker, S. Spin-crossover in Discrete Polynuclear Iron(II) Complexes. *Chem. Soc. Rev.* **2018**, *47*, 7303–7338. [CrossRef] [PubMed]

19. Scott, H.S.; Staniland, R.W.; Kruger, P.E. Spin-crossover in Homoleptic Fe(II) Imidazolylimine Complexes. *Coord. Chem. Rev.* **2018**, *362*, 24–43. [CrossRef]

20. Kumar, K.S.; Bayeh, Y.; Gebretsadik, T.; Elemo, F.; Gebrezgiabher, M.; Thomas, M.; Ruben, M. Spin-Crossover in Iron(II)-Schiff Base Complexes. *Dalton Trans.* **2019**, *48*, 15321–15337. [CrossRef]

21. Sorai, M. Heat Capacity Studies of Spin-crossover Systems. *Top. Curr. Chem.* **2004**, *235*, 153–170.

22. Halcrow, M.A. The Foundation of Modern Spin-Crossover. *Chem. Commun.* **2013**, *49*, 10890–10892. [CrossRef]

23. Ksenofontov, V.; Gaspar, A.B.; Gutlich, P. Pressure Effect Studies on Spin-crossover and Valence Tautomeric Systems. *Top. Curr. Chem.* **2004**, *235*, 23–64.

24. Bousseksou, A.; Varret, F.; Goiran, M.; Boukheddaden, K.; Tuchagues, J.P. The Spin-crossover Phenomenon Under High Magnetic Field. *Top. Curr. Chem.* **2004**, *235*, 65–84.

25. Kamebuchi, H.; Nakamoto, A.; Yokoyama, T.; Kojima, N. Fastener Effect on Uniaxial Chemical Pressure for One-dimensional Spin-crossover System, Magnetostructural Correlation and Ligand Field analysis. *Bull. Chem. Soc. Jpn.* **2015**, *88*, 419–430. [CrossRef]

26. Hauser, A. Light-induced Spin-crossover and the High-spin to Low-spin Relaxation. *Top. Curr. Chem.* **2004**, *234*, 155–198.

27. Hauser, A.; Reber, C. Spectroscopy and Chemical Bonding in Transition Metal Complexes. *Struct. Bond.* **2017**, *172*, 291–312.

28. Bousseksou, A.; Molnar, G.; Matouzenko, G. Switching of Molecular Spin States in Inorganic Complexes by Temperature, Pressure, Magnetic Field and Light: Towards Molecular Devices. *Eur. J. Inorg. Chem.* **2004**, *2004*, 4353–4369. [CrossRef]

29. Mikolasek, M.; Félix, G.; Nicolazzi, W.; Molnar, G.; Salmon, L.; Bousseksou, A. Finite Size Effects in Molecular Spin-crossover Materials. *New J. Chem.* **2014**, *38*, 1834–1839. [CrossRef]

30. Kumar, K.S.; Ruben, M. Emerging Trends in Spin-crossover (SCO) Based Functional Materials and Devices. *Coord. Chem. Rev.* **2017**, *346*, 176–205. [CrossRef]

31. Hauser, A. Ligand Field Theoretical Considerations. *Top. Curr. Chem.* **2004**, *233*, 49–58.

32. Schaffer, C.E.; Jorgensen, C.K. The Nephelauxetic Series of Ligands Corresponding to Increasing Tendency of Partly Covalent Bonding. *J. Inorg. Nuclear Chem.* **1958**, *8*, 143–148. [CrossRef]

33. Phan, H.; Hrudka, J.J.; Igimbayeva, D.; Daku, L.M.L.; Shatruk, M. A Simple Approach for Predicting the Spin State of Homoleptic Fe(II) Tris-diimine Complexes. *J. Am. Chem. Soc.* **2017**, *139*, 6437–6447. [CrossRef]

34. Olguin, J.; Brooker, S. Spin-crossover Active Iron(II) Complexes of Selected Pyrazole-Pyridine/Pyrazine Ligands. *Coord. Chem. Rev.* **2011**, *255*, 203–240. [CrossRef]

35. Lathion, T.; Guénée, L.; Besnard, C.; Bousseksou, A.; Piguet, C. Deciphering the Influence of Meridional versus Facial Isomers in Spin-crossover Complexes. *Chem. Eur. J.* **2018**, *24*, 16873–16888. [CrossRef] [PubMed]

36. Lathion, T.; Furstenberg, A.; Besnard, C.; Hauser, A.; Bousseksou, A.; Piguet, C. Monitoring Fe(II) Spin-State Equilibria via Eu(III) Luminescence in Molecular Complexes: Dream or Reality? *Inorg. Chem.* **2020**, *59*, 1091–1103. [CrossRef]

37. Craig, D.C.; Goodwin, H.A.; Onggo, D. Steric Influences on the Ground-State of Iron(II) in the Tris(3,3′-Dimethyl-2,2′-Bipyridine)Iron(II) Ion. *Aust. J. Chem.* **1988**, *41*, 1157–1169. [CrossRef]

38. Onggo, D.; Hook, J.M.; Rae, A.D.; Goodwin, H.A. The Influence of Steric Effects in Substituted 2,2′-Bipyridine on the Spin State of Iron(II) in [FeN$_6$]$^{2+}$ Systems. *Inorg. Chim. Acta* **1990**, *173*, 19–30. [CrossRef]

39. Goodwin, H.A.; Kucharski, E.S.; White, A.H. Crystal-Structure of Tris (2-Methyl-1,10-Phenanthroline) Iron(II) Tetraphenylborate. *Aust. J. Chem.* **1983**, *36*, 1115–1124. [CrossRef]

40. Goodwin, H.A. Spin-crossover in Iron(II) Tris(diimine) and Bis(terimine) Systems. *Top. Curr. Chem.* **2004**, *233*, 59–90.

41. Edder, C.; Piguet, C.; Bünzli, J.-C.G.; Hopfgartner, G. High-spin Iron(II) as a Semi-Transparent Partner for Tuning Europium(III) Luminescence in Heterodimetallic d-f Complexes. *Chem. Eur. J.* **2001**, *7*, 3014–3024. [CrossRef]

42. Dunitz, J.D. Are Crystal Structures Predictable? *Chem. Commun.* **2003**, 545–548. [CrossRef]

43. Dunitz, J.D.; Gavezzotti, A. How Molecules Stick Together in Organic Crystals: Weak Intermolecular Interactions. *Chem. Soc. Rev.* **2009**, *38*, 2622–2633. [CrossRef]

44. Meng, Y.S.; Liu, T. Manipulating Spin Transition to Achieve Switchable Multifunctions. *Acc. Chem. Res.* **2019**, *52*, 1369–1379. [CrossRef]

45. Kahn, O.; Martinez, C.J. Spin-transition Polymers: From Molecular Materials toward Memory Devices. *Science* **1998**, *279*, 44–48. [CrossRef]

46. Wolsey, W.C. Perchlorate Salts, Their Uses and Alternatives. *J. Chem. Educ.* **1973**, *50*, A335–A337. [CrossRef]

47. Pascal, J.-L.; Favier, F. Inorganic perchlorato complexes. *Coord. Chem. Rev.* **1998**, *178*, 865–902. [CrossRef]

48. Malinowski, E.R.; Howery, D.G. *Factor Analysis in Chemistry*; Wiley: New York, NY, USA, 1980.

49. Gampp, H.; Maeder, M.; Meyer, C.J.; Zuberbuehler, A.D. Calculation of Equilibrium Constants from Multiwavelength Spectroscopic Data—IV. Model-free Least-Squares Refinement by Use of Evolving Factor Analysis. *Talanta* **1986**, *33*, 943–951. [PubMed]

50. Hall, B.R.; Manck, L.E.; Tidmarsh, I.S.; Stephenson, A.; Taylor, B.F.; Blaikie, E.J.; Vander Griend, D.A.; Ward, M.D. Structures, Host-Guest Chemistry and Mechanism of Stepwise Self-Assembly of M$_4$L$_6$ Tetrahedral Cage Complexes. *Dalton Trans.* **2011**, *40*, 12132–12145. [CrossRef]

51. Gampp, H.; Maeder, M.; Meyer, C.J.; Zuberbuehler, A.D. Calculation of Equilibrium Constants from Multiwavelength Spectroscopic Data. III. Model-free Analysis of Spectrophotometric and ESR Titrations. *Talanta* **1985**, *32*, 1133–1139. [CrossRef]

52. Maeder, M.; King, P. Analysis of Chemical Processes, Determination of the Reaction Mechanism and Fitting of Equilibrium and Rate Constants. In *Chemometrics in Practical Applications*; Varmuza, K., Ed.; Intech: Rigeka, Croatia, 2012.

53. *ReactLab^{TM} Equilibria (previously Specfit/32)*; Jplus Consulting Pty. Ltd.: Palmyra, WA, Australia; Available online: https://jplusconsulting.com/products/reactlab-equilibria/ (accessed on 1 April 2020).

54. Bain, G.A.; Berry, J.F. Diamagnetic Corrections and Pascal's Constants. *J. Chem. Educ.* **2008**, *85*, 532–536. [CrossRef]

55. Burla, M.C.; Camalli, M.; Carrozzini, B.; Cascarano, G.L.; Giacovazzo, C.; Polidori, G.; Spagna, R. SIR99, A program for the Automatic Solution of Small and Large Crystal Structures. *Acta Cryst. A* **1999**, *55*, 991–999. [CrossRef]

56. Sheldrick, G.M. A Short History of SHELX. *Acta Cryst. A* **2008**, *64*, 112–122. [CrossRef]

57. Sheldrick, G.M. SHELXT—Integrated Space-Group and Crystal-Structure Determination. *Acta Cryst. A* **2015**, *71*, 3–8. [CrossRef]

58. Sheldrick, G.M. Crystal Structure Refinement with SHELXL. *Acta Cryst. C* **2015**, *71*, 3–8. [CrossRef] [PubMed]

59. Piguet, C.; Bocquet, B.; Hopfgartner, G. Syntheses of Segmental Heteroleptic Ligands for the Self-assembly of Heteronuclear Helical Supramolecular Complexes. *Helv. Chim. Acta* **1994**, *77*, 931–942. [CrossRef]

60. Aboshyan-Sorgho, L.; Lathion, T.; Guénée, L.; Besnard, C.; Piguet, C. Thermodynamic N-donor Trans Influence in Labile Pseudo-Octahedral Zinc Complexes: A Delusion? *Inorg. Chem.* **2014**, *53*, 13093–13104. [CrossRef] [PubMed]

61. Allinger, N.L. Conformational-Analysis.130. MM2—Hydrocarbon Force-Field Utilizing V1 and V2 Torsional Terms. *J. Am. Chem. Soc.* **1977**, *99*, 8127–8134.

62. Llunell, M.; Casanova, D.; Cirera, J.; Alemany, P.; Alvarez, S. SHAPE Is a Free Software. Available online: http://www.ee.ub.edu/ (accessed on 1 April 2020).

63. Pinsky, M.; Avnir, D. Continuous Symmetry Measures. 5. The Classical Polyhedra. *Inorg. Chem.* **1998**, *37*, 5575–5582. [CrossRef] [PubMed]

64. Alvarez, S.; Avnir, D.; Llunell, M.; Pinsky, M. Continuous Symmetry Maps and Shape Classification. The Case of Six-Coordinated Metal Compounds. *New J. Chem.* **2002**, *26*, 996–1009. [CrossRef]

65. Casanova, D.; Cirera, J.; Llunell, M.; Alemany, P.; Avnir, D.; Alvarez, S. Minimal Distortion Pathways in Polyhedral Rearrangements. *J. Am. Chem. Soc.* **2004**, *126*, 1755–1763. [CrossRef]

66. Cirera, J.; Ruiz, E.; Alvarez, S. Shape and Spin State in Four-Coordinate Transition-Metal Complexes: The case of the d(6) Configuration. *Chem. Eur. J.* **2006**, *12*, 3162–3167. [CrossRef]

67. CRC. *Handbook of Chemistry and Physics, Internet Version 2005*; Lide, D.R., Ed.; CRC Press: Boca Raton, FL, USA, 2005.

68. Shannon, R.D. Revised Effective Ionic-Radii and Systematic Studies of Interatomic Distances in Halides and Chalcogenides. *Acta Cryst.* **1976**, *32*, 751–767. [CrossRef]

69. Figgis, B.N. Magnetic Properties of Spin-Free Transition Series Complexes. *Nature* **1958**, *182*, 1568–1570. [CrossRef]

70. Martin, L.L.; Martin, R.L.; Murray, K.S.; Sargeson, A.M. Magnetism and Electronic-Structure of a Series of Encapsulated 1st-Row Transition-Metals. *Inorg. Chem.* **1990**, *29*, 1387–1394. [CrossRef]

71. Sharp, R.; Lohr, L.; Miller, J. Paramagnetic NMR Relaxation Enhancement: Recent Advances in Theory. *Prog. Nucl. Magn. Reson. Spec.* **2001**, *38*, 115–158. [CrossRef]

72. Costes, J.-P.; Clemente-Juan, J.M.; Dahan, F.; Dumestre, F.; Tuchagues, J.-P. Dinuclear (Fe(II), Gd(III)) Complexes Deriving from Hexadentate Schiff Bases: Synthesis, Structure, and Mössbauer and Magnetic Properties. *Inorg. Chem.* **2002**, *41*, 2886–2891. [CrossRef] [PubMed]

73. Boca, R. Zero-field Splitting in Metal Complexes. *Coord. Chem. Rev.* **2004**, *248*, 757–815. [CrossRef]

74. Lever, A.B.P. *Inorganic Electronic Spectroscopy*; Elsevier: Amsterdam, The Netherlands, 1984; pp. 126–127.

75. Triest, N.; Bussiere, G.; Belisle, H.; Reber, C. Why Does the Middle Band in the Absorption Spectrum of $[Ni(H_2O)_6]^{2+}$ Have Two Maxima? *J. Chem. Educ.* **2000**, *77*, 670. [CrossRef]

76. Hart, S.M.; Boeyens, J.C.A.; Hancock, R.D. Mixing of States and the Determination of Ligand Field Parameters for High-Spin Octahedral Complexes of Nickel(II). Electronic Spectrum and Structure of Bis(1,7-diaza-4-thiaheptane)nickel(II) Perchlorate. *Inorg. Chem.* **1983**, *22*, 982–986. [CrossRef]

77. Figgis, B.N.; Hitchman, M.A. *Ligand Field Theory and its Application*; Wiley-VCH: New York, NY, USA, 2000.

78. Robinson, M.A.; Busch, D.H.; Curry, J.D. Complexes Derived from Strong Field Ligands. 17. Electronic Spectra of Octahedral Nickel(II) Complexes with Ligands of α-Diimine and Closley Related Classes. *Inorg. Chem.* **1963**, *2*, 1178–1181. [CrossRef]

79. Nakamoto, K. Ultraviolet Spectra and Structures of 2,2′-Bipyridine and 2,2′,2″-Terpyridine in Aqueous Solution. *J. Phys. Chem.* **1960**, *64*, 1420–1425. [CrossRef]

80. Xu, S.; Smith, J.E.T.; Weber, J.M. UV Spectra of Tris(2,2′-bipyridine)-M(II) Complex Ions in Vacuo (M = Mn, Fe, Co, Ni, Cu, Zn). *Inorg. Chem.* **2016**, *55*, 11937–11943. [CrossRef]

81. Alderighi, L.; Gans, P.; Ienco, A.; Peters, D.; Sabatini, A.; Vacca, A. Hyperquad Simulation and Speciation (HySS): A Utility Program for the Investigation of Equilibria Involving Soluble and Partially Soluble Species. *Coord. Chem. Rev.* **1999**, *184*, 311–318. [CrossRef]

82. Borkovec, M.; Hamacek, J.; Piguet, C. Statistical Mechanical Approach to Competitive Binding of Metal Ions to Multi-center Receptors. *Dalton Trans.* **2004**, 4096–4105. [CrossRef] [PubMed]

83. Piguet, C. Five Thermodynamic Describers for Addressing Serendipity in the Self-assembly of Polynuclear Complexes in Solution. *Chem. Commun.* **2010**, *46*, 6209–6231. [CrossRef] [PubMed]

84. Benson, S.W. Statistical Factors in the Correlation of Rate Constants and Equilibrium Constants. *J. Am. Chem. Soc.* **1958**, *80*, 5151–5154. [CrossRef]

85. Ercolani, G.; Piguet, C.; Borkovec, M.; Hamacek, J. Symmetry Numbers and Statistical Factors in Self-assembly and Multivalency. *J. Phys. Chem. B* **2007**, *111*, 12195–12203. [CrossRef] [PubMed]

86. Ercolani, G.; Schiaffino, L. Allosteric, Chelate, and Intermolecular Cooperativity: A Mise au Point. *Angew. Chem. Int. Ed.* **2011**, *50*, 1762–1768. [CrossRef] [PubMed]

Article

Solid Phase Nitrosylation of Enantiomeric Cobalt(II) Complexes

Mads Sondrup Møller, Morten Czochara Liljedahl, Vickie McKee and Christine J. McKenzie *

Department of Physics, Chemistry and Pharmacy, University of Southern Denmark, Campusvej 55, 5230 Odense M, Denmark; madssm@sdu.dk (M.S.M.); mlilj18@student.sdu.dk (M.C.L.); mckee@sdu.dk (V.M.)
* Correspondence: mckenzie@sdu.dk

Abstract: Accompanied by a change in color from red to black, the enantiomorphic phases of the cobalt complexes of a chiral salen ligand (L^{2-}, Co(L)·CS$_2$, and Co(L) (L = $L^{S,S}$ or $L^{R,R}$)) chemisorb NO (g) at atmospheric pressure and rt over hours for the CS$_2$ solvated phase, and within seconds for the desolvated phase. NO is installed as an axial nitrosyl ligand. Aligned but unconnected voids in the CS$_2$ desorbed Co($L^{R,R}$)·CS$_2$ structure indicate conduits for the directional desorption of CS$_2$ and reversible sorption of NO, which occur without loss of crystallinity. Vibrational circular dichroism (VCD) spectra have been recorded for both hands of LH$_2$, Zn(L), Co(L)·CS$_2$, Co(L), Co(NO)(L), and Co(NO)(L)·CS$_2$, revealing significant differences between the solution-state and solid-state spectra. Chiral induction enables the detection of the v_{NO} band in both condensed states, and surprisingly also the achiral lattice solvent (CS$_2$ (v_{CS} at 1514 cm^{-1})) in the solid-state VCD. Solution-state spectra of the paramagnetic Co(II) complex shows a nearly 10-fold enhancement and more extensive inversion of polarity of the vibrations of dominant VCD bands compared to the spectra of the diamagnetic compounds. This enhancement is less pronounced when there are fewer polarity inversions in the solid state VCD spectra.

Keywords: chirality; vibrational circular dichroism; solid–gas reaction; chemisorption; nitrosyl

Citation: Møller, M.S.; Liljedahl, M.C.; McKee, V.; McKenzie, C.J. Solid Phase Nitrosylation of Enantiomeric Cobalt(II) Complexes. *Chemistry* **2021**, 3, 585–597. https://doi.org/10.3390/chemistry3020041

Academic Editors: Catherine Housecroft and Katharina M. Fromm

Received: 31 March 2021
Accepted: 26 April 2021
Published: 28 April 2021

Publisher's Note: MDPI stays neutral with regard to jurisdictional claims in published maps and institutional affiliations.

1. Introduction

NO is biologically important, but also a highly toxic gas, and materials for its sorptive removal from exhausts are of great interest. A number of diverse materials, from metal oxides, zeolites, and metal-organic frameworks have shown the ability to selectively sorb NO, in some cases reversibly [1–6]. Chemisorptive processes can result in the transformation of the sorbed NO into less toxic compounds and precedence for this was demonstrated recently using the crystalline solid-state of a dicobalt(II) complex, which co-chemisorbs NO and O$_2$. Accompanied by metal oxidation, their conversion to a coordinated nitrite and nitrate counter anion ensues through a series of unidentified in-crystal reactions [7].

We were interested in investigating the chemisorptive reactivity of NO by an isolated mononuclear Co(II) site inside the crystalline lattice of a molecular solid in efforts to understand the mechanism of the host–guest, solid–gas chemistry of the aforementioned molecular dicobalt(II) complexes [7]. In particular, we wanted to investigate whether the in-crystal conversion of NO and O$_2$ to NO$_2^-$ and NO$_3^-$ is dependent on the presence of two closely-located cobalt ions working cooperatively to activate these guest substrates. Mononuclear Co(salen) has been shown to react with NO in both solution- [8] and solid-states [8,9], and offers the opportunity to explore this possibility. Unfortunately, however, no details of the structural or spectroscopic changes in the solids have been reported for the solid–gas reaction. It is unknown whether or not well-defined pores to allow transport through the solid are requisite, whether or not the process is true chemisorption (where bonds are formed), whether the presumed in-solid NO coordination to Co(II) results in its oxidation, and whether the reaction is reversible. We have, therefore, reinvestigated the NO gas–solid reaction for the salen system, however, the parent scaffold has been replaced with N,N'-bis(3,5-di-t-butylsalicylidene)-1,2-cyclohexanediamine (L^{2-}), the derivative used in the manganese(III) chloride complex that is also known as 'Jacobsen's catalyst' [10], and is used for catalyzing asymmetric epoxidation reactions [10–12]. A survey of the Cambridge structural database [13] of the crystal structures of metal complexes of L^{2-} suggests that the peripheral bulky tert-butyl groups of L^{2-} ensure that mononuclear Co sites are isolated from each other by preventing the formation of crystal phases containing dimeric $(M(salen))_2$ [14–18], which form when a phenolato oxygen atom of each salen on adjacent complexes bridge between the two metal ions. Another advantage is that, unlike some phases of the parent Co(salen) and its derivatives, Co(L) does not bind O_2.

Another aspect of this work is the fact that L^{2-} provides chirality by virtue of the aliphatic backbone carbon atoms to give $L^{S,S}$ and $L^{R,R}$ (Scheme 1). With respect to the use of L^{2-} for constructing enantiopure complexes to catalyze asymmetric reactions, we wished also to learn whether a chiral ligand would induce chirality into an achiral axial co-ligand, since this is a proxy for a bound substrate.

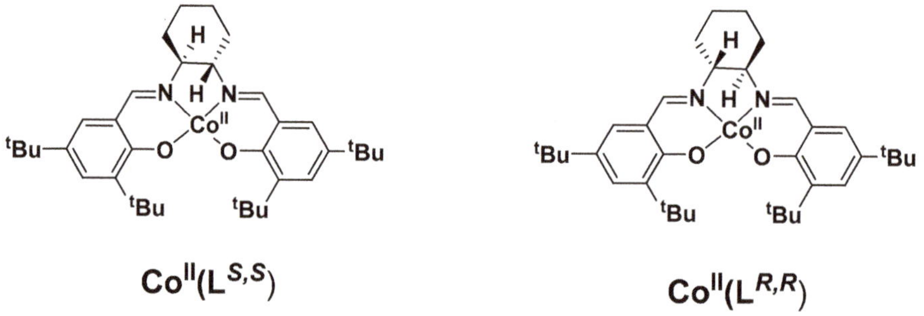

Scheme 1. *S,S* and *R,R* conformations of $Co^{II}(L^{S,S})$ and $Co^{II}(L^{R,R})$.

2. Materials and Methods

Caution NO is a highly toxic gas. CS_2 vapors are also toxic.

2.1. Instrumentation

Mass spectra are recorded with electrospray ionization (ESI) on a Bruker micrOTOF-Q II spectrometer (nanospray, capillary temperature = 180 °C, spray voltage = 3.7 kV). UV-vis spectra were recorded on an Agilent 8453 spectrophotometer in 1 cm quartz cuvettes. VCD and IR spectra were recorded on a CHIRALIR-2X™ spectrometer equipped with a single PEM, a resolution of 4 cm^{-1}, optimized at 1400 cm^{-1}, and in a single block with 50000 scans. All VCD spectra are baseline corrected by half differentiating from the other enantiomer, i.e., subtracting the VCD spectrum of one enantiomer from that of the other enantiomer and dividing the intensity of the resulting spectrum by two, and vice versa. All solution state VCD spectra were recorded in CDCl$_3$ using an ICLSL-4 liquid cell with BaF$_2$ windows and a path length of 75 μm. Solutions were prepared by dissolving 30 mg sample in 0.3 mL CDCl$_3$. All solid phase VCD spectra were recorded as mulls applied between two circular BaF$_2$ windows (Ø25 mm × 4 mm), and the sample holder was rotated at a constant speed throughout the recording to reduce artifacts. Mulls were prepared by grinding, with an agate mortar and pestle, 20 mg sample with 50 μL Nujol oil. Powder X-ray diffraction data were collected on a Synergy, Dualflex, AtlasS2 diffractometer using CuKα radiation (λ = 1.54184 Å) and the CrysAlis PRO 1.171.40.67a suite [19]. Powdered samples were adhered to the mounting loop using Fomblin®Y, and diffractograms were recorded with a detector distance of 120 mm, using Gandolfi scans with a single kappa setting, and an exposure time of 200 s.

2.2. Single Crystal X-ray Diffraction

The crystals used for Single Crystal X-ray Diffraction (SCXRD) were taken directly from the mother liquor and mounted using Fomblin®Y to adhere the crystal to the mounting loop. X-ray crystal diffraction data were collected at 100(2) K on a Synergy, Dualflex, AtlasS2 diffractometer using CuKα radiation (λ = 1.54184 Å) and the CrysAlis PRO 1.171.40.67a suite [19], and corrected for Lorentz-polarization effects and absorption. Using SHELXLE [20], the structure was solved by dual space methods (SHELXT [21]) and refined on F^2 using all the reflections (SHELXL-2018 [22]). All the non-hydrogen atoms were refined using anisotropic atomic displacement parameters, and hydrogen atoms were inserted at calculated positions using a riding model. Parameters for data collection and refinement are summarized in Table 1. The chirality in these complexes is provided by the cyclohexane ring in the ligand backbone, however, the remainder of the structures, including the heavier cobalt and (for Co($L^{R,R}$)·CS$_2$ and Co(NO)($L^{R,R}$)·CS$_2$) sulfur atoms, are arranged almost centrosymmetrically. As a result, the $| E^2 - 1 |$ statistics, cumulative intensity distribution plots, and Wilson plots (SI Figures S6–S8) appear to favor a centrosymmetric structure (as does the initial estimate of the Flack x parameter in SHELXT for Co(NO)($L^{R,R}$)·CS$_2$), and Platon ADDSYM in checkCIF suggests a (pseudo) center of symmetry may be present. Nevertheless, all three structures were successfully refined in the chiral space group $P2_1$, showing the expected chair conformation of the cyclohexane rings, all with R,R chirality. When carbon disulfide is desorbed from Co($L^{R,R}$)·CS$_2$ to produce Co($L^{R,R}$), the crystal quality is reduced slightly, resulting in weak high angle data. The structure of Co($L^{R,R}$), therefore, contains more uncertainties.

2.3. Computational Details

The VCD and IR calculations considered in this work were performed using (unrestricted) density functional theory (DFT) calculations performed using Jaguar [23] through the Maestro graphical interface. [24] All structures have been geometry optimized in isolation using B3LYP/LACVP** [25,26], and vibrational frequencies have also been calculated at this level of theory. All calculations are solution gas phase calculations. For $L^{R,R}H_2$, different conformations were calculated using the OPLS2005 [27] forcefield, and the VCD and IR spectra were based on a Boltzmann average of these conformations. For Zn($L^{R,R}$), the VCD and IR spectra are based on only the most stable conformation.

2.4. Synthesis

The R,R and S,S enantiomers of N,N'-bis(3,5-di-t-butylsalicylidene)-1,2-cyclohexanediamine, $L^{R,R}H_2$ and $L^{S,S}H_2$, were prepared according to literature methods with a reported ee. of >95% [28]. The R,R and S,S enantiomers of N,N'-bis(3,5-di-tert-butylsalicylidene)-1,2-cyclohexanediaminozinc(II) (Zn($L^{R,R}$) and Zn($L^{S,S}$)) [29] and N,N'-bis(3,5-di-tert-butylsalicylidene)-1,2-cyclohexanediaminocobalt(II) (Co(L^{RR}) and Co(L^{SS})) [30] were also synthesized according to literature procedure. All other chemicals were purchased from Sigma-Aldrich and were used without further purification.

2.4.1. Co($L^{R,R}$)·CS$_2$ and Co($L^{S,S}$)·CS$_2$

A degassed solution of Co(L^{RR}) or Co(L^{SS}) (0.250 g) in CS$_2$:*n*-hexane (16 mL, 1:1 *v/v*) was placed in a vial which was then placed inside a large glass jar with an air tight lid, and the solution was allowed to slowly evaporate at 5 °C in the fridge, yielding ruby red block crystals suitable for single crystal X-ray diffraction. Yield 0.227 g, 80.6%.

IR (Nujol) cm^{-1}: 1609 (m, C=C), 1591 (m, C=N), 1524 (vs, C=S), 1254 (m, C–O).

ESI-MS (pos. mode, MeCN): found (calcd) m/z = 603.217 (603.34, [Co(LR,R)]+, $C_{36}H_{52}CoN_2O_2$ 100%).

2.4.2. Co($L^{R,R}$) and Co($L^{S,S}$) (Desolvated from Co($L^{R,R}$)·CS$_2$ and Co($L^{S,S}$)·CS$_2$)

A sample of Co($L^{R,R}$)·CS$_2$ or Co($L^{S,S}$)·CS$_2$ (0.25 g, size 0.05–1 mm^3) was placed inside a 10 mL round bottom flask, and the flask was attached to a rotary evaporator. The crystals were heated to 95 °C in vacuo (~10^{-2} mbar) and rotated for one hour, resulting in a color change from translucent dark red to opaque red/orange. Co($L^{R,R}$) and Co($L^{S,S}$) were isolated in a quantitative yield.

IR (Nujol) cm^{-1}: 1608 (s, C=C), 1596 (vs, C=N), 1524 (m, C=N), 1255 (m, C–O).

ESI-MS (pos. mode, MeCN): found (calcd) m/z = 603.217 (603.34, [Co(L^{R,R})]$^+$, C$_{36}$H$_{52}$CoN$_2$O$_2$ 100%).

UV-vis (DCM) $\lambda_{\max}$/nm (ε/M^{-1} cm^{-1}): 247 (36,012), 428 (11,446).

2.4.3. Co(NO)(L^{R,R}) and Co(NO)(L^{S,S})

Crystals of Co(L^{R,R}) or Co(L^{S,S}) (0.25 g, size 0.05–1 mm^3) were placed in a Schlenk tube (30 mL), and the Schlenk line was evacuated and filled with N$_2$ (3 cycles). NO (1.2 bar) was then admitted into the system, resulting in a color change from opaque red/orange to opaque black within seconds, and the closed tube was then allowed to stand for one hour. Before opening to air, the system was evacuated and flushed with N$_2$ (3 cycles). Co(NO)(L^{R,R}) and Co(NO)(L^{S,S}) were isolated in a quantitative yield. Recrystallization of Co(NO)(L^{R,R}) (0.25 g) from CS$_2$:n-hexane (8 mL, 1:1 v/v) via the same procedure as described above for Co(L^{R,R})·CS$_2$ yielded black needle crystals of Co(NO)(L^{R,R})·CS$_2$ suitable for single crystal X-ray diffraction.

IR (Nujol) cm^{-1}: 1659 (w, N=O) 1638 (vs, C=N), 1608 (s, C=C,), 1524 (w, C=N), 1255 (w, C–O).

ESI-MS (pos. mode, MeCN): found (calcd) m/z = 603.317 (603.34, [Co(L^{R,R})]$^+$, C$_{36}$H$_{52}$CoN$_2$O$_2$, 100%), 634.320 (634.34, [Co(NO)(L^{R,R})H]$^+$, C$_{36}$H$_{53}$CoN$_3$O$_3$, 4.98%)

UV-vis (DCM) $\lambda_{\max}$/nm (ε/M^{-1} cm^{-1}): 267 (18,040), 366 (5391).

3. Results and Discussion

3.1. Reaction of Solid-State Co(II) Complexes with NO

Two new enantiomorphic phases of Co(L) (L = L^{S,S}, L^{R,R}) have been prepared and structurally characterized. Co(L)·CS$_2$, was obtained by recrystallization of Co(L) from CS$_2$/n-hexane. A strong band in the IR spectrum at 1524 cm^{-1} is associated with the ν_{CS} of co-crystallized CS$_2$. This band is very close to that for free CS$_2$ at 1520 cm^{-1}, indicating little interaction with the cobalt atom. Co(L)·CS$_2$ undergoes CS$_2$ loss on heating at 95 °C for 1 h at 10^{-2} mbar to reproduce Co(L), however it is now a new unreported phase that is different to the starting phase. Both Co(L)·CS$_2$ and Co(L) react in the solid state with NO gas (1 atm, rt, unground crystals ranging in size from 0.05 to 1 mm^3) (Scheme 2).

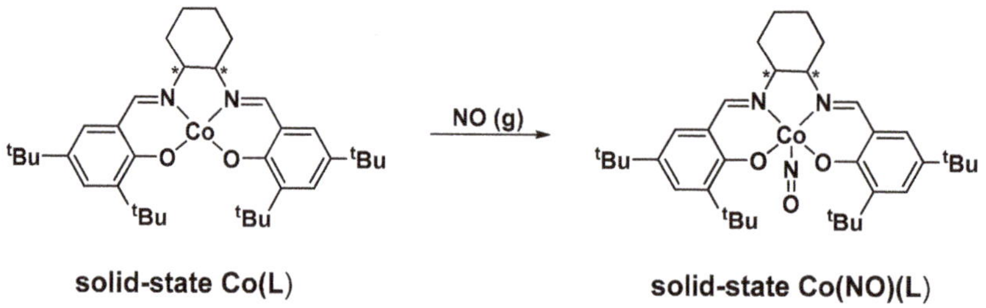

Scheme 2. Preparation of Co(NO)(L) from Co(L) by a solid-state gas reaction with NO (1 atm).

This reaction is accompanied by a color change from red to black (Figure 1, SI film), with the process occurring over several hours for the CS$_2$ solvate and in seconds to minutes for the desolvated phase, depending on sample size. Co(NO)(L) is formed inside the lattice. The IR spectrum of the product shows a low intensity ν_{NO} band at 1657 cm^{-1}.

The NO can be removed stoichiometrically from Co(NO)(L) by heating to 195 °C. This process has been cycled three times without significant decomposition (Figure 2). Powder X-ray diffraction (PXRD) shows that crystallinity is retained after CS$_2$ desorption, NO chemisorption, and subsequent desorption without significant change to the pattern (Figure 3). The pattern for the recrystallized sample of the nitrosyl complex is also similar. These suggest that that the packing is similar throughout these sequential gas (CS$_2$/NO)– solid sorption/chemisorption and desorption processes.

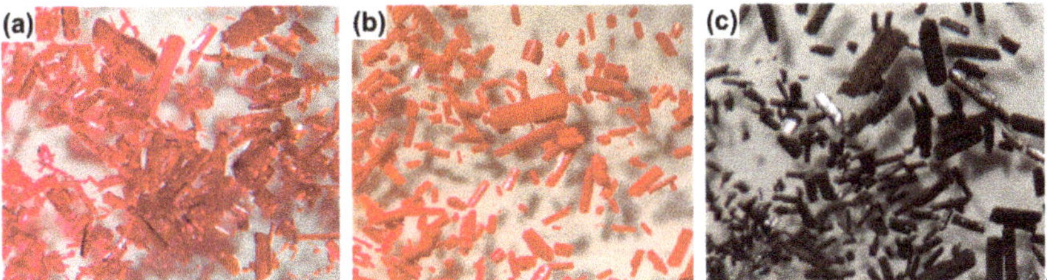

Figure 1. Crystals of (**a**) Co(L)·CS$_2$; (**b**) Co(L), obtained by the desorption of CS$_2$ from Co(L)·CS$_2$; and (**c**) Co(NO)(L) after the NO gas–solid reaction with Co(L).

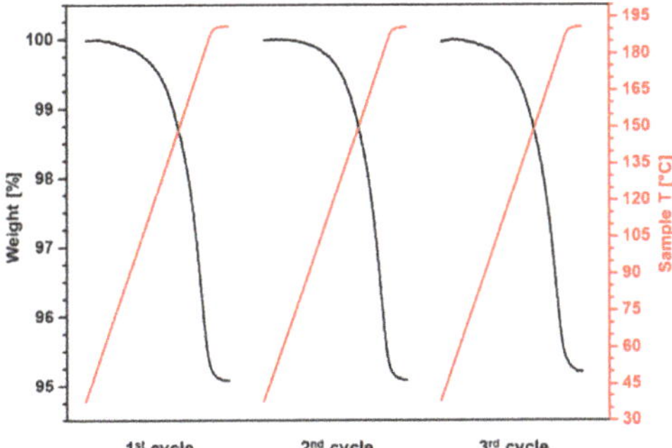

Figure 2. Thermogravimetric analysis showing reversible and ostensibly stoichiometric binding of NO to Co(L) to give Co(NO)(L) over three cycles. The red lines are the heating profile, while the black lines are the weight loss profile.

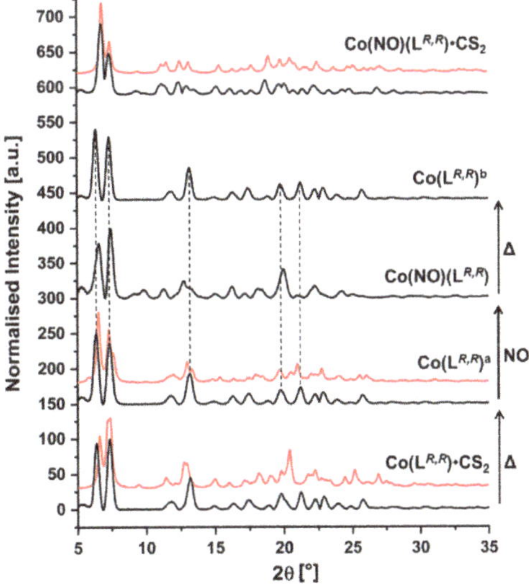

Figure 3. Series of PXRD patterns for an enantiomorphic phase of the CS$_2$ solvate, before and after desorption of CS$_2$, followed by the product of NO chemisorption and subsequent NO desorption. Top pattern is of recrystallized Co(NO)(L) as a CS$_2$ solvate. Note: Black—measured; red—calculated from single crystal structural data.

3.2. Structures of Enantiomorphic Phases

Single crystal X-ray structures were obtained for $Co(L^{R,R}) \cdot CS_2$, $Co(L^{R,R})$ and $Co(NO)(L^{R,R}) \cdot CS_2$. Notably, the data for $Co(L^{R,R})$ were obtained from a crystal of $Co(L^{R,R}) \cdot CS_2$ after desorption of the CS_2, which occurs in a single crystal-to-single crystal (SCSC) transformation for many of the individual crystals, i.e., many of the crystals do not break but retain their morphology in the process. This is not the case for the subsequent NO sorption, where the crystals break into smaller pieces and no crystal of adequate quality for a structure determination by SCXRD was found. PXRD establishes, however, that crystallinity is retained. Recrystallization of the solid $Co(NO)(L^{R,R})$ (from CS_2/n-hexane) was necessary for obtaining a single crystal X-ray structure of the cobalt nitrosyl. Details of the data collections are provided in Table 1. Solid state iron porphyrin complexes have been shown to be capable of NO gas sorption, in this case the process occurs in an SCSC transformation [31,32].

Table 1. Selected crystallographic data for $Co(L^{R,R}) \cdot CS_2$, $Co(L^{R,R})$, and $Co(NO)(L^{R,R}) \cdot CS_2$. All data obtained using Cu K_α radiation.

Compound	$Co(L^{R,R}) \cdot CS_2$	$Co(L^{R,R})$	$Co(NO)(L^{R,R}) \cdot CS_2$
Empirical formula	$C_{37}H_{52}N_2O_2S_2Co$	$C_{36}H_{52}N_2O_2Co$	$C_{37}H_{52}N_3O_3S_2Co$
Formula weight	679.85	603.72	709.86
Temperature/K	100.00(10)	100.00(10)	100.00(10)
Crystal system	monoclinic	monoclinic	monoclinic
Space group	$P2_1$	$P2_1$	$P2_1$
a [Å]	13.55410(10)	26.2740(4)	14.12020(10)
b [Å]	9.96900(10)	9.5866(2)	10.02300(10)
c [Å]	26.59490(10)	29.2826(4)	25.9282(2)
α [°]	90	90	90
β [°]	92.1290(10)	113.379(2)	95.5600(10)
γ [°]	90	90	90
Volume [Å³]	3591.04(5)	6770.1(2)	3652.27(5)
Z	4	8	4
μ [mm⁻¹]	5.085	4.210	5.052
Tmin/Tmax	0.499/0.958	0.688/1.000	0.183/1.000
$(\sin\theta/\lambda)_{max}$	0.749	0.684	0.698
Final R1, wR2 indexes [$I \geq 2\sigma(I)$]	0.0250, 0.0638	0.0952, 0.1822	0.0292, 0.0757
Final R1, wR2 indexes [all data]	0.0267, 0.0649	0.1113, 0.1898	0.0346, 0.0801
Flack parameter [33]	−0.0189(11)	0.082(4)	−0.019(2)
Goof on F^2	1.022	1.198	1.013
Peak/hole [e Å⁻³]	0.28/−0.30	0.58/−0.84	0.42/−0.27
CCDC Numbers	2073657	2073658	2073659

The structures of $Co(L^{R,R}) \cdot CS_2$ and $Co(NO)(L^{R,R}) \cdot CS_2$ are shown in Figure 4. That of $Co(L^{R,R})$ is similar (SI Figure S1). The cobalt ion in $Co(L^{R,R}) \cdot CS_2$ is close to square planar, with the cobalt atom lying 0.023 Å above the O–N–N–O plane of L^{2-}. The sum of angles around the Co(II) is 360.11°, with cis angles of between 86.73° and 94.02°. The average $Co–O_{salen}$ and $Co–N_{salen}$ bond distances are 1.844(2) Å and 1.861(2) Å, respectively. The cyclohexane backbones show the expected chair conformation of the *R,R* enantiomer. The metric parameters for the unsolvated phase of $Co(L^{R,R})$ are similar (Table 2).

The bond lengths and angles for the nitrosyl complex are comparable to those found for its homologue $Co(NO)(salen)$ [34], with the geometry around the cobalt atom being close to square pyramidal. The angles deviate an average of 4.9° from 90°, and the cobalt atom lies even further (0.237 Å) above the O–N–N–O plane of L^{2-} compared to the Co(II) complexes, with no significant interaction to the cobalt ion in the other axial position. The distances and the associated angles are very similar to the cobalt(II) complexes, however, the flat chelating salen ligand can be expected to impose constraints. Analysis of the O/N–Co distances in other cobalt complexes of L^{2-} (SI Figure S2) show that it is typical for square pyramidal Co(III) complexes to display similar values to the Co(II) square planar complexes. Nitrosyl is a well-known non-innocent ligand, and has been assigned formally to the extremes of NO^+, $NO^{.}$ and NO^-. The N–O distance in free radical NO is 1.15 Å, and in $Co(NO)(L^{R,R})·CS_2$ it is less than 0.03 Å greater. With various structural ambiguities, we describe the nitrosyl complex using the Enemark-Feltham notation [35], i.e., $\{Co(NO)\}^8$.

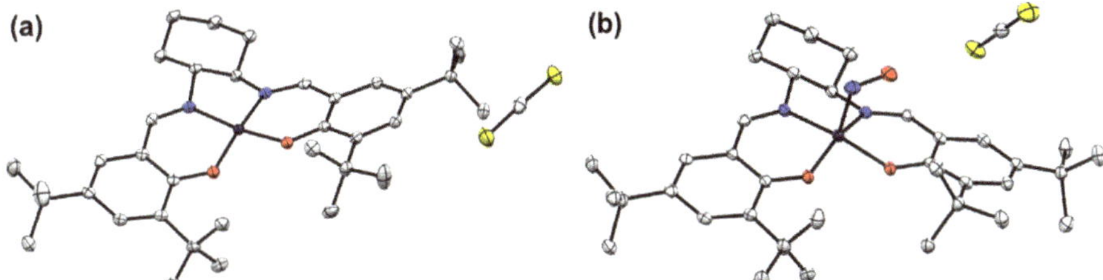

Figure 4. Single crystal X-ray structures of (**a**) $Co(L^{R,R})·CS_2$, and (**b**) $Co(NO)(L^{R,R})·CS_2$. Atom colors: red—oxygen; blue—nitrogen; white—carbon; yellow—sulfur; dark blue—cobalt. Atomic displacement ellipsoids are drawn at 50% probability and hydrogen atoms are omitted for clarity.

Table 2. Selected mean bond distances and angles.

Compound	$Co(L^{R,R})·CS_2$	$Co(L^{R,R})$	$Co(NO)(L^{R,R})·CS_2$
Co–N [Å]	1.861(2)	1.861(9)	1.894(3)
Co–O [Å]	1.844(2)	1.847(8)	1.877(2)
∠O–Co–O [°]	86.93(7)	87.3(3)	84.34(9)
∠N–Co–N [°]	86.54(9)	86.9(4)	85.42(11)
∠N–Co–O [°]	93.78(8)	93.5(4)	93.60(10)
Co–NO [Å]	-	-	1.818(3)
CoN=O [Å]	-	-	1.179(3)
∠Co–N=O [°]	-	-	122.0(2)

The packing in the three structurally characterized compounds show layers of the complexes in a herringbone arrangement, where the tert-butyl groups para to the phenolato O atom reside in columns parallel to the *a* axis (Figure 5, SI Figures S3 and S4). In the solvates, the co-crystallized CS_2 molecules are also located in these columns. Desorption of CS_2 from $Co(L^{R,R})·(CS_2)$, to give $Co(L^{R,R})$, results in a relatively large unit cell change with approximate doubling of the *a* axis from 13.55410(10) Å to 26.2740(4) Å, and an increase in the β-angle from 92.1290(10)° to 113.379(2)°. The packing in $Co(L^{R,R})$ is shown in Figure 5. The voids occur where the CS_2 occupied the lattice of $Co(L^{R,R})·(CS_2)$. They are not connected and have an approximate size of 216.34 Å^3 (3.2% of unit cell volume). The kinetic diameters of CS_2 and NO are approximately 3.6 Å and 3.17 Å, respectively [36], and, on the basis of the structures and the approximate void diameter of 7.4 Å, it is tempting to posit that the exit of CS_2 will be synchronously accompanied by rotation about the C–C bond attaching the tert-butyl groups to the aryl ring. The chemisorption of NO occurs with small shifts in the PXRD pattern (Figure 3), suggesting this unit cell is changed slightly but with related packing. The recrystallized $Co(NO)(L^{R,R})·CS_2$ shows a marginally (1.7%) expanded unit cell compared to that of $Co(L^{R,R})·CS_2$. The two most intense diffraction peaks in the PXRD of $Co(NO)(L^{R,R})$ are moved towards a slightly higher 2θ angle upon chemisorption of NO compared to the parent phase $Co(L^{R,R})$, which corresponds to a change in the interplanar distance (d) from 13.8962 Å and 12.0268 Å in $Co(L^{R,R})$ to 13.4325 Å and 11.8659 Å in $Co(NO)(L^{R,R})$ (Figure 3). This can be attributed to a slight molecular rearrangement, in order to make room for binding of NO at the cobalt atom.

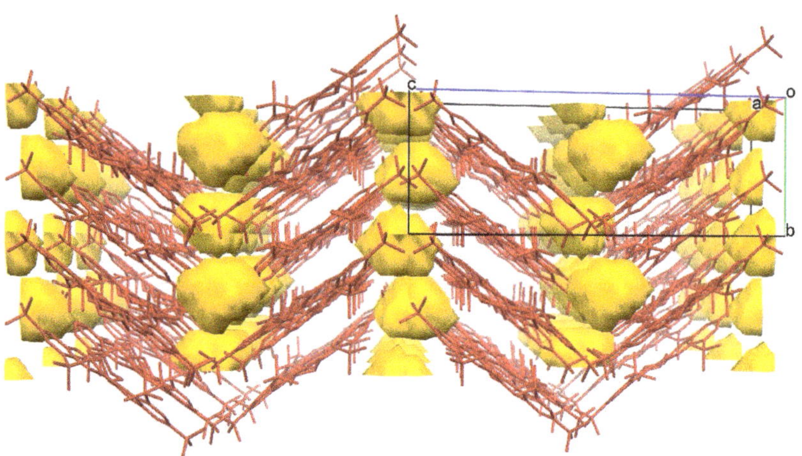

Figure 5. Packing diagram for Co(L^{R,R}) with voids in yellow (1.2 Å probe radius and 0.7 Å grid spacing). Hydrogen atoms are omitted for clarity.

3.3. Solution-State Vibrational Circular Dichroism (VCD) Spectra

VCD allows the characterization of enantiopure compounds through the analysis of IR-active bands. The primary quantity associated with IR absorbance is the dipole strength, however, VCD measures the differential absorbance ($\Delta A(\nu) = A_{\text{left}}(\nu) - A_{\text{right}}(\nu)$) that is proportional to the rotational strength; a quantity which depends both on the electric and magnetic dipole transition moments. Thus, the VCD intensities are not directly correlated with the intensities in the IR spectrum. VCD has predominantly been used to analyze solution and pure liquid phases of organic molecules. In this work, VCD spectra for both hands of LH$_2$ and the Zn(II), Co(II), and Co(III)NO complexes of L^{2-} have been recorded. The IR spectra (purple) and the corresponding VCD spectra for the L^{S,S}- and L^{R,R}-based systems (colored blue and red respectively) are shown in Figure 6. The IR spectra show that the band due to the salen $\nu_{\text{C=N}}$, the most intense in the IR spectra, shifts from 1632 cm^{-1} in LH$_2$ to a lower frequency (1598 cm^{-1}) in coordination with Zn^{2+} in accordance with the decrease in the C=N bond order as a consequence of co-ordinate bond formation by the azomethine nitrogen lone pair. The band due to the $\nu_{\text{C=N}}$, is the most intense in the VCD spectrum of the Zn(II) complexes. Despite being the most intense in the IR spectrum for L^{S,S}H$_2$ and L^{R,R}H$_2$, this is now a relatively moderate signal in the VCD spectra of L^{S,S}H$_2$ and L^{R,R}H$_2$, where, instead, several unassignable C–C and C–H vibrations are the most intense. Computational analysis to aid in the assignment of bands in VCD spectra is common in the study of enantiopure organic compounds, and we have performed such an analysis with reasonable success for L^{R,R} and Zn(L^{R,R}) (SI Figure S5). The introduction of transition metal ions makes calculation a formidable task, perhaps even impossible for the paramagnetic complexes. In this work, we take a fingerprint approach to the characterization. The fact that we have characterized both enantiomers and achieved consistently opposite band polarities demonstrates the reliability of the results.

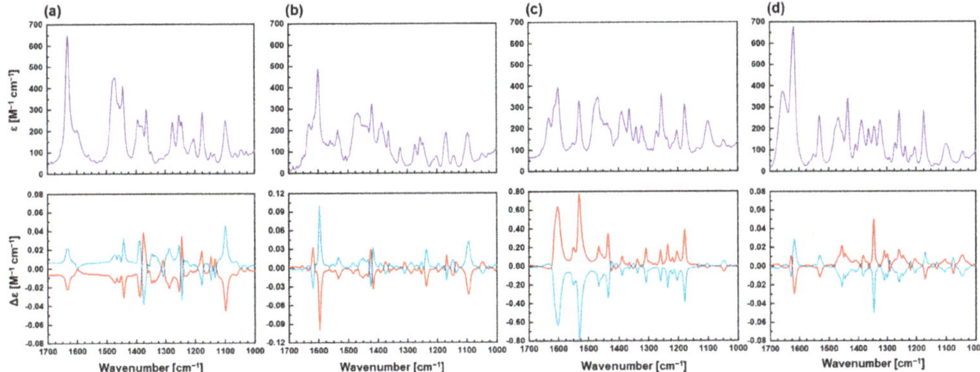

Figure 6. Solution (CDCl$_3$) IR (top, purple line, obtained using equimolar concentrations of both hands) and VCD (bottom) spectra of (**a**) L^{S,S}H$_2$ and L^{R,R}H$_2$, (**b**) Zn(L^{S,S}) and Zn(L^{R,R}), (**c**) Co(L^{S,S}) and Co(L^{R,R}), and (**d**) Co(NO)(L^{S,S}) and Co(NO)(L^{R,R}). The L^{S,S}- and L^{R,R}-based systems are colored blue and red, respectively.

An increase in VCD signal strength is noticeable in the comparison of the VCD spectra of the Zn(II) complexes and those for LH$_2$. This arises because coordination locks the ligand into fewer conformations [37]. The VCD spectra for Co(L^{S,S}) and Co(L^{R,R}) are very different to those of the structurally analogous Zn(II) complexes. Apart from many different positions and relative intensities, the spectrum is close to monosignate (all bands for one hand showing one polarity). A change in the absorbance of left and right circularized IR radiation causes the polarity inversion of some bands. In addition, the signals are approximately nine times as intense as those for all the other compounds. This is associated with the diamagnetism and paramagnetism of low-lying excited electronic states in the Co(II) complexes. Electronic transitions are strongly magnetic-dipole dependent and hence, when present, can induce enhanced VCD by the coupling of the electronic magnetic-dipole transition moments with the smaller vibrational magnetic-dipole moments responsible for normal VCD intensity [37].

Azide is a linear, non-chiral ligand with a track record as a proxy vibrational spectroscopic and structural marker for O$_2$ binding sites. Accordingly, it binds to the iron center at the distal position of the heme group. Bormett et al. [38] have reported the VCD spectra of ferric hemoglobin azide, and show that the azide ligand somehow borrows magnetic dipole intensity from the chiral environment inherent to this protein to produce a VCD signal. The effect we see in this work is analogous, and we have used VCD to assess the chiral induction of the signals for achiral NO in the complexes of L^{2-}. This co-ligand can be regarded as a proxy substrate in a reaction catalyzed by complexes of enantiopure L^{2-}, which, particularly with respect to asymmetric catalytic synthesis, is easily accessible in both hands. Although close to the dominating salen imine ν_{CN} (1616 cm^{-1}) band in Co(NO)(L), the ν_{NO} (1657 cm^{-1}) can be distinguished in the IR spectra as the second most intense band (Figure 6d, top). Notably, a signal for ν_{NO} can also be seen at this position in the VCD spectrum, but its intensity is now very low by comparison to many other bands in the spectrum. This indicates that, while chiral induction occurs, it is not particularly strong. In this context, it is interesting to note that the enantiomeric excess in catalytic reactions with complexes of L^{R,R} and L^{S,S} are sometimes disappointing [39–43].

3.4. Solid State VCD Spectra

While the use of Nujol mull, KBr discs, or even neat powders in sample preparation for recording solid-state IR spectra is commonplace, irreproducibility due to a high sensitivity to inhomogeneity in particle size, especially because the technique is a transmission method, makes classic solid-state sample preparation for VCD spectroscopy less straightforward. This is probably why solid state VCD spectra are rarely reported [37]. In the context of the molecular solid-state chemistry described here, it was interesting for us to develop this technique. We found that, with consistent grinding and Nujol, it was possible to obtain reproducible spectra. The solid-state IR and VCD spectra of Co(L)·CS$_2$ and Co(L) are shown in Figure 7a,b. In contrast to the IR spectra, the VCD spectra of the two phases are easily distinguishable. Obviously, this is information that was lost in the solution state. The differences between the solution-state and solid-state spectra are greater in the VCD spectra compared to the IR spectra: There is far less tendency towards a monosignate spectra for both phases in the solid-state spectra. Fascinatingly, an n$_{CS}$ signal related to co-crystallized CS$_2$ is visible in VCD spectrum at 1514 cm^{-1} (Figure 7a). This indicates supramolecular chiral induction of this achiral molecule. The signal disappears on CS$_2$ desorption to form Co(L) (Figure 7b). The intensity of the IR band at 1524 cm^{-1} also decreases. This band comprises an overlap of the n$_{CS}$ with a band that does not appear in the VCD spectrum.

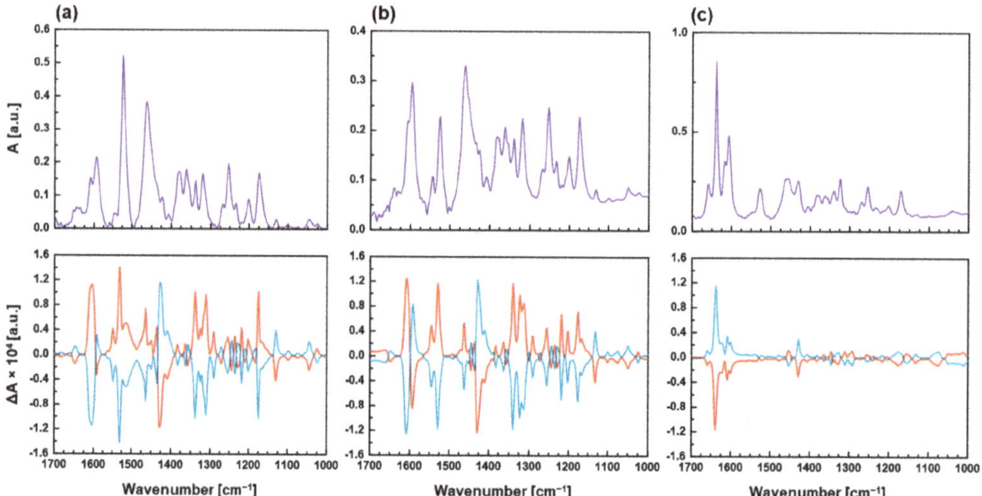

Figure 7. Solid-state IR spectra of a mixture of equal amounts of both hands of the complexes (purple line) and solid-state VCD spectra of (**a**) Co(L)·CS$_2$, (**b**) Co(L), and (**c**) Co(L)(NO) (blue line complexes of L^{S,S}; red line complexes of L^{R,R}).

The solid-state IR and VCD spectra of Co(NO)(L) are shown in Figure 7c. It is difficult to assign the bands in the spectra, however, the band at 1638 cm^{-1}, assigned to the imine v$_{CN}$, is significantly more intense than all the others. It also shows an increase of 22 cm^{-1} compared to the band assigned to this vibration in the solution state spectra of the same complex. This suggests a change in the electronics of the complex due to different supramolecular interactions in the solution-state versus the solid-state. The higher wavenumber for v$_{CN}$ suggests that this bond is stronger in the solid state. Indirectly, this suggests that the NO is a stronger donor to the Co atom in the solid-state. The well-resolved, but low intensity, band at 1659 cm^{-1} is assigned to v$_{NO}$. There is significant polarity inversion between comparable signals in the paramagnetic Co(II) compounds, the diamagnetic {Co(NO)}^{8} phase, and the {Co(NO)}^{8} systems in the solution and solid states. Curiously, an unassigned band at 1344 cm^{-1}, which is most intense in the solution-state spectrum of Co(NO)(L), appears to be absent in the solid-state VCD spectrum.

4. Conclusions

We have demonstrated chemisorption of NO into an ostensibly non-porous crystalline material, but postulate that transient conduits for CS$_2$ desorption and NO chemisorption form because the tert-butyl groups can rotate around the bond between the aryl and the tert-butyl carbon atoms. The gas–solid reactions involved in the transitions between

Co(L)·CS$_2$, Co(L), and Co(L)(NO) occur in crystal to crystal transformations, and in one case an SCSC transformation. The NO binds reversibly, and the materials do not allow O$_2$ to react with the bound NO. This supports the hypothesis that the in-crystal reactivity seen for the crystalline dicobalt(II) complexes described in the introduction, where sorbed NO and O$_2$ were transformed to a nitrite ligand and a nitrate counter anion [7], is dependent on two cobalt(II) ions in close proximity. Accessibility and the specific coordination sphere and geometry, however, also clearly play important roles.

The crystalline solid state offers advantages beyond the solution state, through the provision of tailored cavities that might amplify chirality compared to the solution state. This study shows that in-crystal chemistry for the first time, with the induction of chirality onto both a chemisorbed achiral guest and a physisorbed co-crystallized achiral guest. Crystal phases themselves can occur in chiral space groups without being constructed from enantiopure molecules. Both situations offer feasibility for absolute asymmetric synthesis (AAS) [44] inside solid states. The rare use of VCD spectroscopy to characterize the complexes in this work illustrates its untapped potential in the study of enantiomers of metal coordination complexes in both solution- and solid-states.

Supplementary Materials: The following are available online at https://www.mdpi.com/article/10.3390/chemistry3020041/s1. CCDC 2073657–2073659 contain the supplementary crystallographic data for this paper. These data can be obtained free of charge from The Cambridge Crystallographic Data Centre via www.ccdc.cam.ac.uk/structures.

Author Contributions: M.S.M. synthesized the compounds, performed calculations, and prepared the first manuscript draft; M.S.M. and M.C.L. performed measurements, processed and analyzed the data, and designed the figures; V.M. assisted with X-ray crystallography; C.J.M. conceived and supervised the work. All authors reviewed and edited the manuscript. All authors have read and agreed to the published version of the manuscript.

Funding: The work was supported by the Danish Council for Independent Research (Grant 9041-00170B) and the Carlsberg Foundation grant CF15-0675 for the X-ray diffractometer.

Institutional Review Board Statement: Not applicable.

Informed Consent Statement: Not applicable.

Data Availability Statement: The data presented in this study are available in supplementary material.

Conflicts of Interest: The authors declare no conflict of interest.

References

1. Zhang, W.X.; Yahiro, H.; Mizuno, N.; Izumi, J.; Iwamoto, M. Removal of nitrogen monoxide on copper ion-exchanged zeolites by pressure swing adsorption. *Langmuir* **1993**, *9*, 2337–2343. [CrossRef]
2. Xiao, B.; Byrne, P.J.; Wheatley, P.S.; Wragg, D.S.; Zhao, X.; Fletcher, A.J.; Thomas, K.M.; Peters, L.; Evans, J.S.O.; Warren, J.E.; et al. Chemically blockable transformation and ultraselective low-pressure gas adsorption in a non-porous metal organic framework. *Nat. Chem.* **2009**, *1*, 289–294. [CrossRef] [PubMed]
3. Arai, H.; Machida, M. Removal of NOx through sorption-desorption cycles over metal oxides and zeolites. *Catal. Today* **1994**, *22*, 97–109. [CrossRef]
4. Cruz, W.V.; Leung, P.C.W.; Seff, K. Crystal structure of nitric oxide and nitrogen dioxide sorption complexes of partially cobalt(II)-exchanged zeolite A. *Inorg. Chem.* **1979**, *18*, 1692–1696. [CrossRef]
5. Zhang, J.; Kosaka, W.; Kitagawa, S.; Takata, M.; Miyasaka, H. In Situ Tracking of Dynamic NO Capture through a Crystal-to-Crystal Transformation from a Gate-Open-Type Chain Porous Coordination Polymer to a NO-Adducted Discrete Isomer. *Chem. Eur. J.* **2019**, *25*, 3020–3031. [CrossRef]
6. Zhang, W.-X.; Yahiro, H.; Iwamoto, M.; Izumi, J. Reversible and irreversible adsorption of nitrogen monoxide on cobalt Ion-exchanged ZSM-5 and mordenite zeolites at 273–523 K. *J. Chem. Soc. Faraday Trans.* **1995**, *91*, 767–771. [CrossRef]
7. Møller, M.S.; Haag, A.; McKee, V.; McKenzie, C.J. NO sorption, in-crystal nitrite and nitrate production with arylamine oxidation in gas-solid single crystal to single crystal reactions. *Chem. Commun.* **2019**, *55*, 10551–10554. [CrossRef]
8. Earnshaw, A.; Hewlett, P.C.; Larkworthy, L.F. 877. Transition metal–Schiff's base complexes. Part II. The reaction of nitric oxide with some oxygen-carrying cobalt compounds. *J. Chem. Soc.* **1965**, *0*, 4718–4723. [CrossRef]
9. Diehl, H.; Hach, C.C.; Harrison, G.C.; Liggett, L.M.; Chao, T.S. Studies on oxygen-carrying cobalt compounds; cobalt derivatives of the Schiff's bases of salicylaldehyde with alkylamines. *Iowa State Coll. J. Sci.* **1947**, *21*, 287.

10. Jacobsen, E.N.; Zhang, W.; Muci, A.R.; Ecker, J.R.; Deng, L. Highly enantioselective epoxidation catalysts derived from 1,2-diaminocyclohexane. *J. Am. Chem. Soc.* **1991**, *113*, 7063–7064. [CrossRef]

11. Lee, N.H.; Muci, A.R.; Jacobsen, E.N. Enantiomerically pure epoxychromans via asymmetric catalysis. *Tetrahedron Lett.* **1991**, *32*, 5055–5058. [CrossRef]

12. Hanson, J. Synthesis and use of jacobsen's catalyst: Enantioselective epoxidation in the introductory organic laboratory. *J. Chem. Educ.* **2001**, *78*, 1266. [CrossRef]

13. Groom, C.R.; Bruno, I.J.; Lightfoot, M.P.; Ward, S.C. The cambridge structural database. *Acta Crystallogr. B Struct. Sci. Cryst. Eng. Mater.* **2016**, *72*, 171–179. [CrossRef]

14. Møller, M.S.; Kongsted, J.; McKenzie, C.J. Preparation of organocobalt(iii) complexes via O_2 activation. *Dalton Trans.* **2021**, *50*, 4819–4829. [CrossRef] [PubMed]

15. Brückner, S.; Calligaris, M.; Nardin, G.; Randaccio, L. The crystal structure of the form of N,N'-ethylenebis(salicylaldehydeiminato) cobalt(II) inactive towards oxygenation. *Acta Crystallogr. B Struct. Crystallogr. Cryst. Chem.* **1969**, *25*, 1671–1674. [CrossRef]

16. Hughes, D.L.; Kleinkes, U.; Leigh, G.J.; Maiwald, M.; Sanders, J.R.; Sudbrake, C. New polymeric compounds containing vanadium–oxygen chains. *J. Chem. Soc. Dalton Trans.* **1994**, 2457–2466. [CrossRef]

17. Garcia-Deibe, A.; Sousa, A.; Bermejo, M.R.; Mac Rory, P.P.; McAuliffe, C.A.; Pritchard, R.G.; Helliwell, M. The visible light induced rearrangement of a manganese(III) complex of an unsymmetrical tetradentate Schiff's base ligand, 4-[2-(2-hydroxyphenylmethyleneamino)ethylamino]pent-3-en-2-one, to a manganese(III) complex of the symmetrical ligand salen. *J. Chem. Soc. Chem. Commun.* **1991**, *10*, 728–729. [CrossRef]

18. Bhadbhade, M.M.; Srinivas, D. Effects on molecular association, chelate conformation, and reactivity toward substitution in copper Cu(5-X-salen) complexes, salen2- = N,N'-ethylenebis(salicylidenaminato), X = H, CH3O, and Cl: Synthesis, x-ray structures, and EPR investigations. [Erratum to document cited in CA119(26):285036b]. *Inorg. Chem.* **1993**, *32*, 6122–6130.

19. *CrysAlis*, version 1.171.40.67a; Rigaku Oxford Diffraction: The Woodlands, TX, USA, 2019.

20. Hübschle, C.B.; Sheldrick, G.M.; Dittrich, B. ShelXle: A Qt graphical user interface for SHELXL. *J. Appl. Crystallogr.* **2011**, *44*, 1281–1284. [CrossRef]

21. Sheldrick, G.M. SHELXT—Integrated space-group and crystal-structure determination. *Acta Crystallogr. A Found. Adv.* **2015**, *71*, 3–8. [CrossRef]

22. Sheldrick, G.M. Crystal structure refinement with SHELXL. *Acta Crystallogr. C Struct. Chem.* **2015**, *71*, 3–8. [CrossRef]

23. Bochevarov, A.D.; Harder, E.; Hughes, T.F.; Greenwood, J.R.; Braden, D.A.; Philipp, D.M.; Rinaldo, D.; Halls, M.D.; Zhang, J.; Friesner, R.A. Jaguar: A high-performance quantum chemistry software program with strengths in life and materials sciences. *Int. J. Quantum Chem.* **2013**, *113*, 2110–2142. [CrossRef]

24. *Jaguar*; Schrödinger Release 2019-1; Schrödinger, LLC: New York, NY, USA, 2019.

25. Hay, P.J.; Wadt, W.R. Ab initio effective core potentials for molecular calculations. Potentials for K to Au including the outermost core orbitals. *J. Chem. Phys.* **1985**, *82*, 299–310. [CrossRef]

26. Stephens, P.J.; Devlin, F.J.; Chabalowski, C.F.; Frisch, M.J. Ab Initio Calculation of Vibrational Absorption and Circular Dichroism Spectra Using Density Functional Force Fields. *J. Phys. Chem.* **1994**, *98*, 11623–11627. [CrossRef]

27. Banks, J.L.; Beard, H.S.; Cao, Y.; Cho, A.E.; Damm, W.; Farid, R.; Felts, A.K.; Halgren, T.A.; Mainz, D.T.; Maple, J.R.; et al. Integrated modeling program, applied chemical theory (IMPACT). *J. Comput. Chem.* **2005**, *26*, 1752–1780. [CrossRef] [PubMed]

28. Larrow, J.F.; Jacobsen, E.N. (R,R)-N,N'-Bis(3,5-di-tert-butylsalicylidene)-1,2-cyclohexanediamino manganese(III) chloride, a highly enantioselective epoxidation catalyst. *Org. Synth.* **1998**, *75*, 1.

29. Li, G.; Liu, Y.; Zhang, K.; Shi, J.; Wan, X.; Cao, S. Effect of salen-metal complexes on thermosensitive reversibility of stimuli-responsive water-soluble poly(urethane amine)s. *J. Appl. Polym. Sci.* **2013**, *129*, 3696–3703. [CrossRef]

30. North, M.; Quek, S.C.Z.; Pridmore, N.E.; Whitwood, A.C.; Wu, X. Aluminum(salen) complexes as catalysts for the kinetic resolution of terminal epoxides via co_2 coupling. *ACS Catal.* **2015**, *5*, 3398–3402. [CrossRef]

31. Xu, N.; Powell, D.R.; Cheng, L.; Richter-Addo, G.B. The first structurally characterized nitrosyl heme thiolate model complex. *Chem. Commun.* **2006**, *19*, 2030–2032. [CrossRef] [PubMed]

32. Xu, N.; Powell, D.R.; Richter-Addo, G.B. Nitrosylation in a crystal: Remarkable movements of iron porphyrins upon binding of nitric oxide. *Angew. Chem. Int. Ed. Engl.* **2011**, *50*, 9694–9696. [CrossRef] [PubMed]

33. Parsons, S.; Flack, H.D.; Wagner, T. Use of intensity quotients and differences in absolute structure refinement. *Acta Crystallogr. B Struct. Sci. Cryst. Eng. Mater.* **2013**, *69*, 249–259. [CrossRef]

34. Haller, K.J.; Enemark, J.H. Structural chemistry of the {CoNO}^{8} group. III. The structure of *N, N'*-ethylenebis(salicylideneiminato) nitrosylcobalt(II), Co(NO)(salen). *Acta Crystallogr. B Struct. Sci.* **1978**, *34*, 102–109. [CrossRef]

35. Enemark, J.H.; Feltham, R.D. Principles of structure, bonding, and reactivity for metal nitrosyl complexes. *Coord. Chem. Rev.* **1974**, *13*, 339–406. [CrossRef]

36. Albrecht, E.; Baum, G.; Bellunato, T.; Bressan, A.; Dalla Torre, S.; D'Ambrosio, C.; Davenport, M.; Dragicevic, M.; Duarte Pinto, S.; Fauland, P.; et al. VUV absorbing vapours in n-perfluorocarbons. *Nucl. Instrum. Methods Phys. Res. A Accel. Spectrometers Detect. Assoc. Equip.* **2003**, *510*, 262–272. [CrossRef]

37. Nafie, L.A. *Vibrational Optical Activity: Principles and Applications*; John Wiley & Sons, Ltd.: Chichester, UK, 2011.

38. Bormett, R.W.; Asher, S.A.; Larkin, P.J.; Gustafson, W.G.; Ragunathan, N.; Freedman, T.B.; Nafie, L.A.; Balasubramanian, S.; Boxer, S.G. Selective examination of heme protein azide ligand-distal globin interactions by vibrational circular dichroism. *J. Am. Chem. Soc.* **1992**, *114*, 6864–6867. [CrossRef]
39. Adão, P.; Costa Pessoa, J.; Henriques, R.T.; Kuznetsov, M.L.; Avecilla, F.; Maurya, M.R.; Kumar, U.; Correia, I. Synthesis, characterization, and application of vanadium-salan complexes in oxygen transfer reactions. *Inorg. Chem.* **2009**, *48*, 3542–3561. [CrossRef]
40. Chen, B.-L.; Zhu, H.-W.; Xiao, Y.; Sun, Q.-L.; Wang, H.; Lu, J.-X. Asymmetric electrocarboxylation of 1-phenylethyl chloride catalyzed by electrogenerated chiral [CoI(salen)]−complex. *Electrochem. Commun.* **2014**, *42*, 55–59. [CrossRef]
41. Bam, R.; Pollatos, A.S.; Moser, A.J.; West, J.G. Mild olefin formation via bio-inspired vitamin B$_{12}$ photocatalysis. *Chem. Sci.* **2021**, *12*, 1736–1744. [CrossRef]
42. Chang, S.; Lee, N.H.; Jacobsen, E.N. Regio- and enantioselective catalytic epoxidation of conjugated polyenes. Formal synthesis of LTA4 methyl ester. *J. Org. Chem.* **1993**, *58*, 6939–6941. [CrossRef]
43. Huang, J.; Cai, J.; Feng, H.; Liu, Z.; Fu, X.; Miao, Q. Synthesis of salen Mn(III) immobilized onto the ZnPS-PVPA modified by 1,2,3-triazole and their application for asymmetric epoxidation of olefins. *Tetrahedron* **2013**, *69*, 5460–5467. [CrossRef]
44. Buhse, T.; Cruz, J.-M.; Noble-Terán, M.E.; Hochberg, D.; Ribó, J.M.; Crusats, J.; Micheau, J.-C. Spontaneous Deracemizations. *Chem. Rev.* **2021**, *121*, 2147–2229. [CrossRef] [PubMed]

 *chemistry*

Article

Bis(diphenylphosphino)methane Dioxide Complexes of Lanthanide Trichlorides: Synthesis, Structures and Spectroscopy [†]

Robert D. Bannister, William Levason and Gillian Reid *

School of Chemistry, University of Southampton, Southampton SO17 1BJ, UK;
r.d.bannister@soton.ac.uk (R.D.B.); wxl@soton.ac.uk (W.L.)
* Correspondence: gr@soton.ac.uk
† Dedicated to Dr. Howard Flack (1943–2017).

Received: 18 October 2020; Accepted: 16 November 2020; Published: 19 November 2020

Abstract: Bis(diphenylphosphino)methane dioxide ($dppmO_2$) forms eight-coordinate cations [$M(dppmO_2)_4$]Cl_3 (M = La, Ce, Pr, Nd, Sm, Eu, Gd) on reaction in a 4:1 molar ratio with the appropriate $LnCl_3$ in ethanol. Similar reaction in a 3:1 ratio produced seven-coordinate [$M(dppmO_2)_3Cl$]Cl_2 (M = Sm, Eu, Gd, Tb, Dy, Ho, Er, Tm, Yb), whilst $LuCl_3$ alone produced six-coordinate [$Lu(dppmO_2)_2Cl_2$]Cl. The complexes have been characterised by IR, ^{1}H and ^{31}P{^{1}H}-NMR spectroscopy. X-ray structures show that [$M(dppmO_2)_4$]Cl_3 (M = Ce, Sm, Gd) contain square antiprismatic cations, whilst [$M(dppmO_2)_3Cl$]Cl_2 (M = Yb, Dy, Lu) have distorted pentagonal bipyramidal structures with apical Cl. The [$Lu(dppmO_2)_2Cl_2$]Cl has a *cis* octahedral cation. The structure of [$Yb(dppmO_2)_3(H_2O)$]Cl_3·$dppmO_2$ is also reported. The change in coordination numbers and geometry along the series is driven by the decreasing lanthanide cation radii, but the chloride counter anions also play a role.

Keywords: lanthanide trichloride complexes; diphosphine dioxide; coordination complexes; X-ray structures

---

1. Introduction

Early work viewed the chemistry of the lanthanides (Ln) (Ln = La–Lu, ≠ Pm unless otherwise indicated) in oxidation state III as very similar and often only two or three elements were examined, and the results were assumed to apply to all. More recent work has shown this to be a very unreliable approach and detailed studies of all fourteen elements (excluding only the radioactive Pm) are required to establish properties and trends [1,2]. Sometimes yttrium is also included since it is similar in size to holmium. The main changes along the series are due to the lanthanide contraction, the reduction in the radius of the M^{3+} ions between La (1.22 Å) and Lu (0.85 Å), and at some point a reduction in coordination number may be driven by steric effects, especially with bulky ligands. However, the decrease in radius also results in an increase in the charge/radius ratio along the series and this can lead to significant electronic effects on the ligand preferences. This interplay of steric and electronic effects means that changes in coordination number or ligand donor set can occur at different points along the series with different ligands. The effects are very nicely demonstrated in a recent article, which examined the changes which occurred in the series of lanthanide nitrates with complexes of 2,2′-bipyridyl, 2,4,6-tri-α-pyridyl-1,3,5-triazine and 2,2′; 6′,2″-terpyridine [2]. Tertiary phosphine oxides have proved popular ligands to explore lanthanide chemistry and the area has been the subject of a comprehensive review [3], and several detailed studies of trends along the series La-Lu have been reported [4–7]. We reported bis(diphenylphosphino)methane dioxide ($dppmO_2$) formed square-antiprismatic cations [$La(dppmO_2)_4$]$^{3+}$ with Cl, I or [PF_6] counter ions, but lutetium gave only octahedral

$[Lu(dppmO_2)_2X_2]^+$ (X = Cl, I) and $[Lu(dppmO_2)_2Cl(H_2O)]^{2+}$ [8]. Other $dppmO_2$ complexes reported include several types with $Ln(NO_3)_3$ [4], $[Dy(dppmO_2)_4][CF_3SO_3]_3$ [9], $[Eu(dppmO_2)_4][ClO_4]_3$ [10], $[La(dppmO_2)_4][CF_3SO_3]_3$ and $[Lu(dppmO_2)_3(H_2O)][CF_3SO_3]_3$ [11]. Here, we report a systematic study of the systems $LnCl_3$-$dppmO_2$ for all fourteen accessible lanthanides.

2. Materials and Methods

Infrared spectra were recorded as Nujol mulls between CsI plates using a Perkin-Elmer Spectrum 100 spectrometer over the range 4000–200 cm^{-1}. ^{1}H and $^{31}P\{^1H\}$-NMR spectra were recorded using a Bruker AV–II 400 spectrometer and are referenced to the protio resonance of the solvent and 85% H_3PO_4, respectively. Microanalyses were undertaken by London Metropolitan University or Medac. Hydrated lanthanide trichlorides and anhydrous $LnCl_3$ (Ln–Nd, Pr, Gd, Ho) were from Sigma-Aldrich and used as received. The $Ph_2PCH_2PPh_2$ (Sigma-Aldrich) in anhydrous CH_2Cl_2 was converted to $Ph_2P(O)CH_2P(O)Ph_2$ by air oxidation catalysed by SnI_4 [12].

X-Ray Experimental. Details of the crystallographic data collection and refinement parameters are given in Table 1. Many attempts were made to grow crystals for X-ray examination from a variety of solvents including EtOH and CH_2Cl_2, either by slow evaporation or layering with hexane or pentane. The crystal quality was often rather poor, and all of the structures have disordered co-solvent, either water or ethanol. No attempt was made to locate the protons on the co-solvent. Several showed disorder in one or more of the phenyl rings. Good-quality crystals used for single crystal X-ray analysis were grown from $[Lu(dppmO_2)_4]Cl_2]Cl$ (CH_2Cl_2/hexane), $[Ce(dppmO_2)_4]Cl_3$, $[Sm(dppmO_2)_4]Cl_3$, $[Gd(dppmO_2)_4]Cl_3$ (EtOH), $[Yb(dppmO_2)_3Cl]Cl_2$, $[Yb(dppmO_2)_3(H_2O)]Cl_3 \cdot dppmO_2$ (EtOH), $[Lu(dppmO_2)_3Cl]Cl_2$ (CH_2Cl_2).

Data collections used a Rigaku AFC12 goniometer equipped with a HyPix-600HE detector mounted at the window of an FR-E+ SuperBright molybdenum (λ = 0.71073 Å) rotating anode generator with VHF Varimax optics (70 μm focus) with the crystal held at 100 K (N_2 cryostream). Structure solution and refinements were performed with either SHELX(S/L)97 or SHELX(S/L)2013 [13,14]. The crystallographic data in cif format have been deposited as CCDC 2033611-2033618.

All samples were dried in high vacuum at room temperature for several hours, but this treatment does not remove lattice water or alcohol. Heating the samples in vacuo is likely to cause some decomposition of the complexes [7] and was not applied.

$[La(dppmO_2)_4]Cl_3 \cdot 4H_2O$ and $[Lu(dppmO_2)_2Cl_2]Cl \cdot H_2O$ were made as described [8]. The individual new complexes were isolated as described below, with yields of 50–80%.

$[Ce(dppmO_2)_4]Cl_3 \cdot 6H_2O$—$CeCl_3 \cdot 7H_2O$ (0.025 g, 0.067 mmol) and $dppmO_2$ (0.112 g, 0.268 mmol) afforded colourless crystals of $[Ce(dppmO_2)_4]Cl_3 \cdot 4H_2O$, by concentrating the ethanolic solution and layering with *n*-hexane (1 mL). Required for $C_{100}H_{100}CeCl_3O_{12}P_8$ (2020.1): C, 59.46; H, 4.99%. Found: C, 59.50; H, 4.50%. ^{1}H-NMR (CD_2Cl_2): δ = 1.52 (s, H_2O) 3.60 (vbr, [8H], PCH_2P), 7.10 (s, [32H], Ph), 7.35 (m, [16H], Ph), 7.70 (m, [32H], Ph). $^{31}P\{^1H\}$-NMR (CD_2Cl_2): δ = 48.6 (s). IR (Nujol mull)/cm^{-1}: 3500 br, 1630 (H_2O), 1158, 1099s (P=O).

$[Pr(dppmO_2)_4]Cl_3 \cdot 6H_2O$—To a solution of $PrCl_3 \cdot 6H_2O$ (0.025 g, 0.070 mmol) in ethanol (5 mL) was added a solution of $dppmO_2$ (0.117 g, 0.281 mmol) in ethanol (10 mL). A white powdered solid formed on slow evaporation of the ethanol. Required for $C_{100}H_{100}Cl_3O_{14}P_8Pr$ (2020.9): C, 59.43; H, 4.99%. Found: C, 59.06; H, 4.62% ^{1}H-NMR (CD_2Cl_2): δ = 4.63 (m, [8H], PCH_2P), 7.19 (s, [32H], Ph), 7.44 (m, [16H], Ph), 8.19 (m, [32H], Ph). $^{31}P\{^1H\}$-NMR (CD_2Cl_2): δ = 64.0 (s). IR (Nujol mull)/cm^{-1}: 3500 br, 1630 (H_2O), 1161, 1102 (P=O).

Table 1. X-ray crystallographic data [a].

Compound	[Ce(dppmO$_2$)$_4$] Cl$_3$·9EtOH	[Sm(dppmO$_2$)$_4$] Cl$_3$·9.5EtOH	[Gd(dppmO$_2$)$_4$] Cl$_3$·7EtOH	[Yb(dppmO$_2$)$_3$Cl] Cl$_2$·5EtOH	[Lu(dppmO$_2$)$_3$Cl] Cl$_2$·3.5CH$_2$Cl$_2$·10H$_2$O	[Lu(dppmO$_2$)$_2$Cl$_2$] Cl·CH$_2$Cl$_2$·0.5H$_2$O	[Yb(dppmO$_2$)$_3$(H$_2$O)] Cl$_3$·dppmO$_2$·12H$_2$O
Formula	C$_{118}$H$_{142}$ CeCl$_3$O$_{17}$P$_8$	C$_{119}$H$_{145}$Cl$_3$O$_{17.5}$P$_8$Sm	C$_{114}$H$_{130}$Cl$_3$GdO$_{15}$P$_8$	C$_{87}$H$_{102}$Cl$_3$O$_{12}$P$_6$Yb$_1$	C$_{78.5}$H$_{93}$Cl$_{10}$LuO$_{16}$P$_6$	C$_{51}$H$_{47}$ Cl$_5$LuO$_{4.5}$P$_4$	C$_{100}$H$_{114}$Cl$_3$O$_{21}$P$_8$Yb
M	2326.54	2359.80	2251.53	1805.010	2007.81	1208.49	2179.06
Crystal system	monoclinic	monoclinic	monoclinic	monoclinic	orthorhombic	orthorhombic	orthorhombic
Space group (no.)	P2/c (13)	P2/c (13)	P2/c (13)	Pc (7)	Pcca (54)	Pbca (61)	Pbca (61)
a/Å	29.5926(4	29.7348(4)	29.2352(5)	14.1964(2)	47.7209(4)	21.1303(3)	26.1035(2)
b/Å	23.2600(2)	23.1120(2)	23.1885(3)	12.9572(2)	12.7431(1)	21.7424(5)	27.6790(2)
c/Å	18.0187(2)	17.9915(3)	17.7500(3)	24.0141(3)	28.3698(2)	22.1612(3)	29.1187(2)
α/°	90	90	90	90	90	90	90
β/°	107.4810(10)	106.988(2)	107.116(2)	95.880(1)	90	90	90
γ/°	90	90	90	90	90	90	90
U/Å^3	11829.9(2)	11824.8(3)	11500.1(3)	4394.05(11)	17252.0(2)	10181.4(3)	21038.8(3)
Z	4	4	4	2	8	8	8
μ(Mo-K$_\alpha$)/mm^{-1}	0.613	0.724	0.817	1.322	1.629	2.372	1.153
F(000)	4340	4348	4676	1864	8184	4844	8984
Total number reflns	183494	181022	169524	66370	22287	75026	423866
R_{int}	0.0372	0.0393	0.0561	0.0354	0.0642	0.0558	0.0323
Unique reflns	30589	30564	24123	21530	22287	13145	27176
No. of params, restraints	1253, 132	1261, 35	1143, 0	849, 65	937,264	621, 5	1240, 1
R$_1$,wR$_2$ [I > 2σ(I)] [b]	0.0396, 0.0831	0.0371, 0.0788	0.0526, 0.1302	0.0346, 0.0798	0.0906, 0.1916	0.0337, 0.0708	0.0277, 0.0755
R$_1$, wR$_2$ (all data)	0.0516, 0.0870	0.0532, 0.0846	0.0650,0.1361	0.0387, 0.0814	0.0935, 0.1926	0.0552, 0.0774	0.0331, 0.0783

[a] common data: T = 100 K; wavelength (Mo-K$_\alpha$) = 0.71073 Å; θ(max) = 27.5°; [b] R$_1$ = Σ||Fo| − |Fc||/Σ|Fo|; wR$_2$ = [Σw(Fo^2 − Fc^2)^2/ΣwFo^4]$^{1/2}$.

[Nd(dppmO$_2$)$_4$]Cl$_3$·4H$_2$O—To a solution of NdCl$_3$·6H$_2$O (0.025 g, 0.070 mmol) in ethanol (5 mL) was added a solution of dppmO$_2$ (0.116 g, 0.279 mmol) in ethanol (10 mL). A white powdered solid formed on slow evaporation of the ethanol. Required for C$_{100}$H$_{96}$Cl$_3$NdO$_{12}$P$_8$ (1988.2): C, 60.41; H, 4.87%. Found: C, 60.41; H, 4.62%. ^{1}H-NMR (CD$_2$Cl$_2$): δ = 1.52 (s, H$_2$O) 3.66 (m, [8H], PCH$_2$P), 7.14 (s, [32H], Ph), 7.35 (m, [16H], Ph), 7.76 (m, [32H], Ph). ^{31}P{^{1}H}-NMR (CD$_2$Cl$_2$): δ = 62.9 (s). IR (Nujol mull)/cm^{-1}: 3500 br, 1630 (H$_2$O), 1159 s, 1101 s (P=O).

[Sm(dppmO$_2$)$_4$]Cl$_3$·4H$_2$O—To a solution of SmCl$_3$·6H$_2$O (0.025 g, 0.069 mmol) in ethanol (5 mL) was added a solution of dppmO$_2$ (0.114 g, 0.274 mmol) in ethanol (10 mL). Colourless crystals were formed via slow evaporation of the ethanol. Required for C$_{100}$H$_{96}$Cl$_3$O$_{12}$P$_8$Sm (1994.3): C, 60.22; H, 4.85%. Found: C, 60.05; H, 4.50%. ^{1}H-NMR (CD$_2$Cl$_2$): δ = 2.10 (s, H$_2$O), 5.08 (br, [8H], PCH$_2$P), 7.20 (s, [32H], Ph), 7.39 (m, [16H], Ph), 7.83 (m, [32H], Ph). ^{31}P{^{1}H}-NMR (CD$_2$Cl$_2$): δ = 35.6. IR (Nujol mull)/cm^{-1}: 3500 br, 1630 (H$_2$O), 1162 s, 1101 (P=O).

[Eu(dppmO$_2$)$_4$]Cl$_3$·4H$_2$O—To a solution of EuCl$_3$·6H$_2$O (0.025 g, 0.068 mmol) in ethanol (5 mL) was added a solution of dppmO$_2$ (0.114 g, 0.274 mmol) in ethanol (10 mL) and the solution was stirred for 20 min. The solution was then concentrated, and colourless crystals were formed through layering with *n*-hexane (1 mL). Required for C$_{100}$H$_{96}$Cl$_3$EuO$_{12}$P$_8$ (1995.9): C, 60.41; H, 4.87%. Found: C, 60.73; H, 4.71%. ^{1}H-NMR (CD$_2$Cl$_2$): δ = 2.15 (s, H$_2$O) 3.12 (br, [8H] PCH$_2$P), 7.18 (s, [32H], Ph), 7.38 (m, [16H], Ph), 7.83 (m, [32H], Ph). ^{31}P{^{1}H}-NMR (CD$_2$C$_2$): δ = 25.0 (br, "free" dppmO$_2$), −13.4. IR (Nujol mull)/cm^{-1}: 3500 br, 1630 (H$_2$O), 1159, 1099 (P=O).

[Gd(dppmO$_2$)$_4$]Cl$_3$·4H$_2$O—To a solution of GdCl$_3$·6H$_2$O (0.025 g, 0.067 mmol) in ethanol (5 mL) was added a solution of dppmO$_2$ (0.112 g, 0.269 mmol) in ethanol (10 mL). Colourless crystals were formed through slow evaporation of the solvent. Required for C$_{100}$H$_{96}$Cl$_3$GdO$_{12}$P$_8$ (2001.2): C, 60.02; H, 4.83%. Found: C, 60.05; H, 4.86%. ^{1}H-NMR (CD$_2$Cl$_2$): δ = no resonance. ^{31}P{^{1}H}-NMR (CD$_2$Cl$_2$): δ = no resonance. IR (Nujol mull)/cm^{-1}: 3500 br, 1630 (H$_2$O), 1160, 1099 (P=O).

[Sm(dppmO$_2$)$_3$Cl]Cl$_2$—To a solution of SmCl$_3$·6H$_2$O (0.025 g, 0.069 mmol) in ethanol (5 mL) was added a solution of dppmO$_2$ (0.086 g, 0.206 mmol) in ethanol (10 mL). The solvent was removed in vacuo and the resulting white solid was washed with cold ethanol. Colourless crystals were obtained via slow evaporation of an ethanolic solution of the product. Required for C$_{75}$H$_{66}$Cl$_3$O$_6$P$_6$Sm (1505.9): C, 59.80; H, 4.42%. Found: C, 59.62; H, 4.55%. ^{1}H-NMR (CD$_2$Cl$_2$): δ = 3.67 (br m, [6H], PCH$_2$P), 7.15 (br, [24H], Ph), 7.35 (m, [12H], Ph), 8.05 (m, [24H], Ph). ^{31}P{^{1}H}-NMR (CD$_2$Cl$_2$): δ = 38.15 (s). IR (Nujol mull)/cm^{-1}: 1153 s, 1097 s (P=O).

[Eu(dppmO$_2$)$_3$Cl]Cl$_2$—To a solution of EuCl$_3$·6H$_2$O (0.025 g, 0.068 mmol) in ethanol (5 mL) was added a solution of dppmO$_2$ (0.085 g, 0.205 mmol) in ethanol (10 mL). The solvent was removed in vacuo and the resulting white solid was washed with cold ethanol. Required for C$_{75}$H$_{66}$EuCl$_3$O$_6$P$_6$ (1507.49): C, 59.76; H, 4.41%. Found: C, 59.71; H, 4.56%. ^{1}H-NMR (CDCl$_3$): δ = 3.66 (br, [6H], PCH$_2$P), 7.03 (br m, [36H], Ph), 7.87 (br, [24H], Ph). ^{31}P{^{1}H}-NMR (CDCl$_3$): δ = −14.8 (s). IR (Nujol mull)/cm^{-1}: 1153 s, 1098 s (P=O).

[Gd(dppmO$_2$)$_3$Cl]Cl$_2$·3H$_2$O—To a solution of GdCl$_3$·6H$_2$O (0.025 g, 0.067 mmol) in ethanol (5 mL) was added a solution of dppmO$_2$ (0.084 g, 0.201 mmol) in ethanol (10 mL). The solvent was removed in vacuo and the resulting white solid was washed with cold ethanol. Required for C$_{75}$H$_{66}$Cl$_3$O$_6$P$_6$Gd (166.8): C, 57.49, H, 4.63%; Found: C, 57.17; H, 4.43%. ^{1}H-NMR (CD$_2$Cl$_2$): no resonance. ^{31}P{^{1}H}-NMR (CD$_2$Cl$_2$): no resonance. IR (Nujol mull)/cm^{-1}: 3500 br, 1630 (H$_2$O), 1155 s, 1098 s (P=O).

[Tb(dppmO$_2$)$_3$Cl]Cl$_2$·H$_2$O—To a solution of TbCl$_3$·6H$_2$O (0.025 g, 0.067 mmol) in ethanol (5 mL) was added a solution of dppmO$_2$ (0.084 g, 0.201 mmol) in ethanol (10 cm^3). The solvent was removed in vacuo and the resulting white solid was washed with cold ethanol. Required for C$_{75}$H$_{68}$Cl$_3$O$_7$P$_6$Tb (1532.5): C, 58.78; H, 4.47%. Found: C, 59.41; H, 4.54%. ^{1}H-NMR (CD$_2$Cl$_2$): δ = 1.9 (br H$_2$O), 3.50 (br m, [6H], PCH$_2$P), 5.89 (br, [36H], Ph), 7.46 (br, [24H], Ph). ^{31}P{^{1}H}-NMR (CD$_2$Cl$_2$): δ = −29.2 (s). IR (Nujol mull)/cm^{-1}: 3500 br, 1630 (H$_2$O), 1153 s, 1097 s (P=O).

[Dy(dppmO$_2$)$_3$Cl]Cl$_2$·H$_2$O—To a solution of TbCl$_3$·6H$_2$O (0.025 g, 0.066 mmol) in ethanol (5 mL) was added a solution of dppmO$_2$ (0.083 g, 0.199 mmol) in ethanol (10 cm^3). The solution was filtered then concentrated and layered with hexane (1 mL) yielding a white powdered product. Colourless crystals were formed by layering a CH_2Cl_2 solution of the product with hexane. Required for $C_{75}H_{68}Cl_3DyO_7P_6$ (1536.0): C, 58.64; H, 4.46%. Found: C, 58.21; H, 4.63%. ^{1}H-NMR (CD_2Cl_2): δ = 1.9 (vbr H_2O), 3.66 (br m, [6H], PCH_2P), 7.33 (br, [36H], Ph), 8.66 (br, [24H], Ph). ^{31}P{^{1}H}-NMR (CD_2Cl_2): δ = 18 (vbr, s). IR (Nujol mull)/cm^{-1}: 3500 br, 1630 (H_2O), 1156 s, 1099 s (P=O).

[Ho(dppmO$_2$)$_3$Cl]Cl$_2$·H_2O—To a solution of $HoCl_3$ (0.050 g, 0.124 mmol) in ethanol (5 mL) was added a solution of dppmO$_2$ (0.230 g, 0.55 mmol) in ethanol (10 mL). The solvent was removed in vacuo and the resulting pale pink solid was washed with cold ethanol. Required for $C_{75}H_{68}Cl_3HoO_7P_6$ (1538.5): C, 58.66; H, 4.55%. Found: C, 59.41; H, 4.52%. ^{1}H-NMR (CD_2Cl_2): δ = 2.1 (br, H_2O), 3.72 (br s, [6H], PCH_2P), 6.78 (br, [36H], Ph), 7.68 (br, [24H], Ph)]. ^{31}P{^{1}H}-NMR (CD_2Cl_2): δ = −13.5 (s). IR (Nujol mull)/cm^{-1}: 3500 br, 1630 (H_2O), 1154 s, 1097 s (P=O).

[Er(dppmO$_2$)$_3$Cl]Cl$_2$·3H_2O—To a solution of $ErCl_3$·6H_2O (0.025 g, 0.065 mmol) in ethanol (5 mL) was added a solution of dppmO$_2$ (0.082 g, 0.196 mmol) in ethanol (10 mL). The solvent was removed in vacuo and the resulting white solid was washed with cold ethanol. Required for $C_{75}H_{72}Cl_3ErO_9P_6$ (1576.8): C, 57.13; H, 4.60%. Found: C, 57.08; H, 4.54%. ^{1}H-NMR (CD_2Cl_2): δ = δ = 1.2 (br, H_2O), 3.25 (br s, [6H], PCH_2P), 5.52 (vbr, [12H], Ph), 7.15 (br s, [24H], Ph)], 7.28 (br s, [24H], Ph)]. ^{31}P{^{1}H}-NMR (CD_2Cl_2): δ = −60.8 (s). IR (Nujol mull)/cm^{-1}: 3500 br, 1630 (H_2O), 1155 s, 1097 s (P=O).

[Tm(dppmO$_2$)$_3$Cl]Cl$_2$·3H_2O—To a solution of $TmCl_3$·6H_2O (0.025 g, 0.065 mmol) in ethanol (5 mL) was added a solution of dppmO$_2$ (0.081 g, 0.195 mmol) in ethanol (10 mL). The solvent was removed in vacuo and the resulting white solid was washed with cold ethanol. Required for $C_{75}H_{72}Cl_3O_9P_6Tm$ (1578.5): C, 57.07; H, 4.60%. Found: C, 56.61; H, 4.45%. ^{1}H-NMR (CD_2Cl_2): δ = 3.48 (m, [6H], PCH_2P), 7.11 (br, [24H], Ph), 7.68 (br, [36H], Ph). ^{31}P{^{1}H}-NMR (CD_2Cl_2): δ = −54.8 (s). IR (Nujol mull)/cm^{-1}: 3500 br, 1630 (H_2O), 1156 s, 1096 s (P=O).

[Yb(dppmO$_2$)$_3$Cl]Cl$_2$·H_2O—To a solution of $YbCl_3$·6H_2O (0.025 g, 0.065 mmol) in ethanol (5 mL) was added a solution of dppmO$_2$ (0.080 g, 0.194 mmol) in ethanol (10 mL). The solvent was removed in vacuo and the resulting white powder was washed with cold ethanol. Required for $C_{75}H_{68}Cl_3O_7P_6Yb$ (1546.58): C, 58.24; H, 4.43%. Found: C, 58.73; H, 4.45%. ^{1}H-NMR (CD_2Cl_2): δ = 3.50 (m, [6H], PCH_2P), 6.64 (br, [24H], Ph), 7.15 (br, [36H], Ph). ^{31}P{^{1}H}-NMR (CD_2Cl_2): δ = +9.2 (s). IR (Nujol mull)/cm^{-1}: 3500 br, 1630 (H_2O), 1154 s, 1097 s (P=O).

3. Results

The reaction of $LnCl_3$·nH_2O (Ln = La [8], Ce, Pr, Nd, Sm, Eu or Gd; n = 6 or 7) with four mol. equivalents of dppmO$_2$ in ethanol gave good yields of tetrakis-dppmO$_2$ complexes, [Ln(dppmO$_2$)$_4$]Cl$_3$. The IR and ^{1}H-NMR spectra show the the isolated complexes retain significant amounts of lattice water, and sometimes EtOH, which is not removed by prolonged drying of the bulk powders in vacuo. The high molecular weights make the microanalyses rather insensitive to the amount of water, but are generally consistent with a formulation [Ln(dppmO$_2$)$_4$]Cl$_3$·nH_2O (n = 6: Ce, Pr; n = 4: Nd, Sm, Eu, Gd), although the amount of lattice solvent probably varies with the sample and is unlikely to be stoichiometric. The presence of significant amounts of lattice solvent is common in lanthanide phosphine oxide systems [7–10], and although evident in X-ray crystal structures, it is often disordered and difficult to model. Obtaining good quality crystals of the complexes proved difficult, but crystals of the Ce, Sm and Gd salts were obtained from various organic solvents and the compositions are shown in Table 1. The crystals contain different amounts of solvent of crystallisation to the bulk samples as they were grown from different media (and crystals were not dried in vacuo). The IR spectra (Table 2) show that the υ(PO) stretch in dppmO$_2$ at 1187 cm^{-1} has been lost and replaced by a new very strong and broad band ~1160 cm^{-1} and a second band at ~ 1100 cm^{-1}, which are due to the coordinated phosphine oxide groups. The frequencies appear invariant with the lanthanide present, which may be due to small differences being obscured by the width of the bands. In [LnCl$_3$(OPPh$_3$)$_3$] and [LnCl$_2$(OPPh$_3$)$_4$]$^+$ the frequency of the υ(PO) stretch increases by ~ 10 cm^{-1} between La and Lu [7]. The ^{31}P{^{1}H}-NMR chemical shift of dppmO$_2$ at δ = +25.3 shows a high frequency shift to +33.1

in [La(dppmO$_2$)$_4$]Cl$_3$, whilst the corresponding spectra of the Ce, Pr, Nd and Sm complexes show larger shifts due to the presence of the paramagnetic lanthanide ion (Table 2). In contrast, although the solid [Eu(dppmO$_2$)$_4$]Cl$_3$ complex was isolated without difficulty, the ^{31}P{^{1}H}-NMR spectrum shows a strong feature at $\delta \sim$ +25 ("free" dppmO$_2$), along with a second resonance at δ = −13.4, which may be assigned to [Eu(dppmO$_2$)$_3$Cl]Cl$_2$ (see below), indicating substantial dissociation of one dppmO$_2$ in solution; the broad resonance of the free dppmO$_2$ is indicative of exchange on the NMR timescale. [Gd(dppmO$_2$)$_4$]Cl$_3$ was isolated, and its constitution confirmed by its X-ray crystal structure, but no ^{1}H or ^{31}P{^{1}H}-NMR resonances were observed, an effect seen in other gadolinium systems [6,7] and ascribed to fast relaxation by the f^7 configuration of the metal. Attempts to isolate [Ln(dppmO$_2$)$_4$]Cl$_3$ complexes for Ln = Dy-Lu were unsuccessful. We note that [Dy(dppmO$_2$)$_4$][CF$_3$SO$_3$]$_3$ [9] was isolated with triflate counter ions, but with chloride only [Dy(dppmO$_2$)$_3$Cl]Cl$_2$ was produced (below). An in situ ^{31}P{^{1}H}-NMR spectrum of CeCl$_3$·7H$_2$O + 2 dppmO$_2$ in CH$_2$Cl$_2$ showed a single resonance at δ = +48, which is consistent with formation of [Ce(dppmO$_2$)$_4$]$^{3+}$, confirming the preference for formation of the tetrakis complexes early in the series, even when there is a deficit of ligand.

Table 2. IR and ^{31}P{^{1}H}-NMR spectroscopic data.

Complex	$\delta(^{31}$P) ^{a}	υ(P=O) cm^{-1} ^{b}
dppmO$_2$	+25.3	1187
[La(dppmO$_2$)$_4$]Cl$_3$ ^{c}	+33.1	1159, 1100
[Ce(dppmO$_2$)$_4$]Cl$_3$	+48.6	1158, 1099
[Pr(dppmO$_2$)$_4$]Cl$_3$	+64.0	1161, 1102
[Nd(dppmO$_2$)$_4$]Cl$_3$	+62.9	1159, 1101
[Sm(dppmO$_2$)$_4$]Cl$_3$	+35.6	1162, 1101
[Eu(dppmO$_2$)$_4$]Cl$_3$	−13.4 (+25 dppmO$_2$)	1159, 1099
[Gd(dppmO$_2$)$_4$]Cl$_3$	Not observed	1160, 1099
[Sm(dppmO$_2$)$_3$Cl]Cl$_2$	+38.0	1153, 1097
[Eu(dppmO$_2$)$_3$Cl]Cl$_2$	−14.8	1153, 1098
[Gd(dppmO$_2$)$_3$Cl]Cl$_2$	Not observed	1155, 1099
[Tb(dppmO$_2$)$_3$Cl]Cl$_2$	−29.2	1153, 1097
[Dy(dppmO$_2$)$_3$Cl]Cl$_2$	+18.0	1156, 1099
[Ho(dppmO$_2$)$_3$Cl]Cl$_2$	−13.5	1154, 1095
[Er(dppmO$_2$)$_3$Cl]Cl$_2$	−60.75	1155, 1097
[Tm(dppmO$_2$)$_3$Cl]Cl$_2$	−54.8	1156, 1096
[Yb(dppmO$_2$)$_3$Cl]Cl$_2$	+9.2	1154, 1097
[Lu(dppmO$_2$)$_2$Cl$_2$]Cl ^{c}	+40.0	1158, 1098
[Lu(dppmO$_2$)$_3$Cl]Cl$_2$	+38.3	

^{a} In CD$_2$Cl$_2$ solution 298 K; ^{b} Nujol mull; ^{c} Ref. [8].

The X-ray structures of [Ce(dppmO$_2$)$_4$]Cl$_3$ (Figure 1), [Sm(dppmO$_2$)$_4$]Cl$_3$ (Figure 2) and [Gd(dppmO$_2$)$_4$]Cl$_3$ (Figure 3) show distorted square antiprismatic cations, very similar to those in [La(dppmO$_2$)$_4$][PF$_6$]$_3$ [8] and [Nd(dppmO$_2$)$_4$]Cl$_3$ [15]. The average Ln-O distances in this series are: La = 2.514 Å, Ce = 2.486 Å, Nd = 2.465 Å, Sm = 2.429 Å and Gd = 2.420 Å, correlating well with the decreasing Ln^{3+} radii (La = 1.216 Å, Ce =1.196 Å, Nd = 1.163 Å, Sm = 1.132 Å, Gd = 1.107 Å). The P = O bond lengths and the O-Ln-O chelate angles do not vary significantly along the series. The Ce-O(P) distances in [Ce(dppmO$_2$)$_4$]Cl$_3$ are markedly longer than those in [Ce(Me$_3$PO)$_4$(H$_2$O)$_4$]Cl$_3$ (2.372(2)-2.423(2) Å) [16], which has a distorted dodecahedral geometry with a CeO$_8$ donor set.

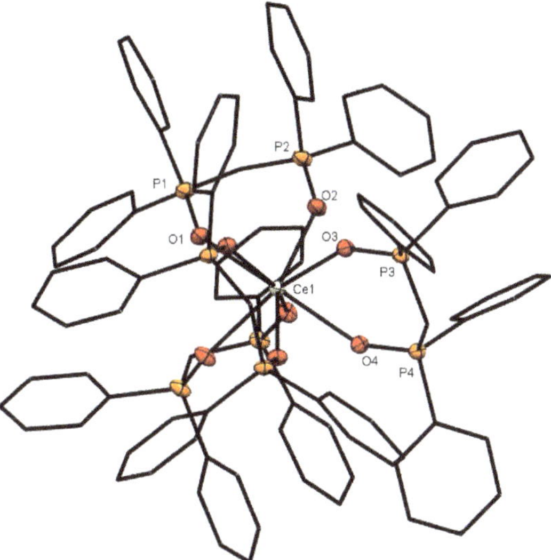

Figure 1. The cation in [Ce(dppmO$_2$)$_4$]Cl$_3$. The chloride anions and solvate molecules are omitted. Selected bond lengths (Å): Ce1–O1 = 2.4874(14), Ce1–O2 = 2.4790(14), Ce1–O3 = 2.4967(13), Ce1–O4 = 2.4803(14), P1–O1 = 1.5031(14), P2–O2 = 1.5021(14), P3–O3 = 1.5018(14), P4–O4 = 1.5031(14). Chelate angle O-Ce-O = 73.1° (av).

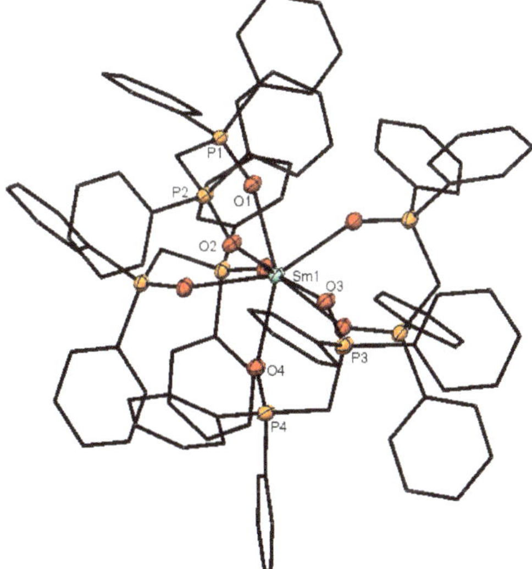

Figure 2. The cation in [Sm(dppmO$_2$)$_4$]Cl$_3$. The chloride anions and solvate molecules are omitted. Selected bond lengths (Å): Sm1–O1 = 2.4160(14), Sm1–O2 = 2.4400(15), Sm1–O3 = 2.4358(14), Sm1–O4 = 2.4268(15), P1–O1 = 1.5025(15), P2–O2 = 1.5019(15), P3–O3 = 1.4961(15), P4–O4 = 1.4961(16). Chelate angle O-Sm-O = 72.9° (av).

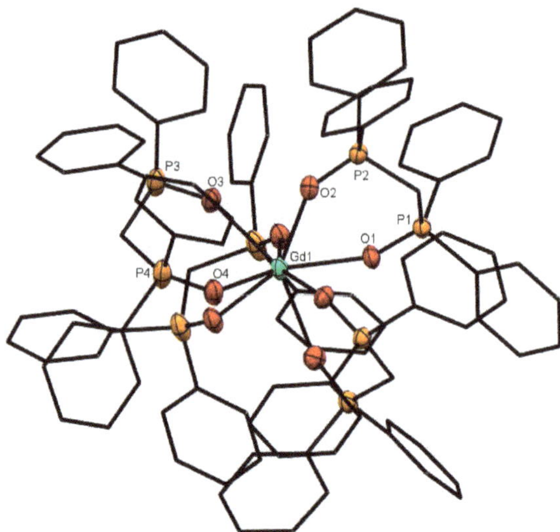

Figure 3. The cation in [Gd(dppmO$_2$)$_4$]Cl$_3$. The chloride anions and solvate molecules are omitted. Selected bond lengths (Å): Gd1–O1 = 2.420(2), Gd1–O2 = 2.409(3), Gd1–O3 = 2.415(2), Gd1–O4 = 2.398(2), P1–O1 = 1.504(2), P2–O2 = 1.501(3), P3–O3 = 1.501(3), P4–O4 = 1.501(3). Chelate angle O-Sm-O = 73.1° (av).

The reaction of LnCl$_3$·6H$_2$O (Ln = Sm, Eu, Gd, Tb, Dy, Ho, Er, Tm, Yb) with 3 mol. equivalents of dppmO$_2$ in EtOH, followed by concentration of the solution or precipitation with hexane, afforded [Ln(dppmO$_2$)$_3$Cl]Cl$_2$ complexes. Examination of the IR and ^{1}H-NMR spectra indicated these incorporated less water or ethanol lattice solvent molecules than the [Ln(dppmO$_2$)$_4$]Cl$_3$, and this was confirmed by the microanalyses. The Sm and Eu complexes appear largely free of solvent of crystallisation, whilst the Tb, Ho and Yb approximate to [Ln(dppmO$_2$)$_3$Cl]Cl$_2$·H$_2$O, and the Gd, Er and Tm complexes are [Ln(dppmO$_2$)$_3$Cl]Cl$_2$·3H$_2$O; again, this is likely to vary from sample to sample and with the isolation method. The IR spectra (Table 2) show the two υ(PO) bands as in the tetrakis complexes, but the higher energy bands of the tris complexes are ~ 5–10 cm^{-1} lower in frequency than in the former. We were unable to identify υ(Ln-Cl) vibrations in the far IR spectra. The ^{31}P{^{1}H}-NMR spectra of the [Ln(dppmO$_2$)$_3$Cl]Cl$_2$ show single resonances to high or low frequency of dppmO$_2$ depending on the f^n configuration of the Ln ion present (Table 2) and are generally similar to those found in other systems [5–7], although the magnitude of the shifts varies widely with the specific f^n configuration. The line broadening is also highly variable between complexes of different Ln ions. The addition of dppmO$_2$ to a solution of [Ln(dppmO$_2$)$_3$Cl]Cl$_2$ (Ln = Eu, Gd, Tb, Dy, Ho, Er, Tm, Yb) in CH$_2$Cl$_2$ showed ^{31}P{^{1}H}-NMR resonances assignable to "free" dppmO$_2$ and [Ln(dppmO$_2$)$_3$Cl]Cl$_2$, but no new resonances that could be attributed to the formation of significant amounts of [Ln(dppmO$_2$)$_4$]$^{3+}$. Although the resonances are broad in some cases, the observed chemical shifts are identical to those in pure [Ln(dppmO$_2$)$_3$Cl]Cl$_2$. For [Sm(dppmO$_2$)$_3$Cl]Cl$_2$ δ(^{31}P{^{1}H}) = 38, the resonance shifts to δ = 35.6 upon addition of dppmO$_2$, attributable to the formation of [Sm(dppmO$_2$)$_4$]Cl$_3$, showing that both tris- and tetrakis-dppmO$_2$ complexes exist in solution for samarium in the presence of the appropriate amount of ligand.

The X-ray structures of [Er(dppmO$_2$)$_3$Cl]Cl$_2$ (Er-O = 2.28 Å av.) [17], [Yb(dppmO$_2$)$_3$Cl]Cl$_2$ (Figure 4; Yb-O = 2.28 Å av.) and [Dy(dppmO$_2$)$_3$Cl]Cl$_2$ (Figure S43) show pentagonal bipyramidal cations with an apical chloride. The Ln-O distances are rather variable (Er-O = 2.244(6)–2.328(6) Å; Yb-O = 2.250(2)–2.269(3) Å), but are shorter than those in the tetrakis-dppmO$_2$ cations, reflecting both the reduced coordination number and the smaller metal ion radii (Er = 1.062, Yb = 1.042 Å). The contraction in ionic radii is also evident in the Ln-Cl distances of 2.598(2) Å (Er) and 2.5829(9) Å (Yb). Crystals of [Dy(dppmO$_2$)$_3$Cl]Cl$_2$ were also obtained and show the same cation type, but during refinement, several of the phenyl rings exhibited severe disorder and the data are therefore not included here (Figure S43).

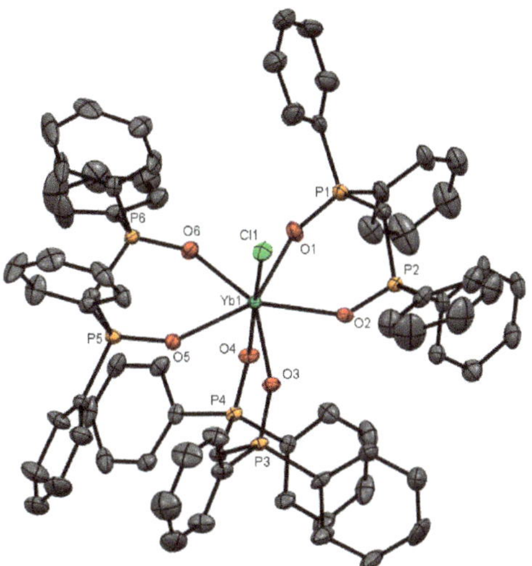

Figure 4. The X-ray structure of [Yb(dppmO$_2$)$_3$Cl]Cl$_2$. The chloride anions and solvate molecules are omitted. Selected bond lengths (Å) and angles (°): Yb1–Cl1 = 2.5834(9), Yb1–O1 = 2.298(3), Yb1–O2 = 2.282(3), Yb1–O3 = 2.250(2), Yb1–O4 = 2.248(2), Yb1–O5 = 2.338(2), Yb1–O6 = 2.269(3), P–O = 1.494(3)-1.509(3), Cl1–Yb1–O4 = 173.89(9), O1–Yb1–O2 = 73.79(9), O3–Yb1–O4 = 80.64(9), O5–Yb1–O6 = 73.95(9).

Lutetium was previously reported to form the only bis-dppmO$_2$ complex, [Lu(dppmO$_2$)$_2$Cl$_2$]Cl, in this series [8], and this has now been confirmed by the X-ray crystal structure which shows a *cis*-octahedral geometry (Figure 5). The Lu-O distance of 2.230 Å (av) is shorter than the Ln-O distances in the seven- or eigth-coordinate complexes, and correlates both with the reduced coordination number and the smaller radius of Lu^{3+} (1.032 Å). Treatment of a CH$_2$Cl$_2$ solution of [Lu(dppmO$_2$)$_2$Cl$_2$]Cl with dppmO$_2$ caused the ^{31}P{^{1}H}-NMR resonance to shift from +40 to +38.3, which suggests that [Lu(dppmO$_2$)$_3$Cl]Cl$_2$ forms in solution. A few crystals of this product were isolated from a mixture containing excess dppmO$_2$. These showed a pentagonal bipyramidal dication (Figure 6). As expected, the Lu-Cl and Lu-O bond lengths are slightly longer than in the six-coordinate cation, but are shorter than the corresponding bonds in [Yb(dppmO$_2$)$_3$Cl]Cl$_2$, showing that the expected contraction continues along the series. The complex, [Lu(dppmO$_2$)$_3$(H$_2$O)][CF$_3$SO$_3$]$_3$, is known and its X-ray crystal structure showed seven-coordinate lutetium [11]. Although not confirmed by an X-ray structure, yttrium is reported to form a six-coordinate complex, [Y(dppmO$_2$)$_2$Cl$_2$]Cl [18].

A different crystal isolated from the YbCl$_3$-dppmO$_2$ reaction proved, on structure solution, to be [Yb(dppmO$_2$)$_3$(H$_2$O)]Cl$_3$·dppmO$_2$·12H$_2$O (Figure 7), which contains a seven-coordinate Yb centre coordinated to three dppmO$_2$ and a water molecule, with the Lu-coordinated water hydrogen-bonded to an adjacent uncoordinated dppmO$_2$ molecule. The geometry is best described as a very distorted pentagonal bipyramid with the water occupying an equatorial position and is similar to the geometry found in [Lu(dppmO$_2$)$_3$(H$_2$O)][CF$_3$SO$_3$]$_3$ [11]. The Yb-OH$_2$ distance of 2.3263(14) Å is ~ 0.05 Å longer than the Yb-O(P).

A large number of disordered solvate water molecules were also present, which proved very difficult to model, but the geometry of the ytterbium cation is clearly defined.

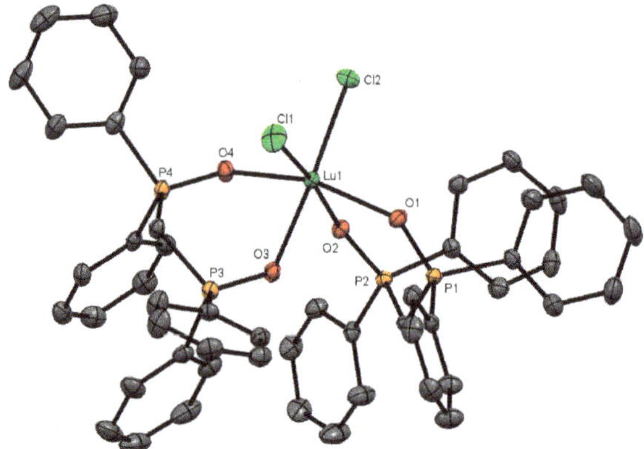

Figure 5. The cation in [Lu(dppmO$_2$)$_2$Cl$_2$]Cl. Selected bond lengths (Å) and angles (°): Lu1–Cl1 = 2.5581(8), Lu1–Cl2 = 2.5163(7), Lu1–O1 = 2.227(2), Lu1–O2 = 2.227(2), Lu1–O3 = 2.274(2), Lu1–O4 = 2.200(2), P1–O1 = 1.510(2), P2–O2 = 1.506(2), P3–O3 = 1.513(2), P4–O4 = 1.507(2), Cl2–Lu1–Cl1 = 95.97(3), O1–Lu1–Cl1 = 97.27(6), O1–Lu1–Cl2 = 99.68(6), O1–Lu1–O2 = 81.56(8), O1–Lu1–O3 = 85.65(7), O2–Lu1–Cl2 = 91.94(5), O2–Lu1–O3 = 84.89(7), O3–Lu1–Cl1 = 87.22(6), O4–Lu1–Cl1 = 94.66(6), O4–Lu1–Cl2 = 92.55(6), O4–Lu1–O2 = 84.74(8), O4–Lu1–O3 = 81.36(7).

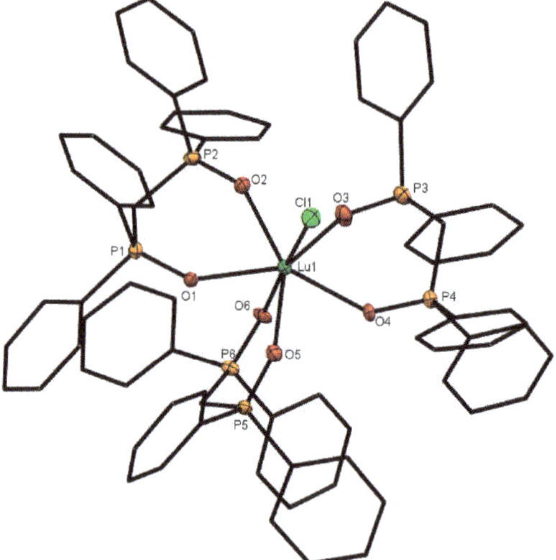

Figure 6. The X-ray structure of [Lu(dppmO$_2$)$_3$Cl]Cl$_2$. The chloride anions and solvate molecules are omitted. Selected bond lengths (Å) and angles (°): Lu1–Cl1 = 2.5604(7), Lu1–O1 = 2.341(5), Lu1–O2 = 2.268(5), Lu1–O3 = 2.268(5), Lu1–O4 = 2.297(5), Lu1–O5 = 2.354(5), Lu1–O6 = 2.227(5), P–O = 1.497(5)-1.510(5), Cl1–Lu1–O6 = 176.35(14), O1–Lu1–O2 = 73.37(17), O3–Lu1–O4 = 73.37(17), O5–Lu1–O6 = 83.45(18).

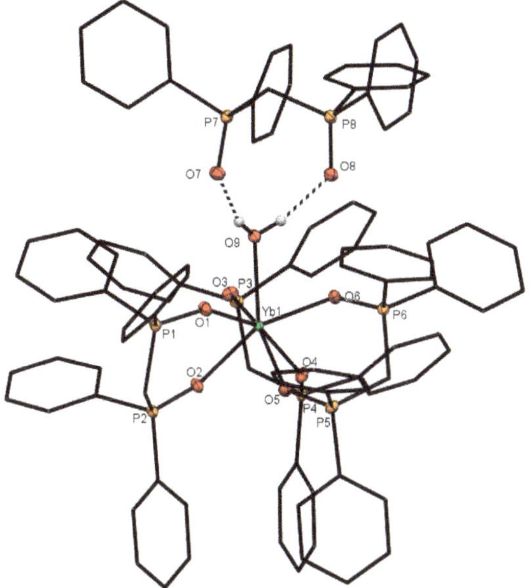

Figure 7. The cation in [Yb(dppmO$_2$)$_3$(H$_2$O)]Cl$_3$·dppmO$_2$·12H$_2$O also showing the hydrogen-bonded dppmO$_2$ molecule. Selected bond lengths (Å): Yb1–O3 = 2.2341(14), Yb1–O2 = 2.2899(13), Yb1–O9 = 2.3263(14), Yb1–O4 = 2.2683(13), Yb1–O6 = 2.2208(13), Yb1–O1 = 2.2328(13), Yb1–O5 = 2.2696(14), P$_n$–O$_n$ (n = 1–6) = 1.5034(14)–1.5072(14), P7–O7 = 1.4924(15), P8–O8 = 1.4926(15).

4. Discussion

The chemistry of dppmO$_2$ with lanthanides described in the previous section proves to be very systematic along the series La–Lu. For La–Gd, it was possible to isolate [Ln(dppmO$_2$)$_4$]Cl$_3$. Although it could be isolated in the solid state, the solution ^{31}P-NMR spectroscopic data indicate that [Eu(dppmO$_2$)$_4$]Cl$_3$ was largely dissociated in CH$_2$Cl$_2$ solution into [Eu(dppmO$_2$)$_3$Cl]$^{2+}$ and dppmO$_2$; the isolation of the tetrakis-dppmO$_2$ complex no doubt resulting from it being the least soluble species in an exchanging mixture in solution, although present in very minor amounts. The case of [Gd(dppmO$_2$)$_4$]Cl$_3$ is likely to be similar, although the fast relaxation of the f^7 ion precluded ^{31}P-NMR study. For the elements Sm-Yb, the complexes [Ln(dppmO$_2$)$_3$Cl]Cl$_2$ were readily isolated, but only for samarium was it possible to convert [Ln(dppmO$_2$)$_3$Cl]Cl$_2$ to [Ln(dppmO$_2$)$_4$]Cl$_3$ in CH$_2$Cl$_2$ solution by treatment with more dppmO$_2$. Similarly, at the end of the series, the complex isolated was [Lu(dppmO$_2$)$_2$Cl$_2$]Cl, for which treatment with dppmO$_2$ afforded a new species in solution, identified as [Lu(dppmO$_2$)$_3$Cl]Cl$_2$ by a structure determination from a few crystals obtained in the presence of excess dppmO$_2$, although a bulk sample could not be isolated [8]. The change from eight-coordination in [Ln(dppmO$_2$)$_4$]Cl$_3$ at the beginning of the series, to seven-coordination from Sm onwards, and finally to six-coordination at Lu, parallels the reduction in Ln^{3+} radii. Isolation of both the eight- and seven-coordinate complexes was possible only for Sm, Eu and Gd. However, one should note that the chloride counter ions also have some role, in that whilst in the LnCl$_3$/dppmO$_2$ series tetrakis-dppmO$_2$ species did not form beyond Gd, the complex [Dy(dppmO$_2$)$_4$][CF$_3$SO$_3$]$_3$ [9] has been isolated from dmf solution with triflate counter ions. The role that anions and solvents play in lanthanide chemistry is often overlooked [2], but can be critical in determining which complex is isolated from solution. For example, the reaction of LnCl$_3$ with Ph$_3$PO results in isolation of [Ln(Ph$_3$PO)$_3$Cl$_3$] from acetone, but [Ln(Ph$_3$PO)$_4$Cl$_2$]Cl from ethanol [7]. On further examination by ^{31}P-NMR spectroscopy, both species were found to be present in either solvent (in varying amounts), and the form isolated reflected the least soluble complex in the particular solvent, which then precipitated from the mixture of rapidly interconverting species.

5. Conclusions

Through this synthetic, structural and spectroscopic study of the coordination of dppmO$_2$ to the lanthanide trichlorides, we have established where the switch from eight-, to seven-, to six-coordination at the Ln(III) centre occurs along the lanthanide series, with X-ray crystallographic authentication for representative examples. The data also reveal subtle, but systematic, variations in the spectroscopic (e.g., ν(PO)) and structural parameters across the series, reflecting the change in ionic radii, the charge:radius ratio and also the influence of the presence of the competitive chloride ions.

Supplementary Materials: The following are available online at http://www.mdpi.com/2624-8549/2/4/60/s1; Figure S1–^{1}H-NMR spectrum of [Ce(dppmO$_2$)$_4$]Cl$_3$ in CD$_2$Cl$_2$; Figure S2–^{31}P{^{1}H} spectrum of Ce(dppmO$_2$)$_4$]Cl$_3$ in CD$_2$Cl$_2$; Figure S3–Infrared spectrum of [Ce(dppmO$_2$)$_4$]Cl$_3$ (Nujol mull); Figure S4–^{1}H-NMR spectrum of [Pr(dppmO$_2$)$_4$]Cl$_3$ in CD$_2$Cl$_2$; Figure S5–^{31}P{^{1}H}-NMR spectrum of [Pr(dppmO$_2$)$_4$]Cl$_3$ in CD$_2$Cl$_2$; Figure S6–Infrared spectrum of [Pr(dppmO$_2$)$_4$]Cl$_3$ (Nujol mull); Figure S7—^{1}H-NMR spectrum of [Nd(dppmO$_2$)$_4$]Cl$_3$ in CD$_2$Cl$_2$; Figure S8—^{31}P{^{1}H}-NMR spectrum of [Nd(dppmO$_2$)$_4$]Cl$_3$ in CD$_2$Cl$_2$; Figure S9—Infrared spectrum of [Nd(dppmO$_2$)$_4$]Cl$_3$ (Nujol mull); Figure S10—^{1}H-NMR spectrum of [Sm(dppmO$_2$)$_4$]Cl$_3$ in CD$_2$Cl$_2$; Figure S11—^{31}P{^{1}H}-NMR spectrum of [Sm(dppmO$_2$)$_4$]Cl$_3$ in CD$_2$Cl$_2$; Figure S12—Infrared spectrum of [Sm(dppmO$_2$)$_4$]Cl$_3$ (Nujol mull); Figure S13—^{1}H-NMR spectrum of [Eu(dppmO$_2$)$_4$]Cl$_3$ in CD$_2$Cl$_2$; Figure S14—^{31}P{^{1}H}-NMR spectrum of [Eu(dppmO$_2$)$_4$]Cl$_3$ in CD$_2$Cl$_2$; Figure S15—Infrared spectrum of [Eu(dppmO$_2$)$_4$]Cl$_3$ (Nujol mull); Figure S16—Infrared spectrum of [Gd(dppmO$_2$)$_4$]Cl$_3$ (Nujol mull); Figure S17—^{1}H-NMR spectrum of [SmCl(dppmO$_2$)$_3$]Cl$_2$ in CD$_2$Cl$_2$ (* = EtOH); Figure S18—^{31}P{^{1}H}-NMR spectrum of [SmCl(dppmO$_2$)$_3$]Cl$_2$ in CD$_2$Cl$_2$; Figure S19—Infrared spectrum of [SmCl(dppmO$_2$)$_3$]Cl$_2$ (Nujol mull); Figure S20—^{1}H-NMR spectrum of [EuCl(dppmO$_2$)$_3$]Cl$_2$ in CDCl$_3$; Figure S21—^{31}P{^{1}H}-NMR spectrum of [EuCl(dppmO$_2$)$_3$]Cl$_2$ in CDCl$_3$; Figure S22—^{31}P{^{1}H}-NMR spectrum of [EuCl(dppmO$_2$)$_3$]Cl$_2$ + excess dppmO$_2$ in CDCl$_3$; Figure S23—Infrared spectrum of [EuCl(dppmO$_2$)$_3$]Cl$_2$ (Nujol mull); Figure S24—Infrared spectrum of [GdCl(dppmO$_2$)$_3$]Cl$_2$ (Nujol mull); Figure S25—^{1}H-NMR spectrum of [TbCl(dppmO$_2$)$_3$]Cl$_2$ in CD$_2$Cl$_2$ (* = EtOH); Figure S26—^{31}P{^{1}H}-NMR spectrum of [TbCl(dppmO$_2$)$_3$]Cl$_2$ in CD$_2$Cl$_2$; Figure S27—Infrared spectrum of [TbCl(dppmO$_2$)$_3$]Cl$_2$ (Nujol mull); Figure S28—^{1}H-NMR spectrum of [DyCl(dppmO$_2$)$_3$]Cl$_2$ in CD$_2$Cl$_2$; Figure S29—^{31}P{^{1}H}-NMR spectrum of [DyCl(dppmO$_2$)$_3$]Cl$_2$ in CD$_2$Cl$_2$; Figure S30—Infrared spectrum of [DyCl(dppmO$_2$)$_3$]Cl$_2$ (Nujol mull); Figure S31—^{1}H-NMR spectrum of [HoCl(dppmO$_2$)$_3$]Cl$_2$ in CD$_2$Cl$_2$ (* = EtOH); Figure S32—^{31}P{^{1}H}-NMR spectrum of [HoCl(dppmO$_2$)$_3$]Cl$_2$ in CD$_2$Cl$_2$; Figure S33—Infrared spectrum of [HoCl(dppmO$_2$)$_3$]Cl$_2$ (Nujol mull); Figure S34—^{1}H-NMR spectrum of [ErCl(dppmO$_2$)$_3$]Cl$_2$ in CD$_2$Cl$_2$ (* = EtOH); Figure S35—^{31}P{^{1}H}-NMR spectrum of [ErCl(dppmO$_2$)$_3$]Cl$_2$ in CD$_2$Cl$_2$; Figure S36—Infrared spectrum of [ErCl(dppmO$_2$)$_3$]Cl$_2$ (Nujol mull); Figure S37—^{1}H-NMR spectrum of [TmCl(dppmO$_2$)$_3$]Cl$_2$ in CD$_2$Cl$_2$ (* = EtOH); Figure S38—^{31}P{^{1}H}-NMR spectrum of [TmCl(dppmO$_2$)$_3$]Cl$_2$ in CD$_2$Cl$_2$; Figure S39—Infrared spectrum of [TmCl(dppmO$_2$)$_3$]Cl$_2$ (Nujol mull); Figure S40—^{1}H-NMR spectrum of [YbCl(dppmO$_2$)$_3$]Cl$_2$ in CD$_2$Cl$_2$ (* = EtOH); Figure S41—^{31}P{^{1}H}-NMR spectrum of [YbCl(dppmO$_2$)$_3$]Cl$_2$ in CD$_2$Cl$_2$; Figure S42—Infrared spectrum of [YbCl(dppmO$_2$)$_3$]Cl$_2$ (Nujol mull); Figure S43—The cation in [DyCl(dppmO$_2$)$_3$]Cl$_2$. The chloride anions and solvate molecules are omitted.

Author Contributions: Conceptualization, R.D.B., W.L. and G.R.; formal analysis, R.D.B.; investigation, R.D.B.; data curation, R.D.B.; writing—original draft preparation, R.D.B. and W.L.; writing—review and editing, R.D.B., W.L. and G.R.; supervision, W.L. and G.R. All authors have read and agreed to the published version of the manuscript.

Funding: This research received no external funding.

Conflicts of Interest: The authors declare no conflict of interest.

References

1. Cotton, S.A. *Lanthanide and Actinide Chemistry*; John Wiley: Chichester, UK, 2006.
2. Cotton, S.A.; Raithby, P.R. Systematics and surprises in lanthanide coordination chemistry. *Coord. Chem. Rev.* **2017**, *340*, 220–231. [CrossRef]
3. Platt, A.W.G. Lanthanide phosphine oxide complexes. *Coord. Chem. Rev.* **2017**, *340*, 62–78. [CrossRef]
4. Lees, A.M.J.; Platt, A.W.G. Complexes of lanthanide nitrates with bis(diphenylphosphino)methane dioxide. *Inorg. Chem.* **2003**, *42*, 4673–4679. [CrossRef] [PubMed]

5. Levason, W.; Newman, E.H.; Webster, M. Tetrakis(triphenylphosphine oxide) complexes of the lanthanide nitrates; synthesis, characterization and crystal structures of [La(Ph$_3$PO)$_4$(NO$_3$)$_3$] and [Lu(Ph$_3$PO)$_4$(NO$_3$)$_2$]NO$_3$. *Polyhedron* **2000**, *19*, 2697–2705. [CrossRef]

6. Hill, N.J.; Leung, L.-S.; Levason, W.; Webster, M. Homoleptic octahedral hexakis(trimethylphosphine oxide)lanthanide hexafluorophosphates, [Ln(Me$_3$PO)$_6$][PF$_6$]$_3$; synthesis, structures and properties. *Inorg. Chim. Acta* **2003**, *343*, 169–174. [CrossRef]

7. Glazier, M.J.; Levason, W.; Matthews, M.L.; Thornton, P.L.; Webster, M. Synthesis, properties and solution speciation of lanthanide chloride complexes of triphenylphosphine oxide. *Inorg. Chim. Acta* **2004**, *357*, 1083–1091. [CrossRef]

8. Bannister, R.D.; Levason, W.; Reid, G. Diphosphine dioxide complexes of lanthanum and lutetium—The effects of ligand architecture and counter-anion. *Polyhedron* **2017**, *133*, 264–269. [CrossRef]

9. Jin, Q.-H.; Wu, J.-Q.; Zhang, Y.-Y.; Zhang, C.-L. Crystal structure of tetra(bis(diphenylphosphino)methane dioxide-*O,O*]dysprosium(III) tri(trifluoromethanesulfonate)-N,N-dimethylformamide (1:2) [Dy(C$_{25}$H$_{22}$P$_2$O$_2$)$_4$][CF$_3$SO$_3$]$_3$·2C$_3$H$_7$NO. *Z. Krystallogr. NCS* **2009**, *224*, 428–432. [CrossRef]

10. Huang, L.; Ma, B.-Q.; Huang, C.-H.; Makk, T.C.W.; Yao, G.-Q.; Xu, G.-X. Synthesis, characterisayion and crystal structure of a complex of europium perchlorate with methylenebis(diphenylphosphine oxide). *J. Coord. Chem.* **2001**, *54*, 95–103. [CrossRef]

11. Fawcett, J.; Platt, A.W.G. Structures and catalytic properties of complexes of bis(diphenylphosphino)methane dioxide with scandium and lanthanide trifluoromethanesulfonates. *Polyhedron* **2003**, *22*, 967–973. [CrossRef]

12. Levason, W.; Patel, R.; Reid, G. Catalytic air oxidation of tertiary phosphines in the presence of tin(IV) iodide. *J. Organomet. Chem.* **2003**, *688*, 280–282. [CrossRef]

13. Sheldrick, G.M. A short history of SHELX. *Acta Crystallog. Sect. A* **2008**, *64*, 112–122. [CrossRef] [PubMed]

14. Sheldrick, G.M. Crystal structure refinement with SHELX. *Acta Crystallog. Sect. C* **2015**, *71*, 3–8. [CrossRef] [PubMed]

15. Grachova, E.V.; Linti, G.; Vologzhanina, A.V. *Personal communication CCDC 860683: Experimental Crystal Structure Determination*; The Cambridge Crystallographic Data Centre (CCDC): Cambridge, UK, 2016. [CrossRef]

16. Hill, N.J.; Leung, L.-S.; Levason, W.; Webster, M. Tetraaquatetrakis(trimethylphosphineoxide-κ-O)cerium(III) trichloride trihydrate. *Acta Crystallog. Sect. C* **2002**, *58*, m295. [CrossRef] [PubMed]

17. Grachova, E.V.; Linti, G.; Vologzhanina, A.V. *Personal communication CCDC 860682: Experimental Crystal Structure Determination*; The Cambridge Crystallographic Data Centre (CCDC): Cambridge, UK, 2016. [CrossRef]

18. Hill, N.J.; Levason, W.; Popham, M.C.; Reid, G.; Webster, M. Yttrium halide complexes of phosphine- and arsine-oxides: Synthesis, multinuclear NMR and structural studies. *Polyhedron* **2002**, *21*, 445–455. [CrossRef]

Publisher's Note: MDPI stays neutral with regard to jurisdictional claims in published maps and institutional affiliations.

Article

Chloropentaphenyldisiloxane—Model Study on Intermolecular Interactions in the Crystal Structure of a Monofunctionalized Disiloxane [†]

Jonathan O. Bauer * and Tobias Götz

Institut für Anorganische Chemie, Fakultät für Chemie und Pharmazie, Universität Regensburg, Universitätsstraße 31, D-93053 Regensburg, Germany; Tobias.Goetz@chemie.uni-regensburg.de
* Correspondence: Jonathan.Bauer@chemie.uni-regensburg.de
† Dedicated to Dr. Howard Flack (1943–2017).

Abstract: Small functional siloxane units have gained great interest as molecular model systems for mimicking more complex silicate structures both in nature and in materials chemistry. The crystal structure of chloropentaphenyldisiloxane, which was synthesized for the first time, was elucidated by single-crystal X-ray diffraction analysis. The molecular crystal packing was studied in detail using state-of-the-art Hirshfeld surface analysis together with a two-dimensional fingerprint mapping of the intermolecular interactions. It was found that the phenyl C–H bonds act as donors for both weak C–H···π and C–H···Cl hydrogen bond interactions. The influence of intramolecular Si–O–Si bond parameters on the acceptor capability of functional groups in intermolecular hydrogen bond interactions is discussed.

Keywords: disiloxanes; intermolecular interactions; Hirshfeld surface analysis; molecular models; silicon; X-ray crystallography

Citation: Bauer, J.O.; Götz, T. Chloropentaphenyldisiloxane— Model Study on Intermolecular Interactions in the Crystal Structure of a Monofunctionalized Disiloxane . *Chemistry* **2021**, *3*, 444–453. https://doi.org/10.3390/chemistry3020033

Academic Editor: Catherine Housecroft

Received: 14 March 2021
Accepted: 25 March 2021
Published: 29 March 2021

Publisher's Note: MDPI stays neutral with regard to jurisdictional claims in published maps and institutional affiliations.

1. Introduction

Siloxanes are known to be quite resistant towards thermal and chemical decomposition [1]. Their structural motif, the Si–O–Si bond, therefore, not only forms the basis for silicate minerals in nature, which are built from both geological [2] and biosilicification processes [3], but also the backbone for technologically important organic–inorganic hybrid polymers (silicones) [4] and for new synthetic silicate materials [5,6]. Studies on small and defined molecular siloxane models can provide very useful information on structure and reactivity of more complex siloxane-based materials and surfaces [7–17]. Silica-based biomimetic model systems [18] have also gained much interest in order to understand natural coral shapes [19] and shell formation of unicellular organisms such as diatoms [20]. We recently reported on monofunctionalized disiloxane units that served as simplified molecular model systems for investigating the reactivity and chemoselectivity in targeted further transformations [15].

The identification of weak intermolecular interactions in molecular crystals is an interesting undertaking with the aim of gaining knowledge about structure-forming forces and making it usable for the targeted formation of functional crystalline networks [21,22]. Siloxanes are of particular interest, since a large number of three-dimensional architectures can be formed through Si–O bond formation [23]. The assembly of several siloxane units to form complex framework structures therefore requires a more detailed study of intermolecular interactions.

As part of our studies on weak intermolecular interactions in molecular crystals [24–26], we were now interested in taking a closer look at monofunctionalized disiloxanes with regard to their crystal packing and intermolecular contacts. Disiloxanes as the smallest units of oligo- and polysiloxanes are a fascinating class of substances, ideal for model studies [27]. Unymmetrically substituted crystalline disiloxanes with only one heterofunction are extremely rare [15,28–33], but they shed light on the influence of single substituents on the Si–O–Si unit [15]. Chlorosilanes in general are important precursors for the synthesis of other functional silanes [34] and used as silylation reagents for surface modifications [35]. We therefore chose a chlorodisiloxane (**2**) as an appropriate model system, which is only equipped with aryl groups as additional substituents in order to examine the role of C–H$\cdots\pi$ and C–H$\cdots$Cl–Si interactions in the crystalline state more closely by using state-of-the-art analytical methods, Hirshfeld surface analysis [36] along with two-dimensional (2D) fingerprint plots [37]. As already successfully applied in previous work [15,38–40], we took advantage of the good crystallization properties that result when compounds are equipped with triphenylsiloxy groups.

2. Experimental Details

2.1. General Remarks

All experiments were performed under an inert atmosphere of purified nitrogen by using standard Schlenk techniques. Glassware was heated at 140 °C prior to use. Dichloromethane, pentane, tetrahydrofuran, and toluene were dried and degassed with an MBraun SP800 solvent purification system. *n*-Butyllithium (2.5 M solution in hexane, Merck KGaA, Darmstadt, Germany), dichlorodiphenylsilane (98%, Merck KGaA, Darmstadt, Germany), and triphenylsilanol (98%, Merck KGaA, Darmstadt, Germany) were used without further purification. [D_6]-Benzene used for NMR spectroscopy was dried over Na/K amalgam. NMR spectra were recorded on a Bruker Avance 300 (300.13 MHz, Bruker Corporation, Billerica, MA, USA) and a Bruker Avance III HD 400 (400.13 MHz) spectrometer at 25 °C. Chemical shifts (δ) are reported in parts per million (ppm). ^{1}H and ^{13}C{^{1}H} NMR spectra are referenced to tetramethylsilane ($SiMe_4$, δ = 0.0 ppm) as external standard, with the deuterium signal of the solvent serving as internal lock and the residual solvent signal as an additional reference. The ^{29}Si NMR spectrum is referenced to $SiMe_4$ (δ = 0.0 ppm) as the external standard. For the assignment of the multiplicities, the following abbreviations were used: s = singlet, m = multiplet. High resolution mass spectrometry was carried out on a Jeol AccuTOF GCX spectrometer. Elemental analysis was performed on a Vario MICRO cube apparatus. The IR spectrum was recorded on a Bruker ALPHA FT-IR spectrometer equipped with a diamond ATR unit. For the intensities of the bands, the following abbreviations were used: s = strong, m = medium, w = weak.

2.2. Synthesis of Ph$_2$SiCl(OSiPh$_3$) (**2**)

n-Butyllithium (22.0 mL of a 2.5 M solution in hexane, 55.0 mmol, 1.1 equiv.) was added dropwise to a solution of triphenylsilanol (13.82 g, 50.0 mmol, 1.0 equiv.) in tetrahydrofuran (200 mL) at 0 °C. The clear colorless solution was then allowed to slowly warm to room temperature and stirred for 1 h. Then, the reaction mixture was again cooled to 0 °C, dichlorodiphenylsilane (**1**) (10.5 mL, 50.0 mmol, 1.0 equiv.) was added and the mixture was allowed to warm to room temperature. The reaction mixture was refluxed for 5 h. After cooling down to room temperature, all volatiles were removed in vacuo and the product was extracted in dichloromethane (100 mL). Again, all volatiles were removed in vacuo and the crude solid material was recrystallized from hot toluene (30 mL). The crystals were isolated via filtration and washed with pentane to obtain compound **2** as a white crystalline solid (15.09 g, 30.6 mmol, 61%). ^{1}H NMR (400.13 MHz, C$_6$D$_6$): δ = 7.00–7.15 (m, 15H, H_{Ph}), 7.71–7.79 (m, 10H, H_{Ph}). ^{13}C{^{1}H} NMR (75.44 MHz, C$_6$D$_6$): δ = 128.3 (s, C_{Ph}), 130.5 (s, C_{Ph}), 131.1 (s, C_{Ph}), 134.2 (s, C_{Ph}), 134.7 (s, C_{Ph}), 135.0 (s, C_{Ph}), 135.7 (s, C_{Ph}). ^{29}Si NMR (79.49 MHz, C$_6$D$_6$): δ = −19.6 (m, $SiClPh_2$), −15.7 (m, $SiPh_3$). HRMS (EI+): C$_{30}$H$_{25}$ClOSi$_2$ calcd. *m/z* for [M$^+$] 492.1127; found 492.1119. CHN analysis: C$_{30}$H$_{25}$ClOSi$_2$ calcd. C 73.07%, H 5.11%; found C 73.11%, H 4.97%. FT-IR (cm^{-1}): 3070 (w), 3024 (w), 1590 (w), 1486 (w), 1427 (m), 1116 (s, Si–O–Si), 1096 (s), 1026 (m), 997 (m), 711 (s), 696 (s), 540 (s), 507 (s), 491 (s), 475 (s).

2.3. X-Ray Crystallography

Single-crystal X-ray diffraction analysis of chloropentaphenyldisiloxane (**2**) was performed on a GV50 diffractometer equipped with a TitanS2 CCD detector at 123(2) K using graphite-monochromated Cu-Kβ radiation (λ = 1.39222 Å). Data collection and reduction was performed using the CrysAlisPro software system, version 1.171.40.14a [41]. The crystal structure was solved with SHELXT 2018/2 [42,43] and a full-matrix least-squares refinement based on F^2 was carried out with SHELXL-2018/3 [43–45] using Olex2 [46] and the SHELX program package as implemented in WinGX [47]. A multi-scan absorption correction using spherical harmonics as implemented in SCALE3 ABSPACK was employed [41]. The non-hydrogen atoms were refined using anisotropic displacement parameters. The hydrogen atoms were located on the difference Fourier map and refined independently. The Hirshfeld surface was mapped over d_{norm} ranging from -0.0425 to 1.3719 a.u. d_i and d_e in the 2D fingerprint diagrams are the distances from the surface to the nearest atom *interior* and *exterior* to the surface, respectively, and are each given in the range of 0.4 to 3.0 Å. Details on crystal data and structure refinement are summarized in Table 1 (see also Supplementary Materials). The Hirshfeld surface and 2D fingerprint plots including Figures 1–3 and Appendix A Figure A1 were created using CrystalExplorer 17.5 [48].

Table 1. Crystal data and structure refinement of chloropentaphenyldisiloxane (**2**).

Empirical formula	$C_{30}H_{25}ClOSi_2$
Formula weight [g·mol^{-1}]	493.13
Crystal system	Monoclinic
Space group	$P2_1/n$
a [Å]	10.6741(2)
b [Å]	14.2858(2)
c [Å]	17.5012(3)
α [°]	90
β [°]	99.597(2)
γ [°]	90
Volume [Å^3]	2631.37(8)
Z	4
Density (calculated) ρ [g·cm^{-3}]	1.245
Absorption coefficient μ [mm^{-1}]	1.690
$F(000)$	1032
Crystal size [mm^3]	0.161 × 0.100 × 0.084
Theta range for data collection θ [°]	3.627–69.661
Index ranges	$-12 \le h \le 14$
	$-18 \le k \le 19$
	$-22 \le l \le 23$
Reflections collected	22059
Independent reflections	6570 (R_{int} = 0.0209)
Completeness to θ = 56.650°	99.9%
Max. and min. transmission	1.000 and 0.795
Data/restraints/parameters	6570/0/407
Goodness-of-fit on F^2	1.045
Final R indices [$I > 2\sigma(I)$]	$R1$ = 0.0347, w$R2$ = 0.0954
R indices (all data)	$R1$ = 0.0397, w$R2$ = 0.0994
Largest diff. peak and hole [e·Å^{-3}]	0.393 and -0.548

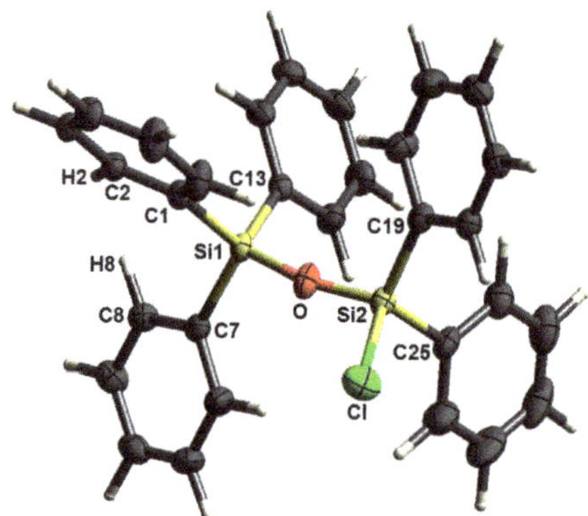

Figure 1. Molecular structure of chloropentaphenyldisiloxane (**2**) (displacement ellipsoids set at the 50% probability level). Selected bond lengths (Å) and angles (°): Si1–C1 1.8649(12), Si1–C7 1.8636(13), Si1–C13 1.8595(13), Si2–C19 1.8543(14), Si2–C25 1.8516(13), Si2–Cl 2.0700(5), Si1–O 1.6305(10), Si2–O 1.6012(10), Si1–O–Si2 165.08(8), O–Si2–C19 109.26(6), O–Si2–C25 111.83(6), O–Si2–Cl 105.62(4). Shortest intramolecular H···H contact: H2···H8 2.423 Å.

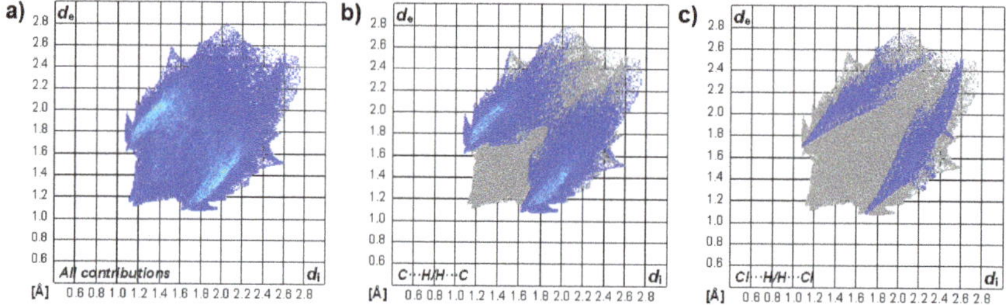

Figure 2. 2D fingerprint plots of chloropentaphenyldisiloxane (**2**) showing (**a**) all contributions of intermolecular contacts, (**b**) C···H/H···C (37.9%), and (**c**) Cl···H/H···Cl (8.9%) contacts.

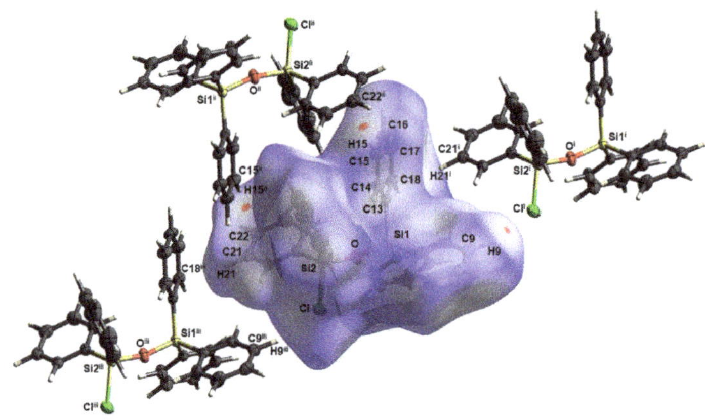

Figure 3. Hirshfeld surface analysis of chloropentaphenyldisiloxane (**2**) highlighting C–H···Cl and C–H···π hydrogen bonds (displacement ellipsoids set at the 50% probability level). Distances (Å) and angle (°) of the C9–H9···Cl contact: C9–H9 0.917, H9···Cl 2.913, C9···Cl 3.669, C9–H9···Cl 140.61. Distances (Å) and angle (°) of the C15–H15···C22 contact: C15–H15 0.973, H15···C22 2.822, C15···C22 3.748, C15–H15···C22 159.46. Symmetry transformations used to generate equivalent atoms: (i) –1 + x, y, z; (ii) 1–x, 1–y, 1–z; (iii) 1 + x, y, z.

3. Results and Discussion

According to a synthetic protocol recently published by us [15], chloropentaphenyldisiloxane (**2**) was easily obtained in 61% isolated yield for the first time after reacting dichlorodiphenylsilane (**1**) with lithium triphenylsiloxide (Scheme 1). The use of metallated siloxide reagents for the stepwise and controlled building of organopolysiloxane polymers was impressively shown by Muzavarov and Rebrov [49]. Recrystallization from hot toluene afforded single-crystals of disiloxane **2** suitable for single-crystal X-ray diffraction analysis (Table 1, Figure 1, see also Supplementary Materials). The asymmetric unit of the monoclinic crystal system, space group $P2_1/n$, contains one molecule of compound **2**. The intramolecular bond parameters of the Si–O–Si backbone show a significantly shortened Si2–O bond [1.6012(10) Å], i.e., the bond that contains the silicon atom attached to the chlorine substituent, in comparison to the Si1–O bond [1.6305(10) Å] (Figure 1). It has already been noticed earlier that the Si–O–Si bond angle in chloro-substituted disiloxanes is remarkably larger than in the respective methoxy- and aminodisiloxanes [15]. In compound **2**, the Si1–O–Si2 bond angle of 165.08(8)° is even larger than in the previously described [15] chlorodisiloxane MesPhSiCl(OSiPh₃) and, together with the short Si2–O bond, may be indicative for a pronounced negative hyperconjugation of the type LP(O)→σ*(Si–R) [50,51]. However, neither the Si2–Cl bond [2.0700(5) Å] nor the Si–C bonds show any appreciable elongation when compared to other aryl-substituted chlorosilanes [52–54]. A thorough analysis of the Si–O–Si bonding parameters is not only important for organosiloxanes, but also in the crystal chemistry of minerals and has contributed significantly to a deeper understanding of mineral properties [55].

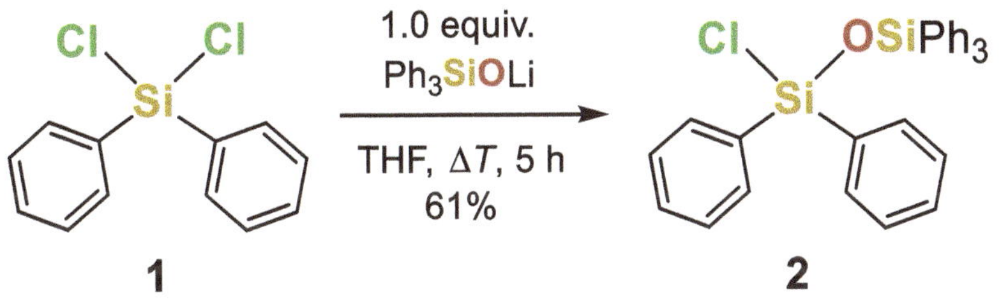

Scheme 1. Synthesis of chloropentaphenyldisiloxane (**2**).

Figure 2 shows the 2D fingerprint diagrams of intermolecular interactions in the crystal structure of disiloxane **2**, all contributions (plot **a**) and subdivided into the individual contributions between atoms inside and outside the Hirshfeld surfaces (plots **b** and **c**). As expected, isotropic H⋯H contacts (52.9%) make the largest percentage contribution to the intermolecular interactions. The point on the Hirshfeld surface where $d_i = d_e \approx 1.2$ Å belongs to the shortest intermolecular H⋯H contact, i.e., H5⋯H14 (2.455 Å), which is not unusually short for H⋯H contacts between phenyl groups [25,37,38,56] and almost as long as the shortest intramolecular H⋯H contact (H2⋯H8 2.423 Å) (Figures 1 and 2, plot **a**). Two types of short C–H⋯π (i.e., H⋯C) contacts can be found in the crystal structure of compound **2** (Figure 2, plot **b**). The closest H⋯C contact amounts to 2.822 Å ($d_i \approx$ 1.65 Å, $d_e \approx$ 1.15 Å), is represented by the spikes, and contains a C–H bond directed towards a single carbon atom (C15–H15⋯C22) (Figure 3). The other of these shortest C–H⋯π contacts, located at $d_i \approx 1.8$ Å and $d_e \approx 1.1$ Å, within the only faintly indicated but typical wing at the lower right of the C⋯H/H⋯C contact plot points almost directly to the center of a phenyl ring and can be identified as the C21–H21⋯π(Ph) interaction with the π-bonded acceptor group containing the carbon atoms C13 to C18 (shortest contact: H21⋯C18 2.901 Å) (Table 2 and Figure 3).

Table 2. H21···C distances of the almost centered C21–H21···π(Ph) contact in compound **2**.

Contact	Distance (Å)
H21···C13	2.997
H21···C14	3.083
H21···C15	3.093
H21···C16	3.001
H21···C17	2.908
H21···C18	2.901

The designation of a C–H···π contact as a hydrogen bond [57,58] applies at least to the most acidic C–H donors such as alkynyl C≡C–H groups [59,60]. C–H···π(Ph) interactions, even with weak $C(sp^2)$–H or even $C(sp^3)$–H donors, generally still have important structure-determining and directing abilities, although they are borderline cases at the weak end of the hydrogen bond classification [26,59]. It was impressively shown by Nishio et al. [61] that C–H···π interactions can play a crucial role in molecular recognition, for the formation of inclusion compounds, and in controlling specificities in organic reactions. Furthermore, due to their weak but still orienting character, they should also play an important role in the dynamic formation of supramolecular structures of biopolymers during the processes in living cells. Recently, the importance of anisotropic C–H···π interactions in the crystal structure formation of arylmethoxysilanes has also been pointed out [26].

There are no intermolecular C–H···O contacts to be found, which, on the one hand, can be explained by the difficult steric accessibility of the effectively shielded siloxane oxygen atom as a consequence of the large Si–O–Si angle of 165.08(8)°. On the other hand, this might also have an electronic reason, as recently pointed out by theoretical investigations on the hydrogen bond interaction energy as a function of the Si–O–Si angle [51]. In this picture, the decreased accessibility of oxygen lone electron pairs due to increased negative hyperconjugation may be the reason for the low basicity of the Si–O–Si linkage in compound **2**. This could be interesting with regard to a siloxane—functional group cooperation and lead to the design of precisely defined functional units in which intramolecular Si–O–Si-specific bond parameters can influence the acceptor capabilities of functional groups or vice versa.

The fingerprint plot for the Cl···H/H···Cl contacts shows distinct spikes that closely resemble that of typical hydrogen bonding pattern (Figure 2, plot c) [26,37]. In the meanwhile, the existence of C–H···Cl hydrogen bonds has been well documented and evidenced [62–65]. The H···Cl contact in disiloxane **2** is represented by the spike where d_i ≈ 1.7 Å and d_e ≈ 1.1 Å (actually found in the crystal structure: 2.913 Å) and belongs to the C9–H9···Cl–Si2 hydrogen bond (C9···Cl 3.669 Å, C9–H9···Cl 140.61°) (Figure 3). It is in the range of the sum of the van der Waals radii for hydrogen (1.2 Å) and chlorine (1.75 Å) [66] and is quite the same as found for H···Cl contacts in chloroform at around 2.95 Å [37]. Since the H9···Cl distance is also in the typical range for chloro-substituted hydrocarbons [63], we therefore anticipate an essentially anisotropic contribution of the C–H···Cl–Si hydrogen bond with a directional influence on the crystal packing. The essential directing structure-forming interactions that were identified from the Hirshfeld surface analysis are also clearly reflected in the crystal packing of disiloxane **2** (Figure A1).

For comparison: In $MesPhSiCl(OSiPh_3)$ [15], the C···H/H···C and Cl···H/H···Cl contacts with 29.8% and 6.4%, respectively, contribute less to the intermolecular interactions. Although, the directionality of these contacts seems to be less pronounced in $MesPhSiCl(OSiPh_3)$, the mesityl CH_3 groups can also participate in intermolecular interactions.

4. Conclusions

Monofunctional disiloxanes are scarce, but helpful model systems in order to provide information on substituent effects on the Si–O–Si structural motif and on the packing in the molecular crystalline state. The present investigation on intermolecular interactions in the crystal structure of a chlorodisiloxane (**2**) was carried out using Hirshfeld surface analysis and 2D fingerprint plots. Two major types of anisotropic short C–H···π contacts and a C–H···Cl–Si hydrogen bond-like interaction were identified to have the strongest directional influence on the packing within the molecular crystal. Although the siloxane unit does not appear to have a pronounced effect on the chlorine substituent in this molecule, it seems worthwhile to address the influence of the siloxane motif on the acceptor capabilities of functional groups directly connected to the Si–O–Si unit in future investigations. The information on intermolecular interactions provided herein may be of particular interest with regard to the design of supramolecular functional polysiloxane architectures.

Supplementary Materials: CCDC-2068445 (compound **2**) contains the supplementary crystallographic data for this paper. These data can be obtained free of charge from the Cambridge Crystallographic Data Centre, CCDC, 12 Union Road, Cambridge CB21EZ, UK (Fax: + 44-1223-336-033; E-mail: deposit@ccdc.cam.ac.uk; https://www.ccdc.cam.ac.uk/structures/).

Author Contributions: Conceptualization, J.O.B.; methodology, J.O.B. and T.G.; validation, J.O.B. and T.G.; formal analysis, J.O.B.; investigation, J.O.B. and T.G.; resources, J.O.B.; data curation, J.O.B. and T.G.; writing—original draft preparation, J.O.B.; writing—review and editing, J.O.B. and T.G.; visualization, J.O.B.; supervision, J.O.B.; project administration, J.O.B.; funding acquisition, J.O.B. All authors have read and agreed to the published version of the manuscript.

Funding: This research was funded by the University of Regensburg, the Elite Network of Bavaria (ENB), and the Bavarian State Ministry of Science and the Arts (StMWK), grant number N-LW-NW-2016-366.

Data Availability Statement: The data presented in this study are available in Supplementary Materials.

Acknowledgments: The authors would like to thank Manfred Scheer and Jörg Heilmann for continuous support and providing laboratory facilities.

Conflicts of Interest: The authors declare no conflict of interest.

Appendix A

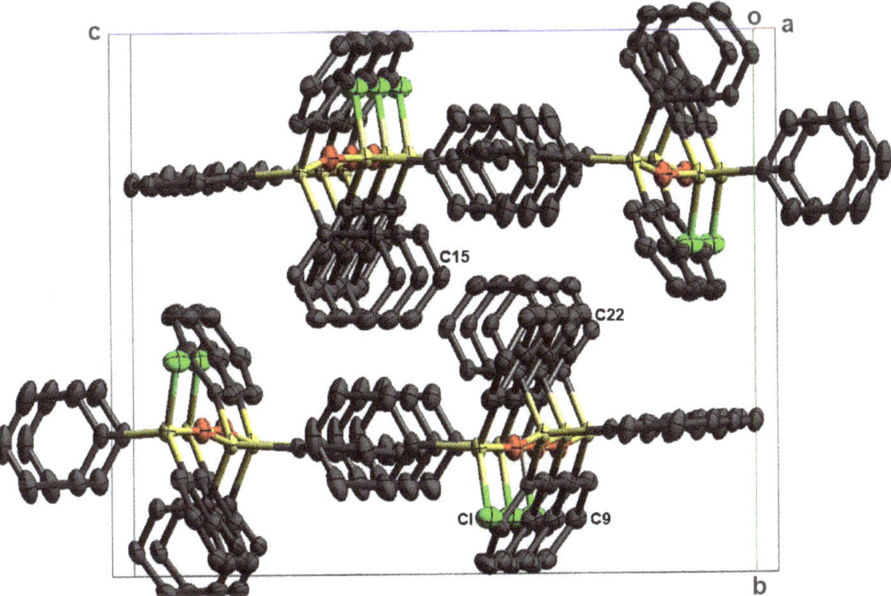

Figure A1. Crystal packing of chloropentaphenyldisiloxane (**2**) along the a axis. Hydrogen atoms are omitted for clarity.

References

1. Liebau, F. *Structural Chemistry of Silicates: Structure, Bonding, and Classification*; Springer: Berlin/Heidelberg, Germany, 1985.
2. Swaddle, T.W.; Salerno, J.; Tregloan, P.A. Aqueous aluminates, silicates, and aluminosilicates. *Chem. Soc. Rev.* **1994**, *23*, 319–325. [CrossRef]
3. Perry, C.C.; Keeling-Tucker, T. Biosilicification: The role of the organic matrix in structure control. *J. Biol. Inorg. Chem.* **2000**, *5*, 537–550. [CrossRef]
4. Ganachaud, S.; Boileau, S.; Boury, B. *Silicon Based Polymers: Advances in Synthesis and Supramolecular Organization*; Springer: Berlin/Heidelberg, Germany, 2008.
5. Schubert, U.; Hüsing, N. *Synthesis of Inorganic Materials*, 4th ed.; Wiley-VCH: Weinheim, Germany, 2019.
6. Höppe, H.A.; Stadler, F.; Oeckler, O.; Schnick, W. Ca[Si$_2$O$_2$N$_2$]—A Novel Layer Silicate. *Angew. Chem. Int. Ed.* **2004**, *43*, 5540–5542. [CrossRef] [PubMed]
7. Däschlein, C.; Bauer, J.O.; Strohmann, C. From the Selective Cleavage of the Si–O–Si Bond in Disiloxanes to Zwitterionic, Water-Stable Zinc Silanolates. *Angew. Chem. Int. Ed.* **2009**, *48*, 8074–8077. [CrossRef]
8. Spirk, S.; Nieger, M.; Belaj, F.; Pietschnig, R. Formation and hydrogen bonding of a novel POSS-trisilanol. *Dalton Trans.* **2008**, 163–167. [CrossRef]
9. Hurkes, N.; Bruhn, C.; Belaj, F.; Pietschnig, R. Silanetriols as Powerful Starting Materials for Selective Condensation to Bulky POSS Cages. *Organometallics* **2014**, *33*, 7299–7306. [CrossRef]
10. Čas, D.; Hurkes, N.; Spirk, S.; Belaj, F.; Bruhn, C.; Rechberger, G.N.; Pietschnig, R. Dimer formation upon deprotonation: Synthesis and structure of a *m*-terphenyl substituted (*R*,*S*)-dilithium disiloxanolate disilanol. *Dalton Trans.* **2015**, *44*, 12818–12823. [CrossRef]
11. Oguri, N.; Egawa, Y.; Takeda, N.; Unno, M. Janus-Cube Octasilsesquioxane: Facile Synthesis and Structure Elucidation. *Angew. Chem. Int. Ed.* **2016**, *55*, 9336–9339. [CrossRef] [PubMed]
12. Lokare, K.S.; Frank, N.; Braun-Cula, B.; Goikoetxea, I.; Sauer, J.; Limberg, C. Trapping Aluminum Hydroxide Clusters with Trisilanols during Speciation in Aluminum(III)–Water Systems: Reproducible, Large Scale Access to Molecular Aluminate Models. *Angew. Chem. Int. Ed.* **2016**, *55*, 12325–12329. [CrossRef]
13. Bauer, J.O.; Strohmann, C. Synthesis and molecular structure of a zwitterionic ZnI$_2$ silanolate. *Inorg. Chim. Acta* **2018**, *469*, 133–135. [CrossRef]
14. Lokare, K.S.; Braun-Cula, B.; Limberg, C.; Jorewitz, M.; Kelly, J.T.; Asmis, K.R.; Leach, S.; Baldauf, C.; Goikoetxea, I.; Sauer, J. Structure and Reactivity of Al−O(H)−Al Moieties in Siloxide Frameworks: Solution and Gas-Phase Model Studies. *Angew. Chem. Int. Ed.* **2019**, *58*, 902–906. [CrossRef]
15. Espinosa-Jalapa, N.A.; Bauer, J.O. Controlled Synthesis and Molecular Structures of Methoxy-, Amino-, and Chloro-Functionalized Disiloxane Building Blocks. *Z. Anorg. Allg. Chem.* **2020**, *646*, 828–834. [CrossRef]
16. Weitkamp, R.F.; Neumann, B.; Stammler, H.; Hoge, B. Synthesis and Reactivity of the First Isolated Hydrogen-Bridged Silanol–Silanolate Anions. *Angew. Chem. Int. Ed.* **2020**, *59*, 5494–5499. [CrossRef] [PubMed]
17. Weitkamp, R.F.; Neumann, B.; Stammler, H.-G.; Hoge, B. The Influence of Weakly Coordinating Cations on the O–H⋯O$^-$ Hydrogen Bond of Silanol–Silanolate Anions. *Chem. Eur. J.* **2021**, *27*, 915–920. [CrossRef] [PubMed]
18. Tacke, R. Milestones in the Biochemistry of Silicon: From Basic Research to Biotechnological Applications. *Angew. Chem. Int. Ed.* **1999**, *38*, 3015–3018. [CrossRef]
19. Voinescu, A.E.; Kellermeier, M.; Bartel, B.; Carnerup, A.M.; Larsson, A.-K.; Touraud, D.; Kunz, W.; Kienle, L.; Pfitzner, A.; Hyde, S.T. Inorganic Self-Organized Silica Aragonite Biomorphic Composites. *Cryst. Growth Des.* **2008**, *8*, 1515–1521. [CrossRef]
20. Volkmer, D.; Tugulu, S.; Fricke, M.; Nielsen, T. Morphosynthesis of Star-Shaped Titania–Silica Shells. *Angew. Chem. Int. Ed.* **2003**, *42*, 58–61. [CrossRef]
21. Desiraju, G.R. Supramolecular Synthons in Crystal Engineering—A New Organic Synthesis. *Angew. Chem. Int. Ed. Engl.* **1995**, *34*, 2311–2327. [CrossRef]
22. Desiraju, G.R. Hydrogen Bridges in Crystal Engineering: Interactions without Borders. *Acc. Chem. Res.* **2002**, *35*, 565–573. [CrossRef]
23. Thompson, D.B.; Brook, M.A. Rapid Assembly of Complex 3D Siloxane Architectures. *J. Am. Chem. Soc.* **2008**, *130*, 32–33. [CrossRef]
24. Bauer, J.O. The crystal structure of the triclinic polymorph of hexameric (trimethylsilyl)methyllithium, C$_{24}$H$_{66}$Li$_6$Si$_6$. *Z. Kristallogr. NCS* **2020**, *235*, 353–356. [CrossRef]
25. Bauer, J.O. The crystal structure of the first ether solvate of hexaphenyldistannane [(Ph$_3$Sn)$_2$ · 2 THF]. *Main Group Met. Chem.* **2020**, *43*, 1–6. [CrossRef]
26. Bauer, J.O. Crystal Structure and Hirshfeld Surface Analysis of Trimethoxy(1-naphthyl)silane—Intermolecular Interactions in a One-Component Single-Crystalline Trimethoxysilane. *Z. Anorg. Allg. Chem.* **2021**, *647*, 1053–1057. [CrossRef]
27. Pietschnig, R.; Merz, K. Selective Formation of Functionalized Disiloxanes from Terphenylfluorosilanes. *Organometallics* **2004**, *23*, 1373–1377. [CrossRef]
28. Wojnowski, W.; Becker, B.; Peters, K.; Peters, E.-M.; von Schnering, H.G. Beiträge zur Chemie der Silicium-Schwefel-Verbindungen. 53. Die Struktur des 1,3-Dimethyl-1,1,3,3-Tetraphenyldisilthians. *Z. Anorg. Allg. Chem.* **1988**, *563*, 48–52. [CrossRef]
29. Coelho, A.C.; Amarante, T.R.; Klinowski, J.; Gonçalves, I.S.; Almeida Paz, F.A. 1-Hydroxy-1,1,3,3,3-pentaphenyldisiloxane, [Si$_2$O(OH)(Ph)$_5$], at 100 K. *Acta Crystallogr. Sect. E* **2008**, *64*, o237–o238. [CrossRef] [PubMed]

30. Amarante, T.R.; Coelho, A.C.; Klinowski, J.; Gonçalves, I.S.; Almeida Paz, F.A. 1-Hydroxy-1,1,3,3,3-pentaphenyldisiloxane, [Si$_2$O(OH)(Ph)$_5$], at 150 K. *Acta Crystallogr. Sect. E* **2008**, *64*, o239. [CrossRef] [PubMed]
31. Bauer, J.O.; Strohmann, C. One-step conversion of methoxysilanes to aminosilanes: A convenient synthetic strategy to *N,O*-functionalised organosilanes. *Chem. Commun.* **2012**, *48*, 7212–7214. [CrossRef]
32. Bauer, J.O.; Strohmann, C. Stereoselective Synthesis of Silicon-Stereogenic Aminomethoxysilanes: Easy Access to Highly Enantiomerically Enriched Siloxanes. *Angew. Chem. Int. Ed.* **2013**, *53*, 720–724. [CrossRef]
33. Woińska, M.; Grabowsky, S.; Dominiak, P.M.; Woźniak, K.; Jayatilaka, D. Hydrogen atoms can be located accurately and precisely by x-ray crystallography. *Sci. Adv.* **2016**, *2*, e1600192. [CrossRef]
34. Wakabayashi, R.; Sugiura, Y.; Shibue, T.; Kuroda, K. Practical Conversion of Chlorosilanes into Alkoxysilanes without Generating HCl. *Angew. Chem. Int. Ed.* **2011**, *50*, 10708–10711. [CrossRef]
35. Deschner, T.; Liang, Y.; Anwander, R. Silylation Efficiency of Chorosilanes, Alkoxysilanes, and Monosilazanes on Periodic Mesoporous Silica. *J. Phys. Chem. C* **2010**, *114*, 22603–22609. [CrossRef]
36. Spackman, M.A.; Jayatilaka, D. Hirshfeld surface analysis. *CrystEngComm* **2009**, *11*, 19–32. [CrossRef]
37. Spackman, M.A.; McKinnon, J.J. Fingerprinting intermolecular interactions in molecular crystals. *CrystEngComm* **2002**, *4*, 378–392. [CrossRef]
38. Bauer, J.O.; Strohmann, C. *tert*-Butoxytriphenylsilane. *Acta Crystallogr. Sect. E* **2010**, *66*, o461–o462. [CrossRef]
39. Bauer, J.O.; Strohmann, C. Hydrogen bonding principles in inclusion compounds of triphenylsilanol and pyrrolidine: Synthesis and structural features of [(Ph$_3$SiOH)$_4$·HN(CH$_2$)$_4$] and [Ph$_3$SiOH·HN(CH$_2$)$_4$·CH$_3$CO$_2$H]. *J. Organomet. Chem.* **2015**, *797*, 52–56. [CrossRef]
40. Lokare, K.S.; Wittwer, P.; Braun-Cula, B.; Frank, N.; Hoof, S.; Braun, T.; Limberg, C. Mimicking Base Interaction with Acidic Sites [Si–O(*H*)–Al] of Zeolites in Molecular Models. *Z. Anorg. Allg. Chem.* **2017**, *643*, 1581–1588. [CrossRef]
41. Rigaku Oxford Diffraction. *CrysAlisPro Software System*; Rigaku Corporation: Oxford, UK, 2018.
42. Sheldrick, G.M. *SHELXT*—Integrated space-group and crystal-structure determination. *Acta Crystallogr. Sect. A* **2015**, *71*, 3–8. [CrossRef]
43. Sheldrick, G.M. A short history of *SHELX*. *Acta Crystallogr. Sect. A* **2008**, *64*, 112–122. [CrossRef] [PubMed]
44. Sheldrick, G.M. Crystal structure refinement with *SHELXL*. *Acta Crystallogr. Sect. C* **2015**, *71*, 3–8. [CrossRef]
45. Sheldrick, G.M. *SHELXL-2018*; Universität Göttingen: Göttingen, Germany, 2018.
46. Dolomanov, O.V.; Bourhis, L.J.; Gildea, R.J.; Howard, J.A.K.; Puschmann, H. *OLEX2*: A complete structure solution, refinement and analysis program. *J. Appl. Crystallogr.* **2009**, *42*, 339–341. [CrossRef]
47. Farrugia, L.J. *WinGX* and *ORTEP for Windows*: An update. *J. Appl. Crystallogr.* **2012**, *45*, 849–854. [CrossRef]
48. Turner, M.J.; McKinnon, J.J.; Wolff, S.K.; Grimwood, D.J.; Spackman, P.R.; Jayatilaka, D.; Spackman, M.A. *CrystalExplorer17*; University of Western Australia: Perth, Australia, 2017.
49. Muzafarov, A.M.; Rebrov, E.A. From the Discovery of Sodiumoxyorganoalkoxysilanes to the Organosilicon Dendrimers and Back. *J. Polym. Sci. Part A: Polym. Chem.* **2008**, *46*, 4935–4948. [CrossRef]
50. Weinhold, F.; West, R. Hyperconjugative Interactions in Permethylated Siloxanes and Ethers: The Nature of the SiO Bond. *J. Am. Chem. Soc.* **2013**, *135*, 5762–5767. [CrossRef]
51. Fugel, M.; Hesse, M.F.; Pal, R.; Beckmann, J.; Jayatilaka, D.; Turner, M.J.; Karton, A.; Bultinck, P.; Chandler, G.S.; Grabowsky, S. Covalency and Ionicity Do Not Oppose Each Other—Relationship Between Si–O Bond Character and Basicity of Siloxanes. *Chem. Eur. J.* **2018**, *24*, 15275–15286. [CrossRef] [PubMed]
52. Liew, S.K.; Al-Rafia, S.M.I.; Goettel, J.T.; Lummis, P.A.; McDonald, S.M.; Miedema, L.J.; Ferguson, M.J.; McDonald, R.; Rivard, E. Expanding the Steric Coverage Offered by Bis(amidosilyl) Chelates: Isolation of Low-Coordinate *N*-Heterocyclic Germylene Complexes. *Inorg. Chem.* **2012**, *51*, 5471–5480. [CrossRef] [PubMed]
53. Reuter, K.; Maas, R.G.M.; Reuter, A.; Kilgenstein, F.; Asfaha, Y.; von Hänisch, C. Synthesis of heteroatomic bridged paracyclophanes. *Dalton Trans.* **2017**, *46*, 4530–4541. [CrossRef]
54. Marin-Luna, M.; Pölloth, B.; Zott, F.; Zipse, H. Size-dependent rate acceleration in the silylation of secondary alcohols: The bigger the faster. *Chem. Sci.* **2018**, *9*, 6509–6515. [CrossRef]
55. Gibbs, G.V.; Downs, R.T.; Cox, D.F.; Ross, N.L.; Prewitt, C.T.; Rosso, K.M.; Lippmann, T.; Kirfel, A. Bonded interactions and the crystal chemistry of minerals: A review. *Z. Kristallogr.* **2008**, *223*, 1–40. [CrossRef]
56. Brendler, E.; Heine, T.; Seichter, W.; Wagler, J.; Witter, R. ^{29}Si NMR Shielding Tensors in Triphenylsilanes—^{29}Si Solid State NMR Experiments and DFT-IGLO Calculations. *Z. Anorg. Allg. Chem.* **2012**, *638*, 935–944. [CrossRef]
57. Arunan, E.; Desiraju, G.R.; Klein, R.A.; Sadlej, J.; Scheiner, S.; Alkorta, I.; Clary, D.C.; Crabtree, R.H.; Dannenberg, J.J.; Hobza, P.; et al. Defining the hydrogen bond: An account (IUPAC Technical Report). *Pure Appl. Chem.* **2011**, *83*, 1619–1636. [CrossRef]
58. Arunan, E.; Desiraju, G.R.; Klein, R.A.; Sadlej, J.; Scheiner, S.; Alkorta, I.; Clary, D.C.; Crabtree, R.H.; Dannenberg, J.J.; Hobza, P.; et al. Definition of the hydrogen bond (IUPAC Recommendations 2011). *Pure Appl. Chem.* **2011**, *83*, 1637–1641. [CrossRef]
59. Steiner, T. The Hydrogen Bond in the Solid State. *Angew. Chem. Int. Ed.* **2002**, *41*, 48–76. [CrossRef]

60. Steiner, T.; Starikov, E.B.; Amado, A.M.; Teixeira-Dias, J.J.C. Weak hydrogen bonding. Part 2. The hydrogen bonding nature of short C–H···π contacts: Crystallographic, spectroscopic and quantum mechanical studies of some terminal alkynes. *J. Chem. Soc. Perkin Trans. 2* **1995**, 1321–1326. [CrossRef]

61. Nishio, M.; Umezawa, Y.; Hirota, M.; Takeuchi, Y. The CH/π Interaction: Significance in Molecular Recognition. *Tetrahedron* **1995**, *51*, 8665–8701. [CrossRef]

62. Taylor, R.; Kennard, O. Crystallographic Evidence for the Existence of C–H···O, C–H···N, and C–H···Cl Hydrogen Bonds. *J. Am. Chem. Soc.* **1982**, *104*, 5063–5070. [CrossRef]

63. Desiraju, G.R.; Parthasarathy, R. The Nature of Halogen···Halogen Interactions: Are Short Halogen Contacts Due to Specific Attractive Forces or Due to Close Packing of Nonspherical Atoms? *J. Am. Chem. Soc.* **1989**, *111*, 8725–8726. [CrossRef]

64. Aakeröy, C.B.; Evans, T.A.; Seddon, K.R.; Pálinkó, I. The C–H···Cl hydrogen bond: Does it exist? *New J. Chem.* **1999**, *23*, 145–152. [CrossRef]

65. Liu, M.; Yin, C.; Chen, P.; Zhang, M.; Parkin, S.; Zhou, P.; Li, T.; Yu, F.; Long, S. $^{sp^2}$CH···Cl hydrogen bond in the conformational polymorphism of 4-chloro-phenylanthranilic acid. *CrystEngComm* **2017**, *19*, 4345–4354. [CrossRef]

66. Bondi, A. van der Waals Volumes and Radii. *J. Phys. Chem.* **1964**, *68*, 441–451. [CrossRef]

Article

Synthesis and Crystallographic Characterization of X-Substituted 2,4-Dinitrophenyl-4′-phenylbenzenesulfonates [†]

Brock A. Stenfors [1], Richard J. Staples [2], Shannon M. Biros [1] and Felix N. Ngassa [1,*]

[1]   Department of Chemistry, Grand Valley State University, 1 Campus Drive., Allendale, MI 49401, USA; stenforb@mail.gvsu.edu (B.A.S.); biross@gvsu.edu (S.M.B.)
[2]   Center for Crystallographic Research, Department of Chemistry, Michigan State University, 578 S. Shaw Lane, East Lansing, MI 48824, USA; staples@Chemistry.msu.edu
*   Correspondence: ngassaf@gvsu.edu
†   Dedicated to Dr. Howard Flack (1943–2017).

Received: 9 May 2020; Accepted: 21 May 2020; Published: 15 June 2020

Abstract: Treatment of 2,4-dinitrophenol with sulfonyl chlorides in the presence of pyridine results in the formation of undesired pyridinium salts. In non-aqueous environments, the formation of the insoluble pyridinium salt greatly affects the formation of the desired product. A facile method of producing the desired sulfonate involves the use of an aqueous base with a water-miscible solvent. Herein, we present the optimization of methods for the formation of sulfonates and its application in the production of desired x-substituted 2,4-dinitrophenyl-4′-phenylbenzenesulfonates. This strategy is environmentally benign and supports a wide range of starting materials. Additionally, the intermolecular interactions of these sulfonate compounds were investigated using single-crystal x-ray diffraction data.

Keywords: arylsulfonates; pyridinium salt formation; single-phase solvent system; sulfonate synthesis; sulfonyl chlorides; X-ray crystallography

1. Introduction

A reliable method for producing arylsulfonates involves the nucleophilic substitution reaction of alcohols and sulfonyl halides. This reaction is highly efficient in creating the sulfonate ester, a synthetically important electrophile in organic chemistry due to its high chemical reactivity [1]. This property has been utilized in the detection and fluorescent imaging of biologically important compounds. In particular, recent research has shown that 2,4-dinitrobenzenesulfonate-functionilized carbon dots (g-CD-DNBS) are significant regarding the detection and fluorescence imaging of biothiols (Scheme 1) [2].

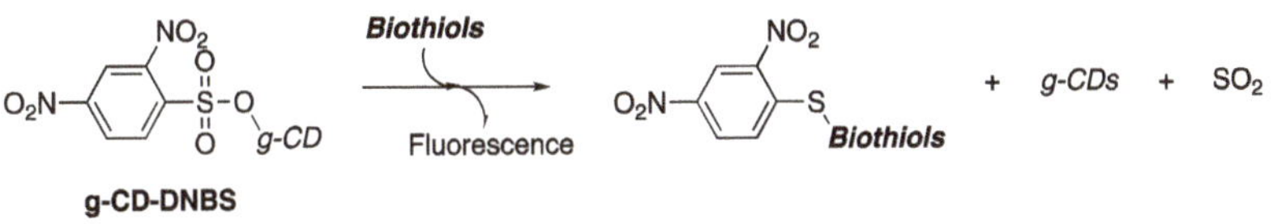

Scheme 1. The fluorescence response from the treatment of g-CD-DNBS with biothiols.

The 2,4-dinitrobenzenesulfonyl (DNBS) group resembles the x-substituted 2,4-dinitrophenyl-4′-phenylbenzenesulfonate compounds. Similarly, this resemblance is found in the fluorescent probe O-hNRSel. O-hNRSel was developed for the detection and imaging of selenocysteine (Sec) in living cells (Scheme 2) [3]. Sec is a selenium-containing amino acid that resides in proteins of organisms and viruses.

Selenium is linked to several health benefits, including the prevention of cancer and cardiovascular diseases [4]. The role of Sec in human health underlines the importance of Sec fluorescence probes.

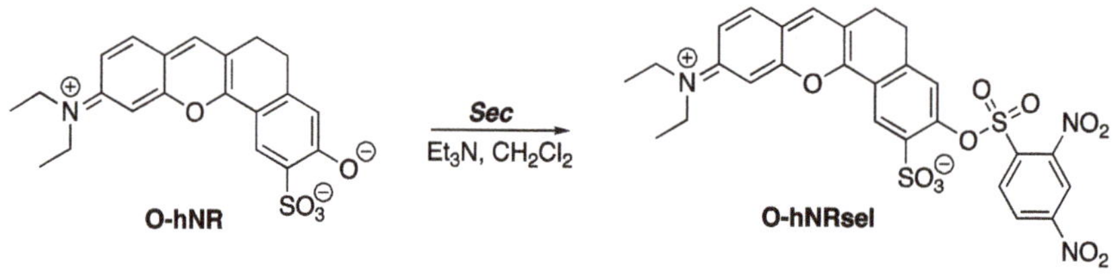

Scheme 2. The formation of the probe O-hNRSel from the treatment of O-hNR with Sec in the presence of triethylamine (Et_3N) and dichloromethane (CH_2Cl_2).

A facile synthesis of sulfonates is necessary to produce these biologically significant compounds. A review of the current literature suggests the use of sulfonic acids or sulfonyl halides for synthesizing sulfonates [5,6]. The compounds synthesized hereafter were done so through the sulfonylation of 2,4-dinitrophenol with sulfonyl chloride in the presence of a base. Basic conditions allow for the neutralization of hydrochloric acid. The presence of hydrochloric acid lowers the pH of the reaction, resulting in poor deprotonation of the intermediary structure. Preliminary experiments for the synthesis of 2,4-dinitrophenyl-4-methylbenzenesulfonate involved the treatment of 4-methylbenzenesulfonyl chloride (**1**) with 2,4-dinitrophenol (**2**) in the presence of pyridine and dichloromethane. This reaction was done under N_2 atmosphere to avoid hydrolysis of 4-methylbenzenesulfonyl chloride. The production of the insoluble pyridinium salt, the pyridine adduct of 2,4-dinitrophenyl 4-methylbenzenesulfonate (**3**), was evidence of an unsuccessful reaction. Compound **3** is insoluble in dichloromethane and pulled out of the solution as a white precipitate. The proposed mechanism for the formation of the pyridinium salt (**3**) in the presence of pyridine is found in Scheme 3. The desired sulfonate (**4**) was not formed as a result of compound **3** being insoluble in solution. A review of the literature supports the theory that a pyridinium salt was formed [7].

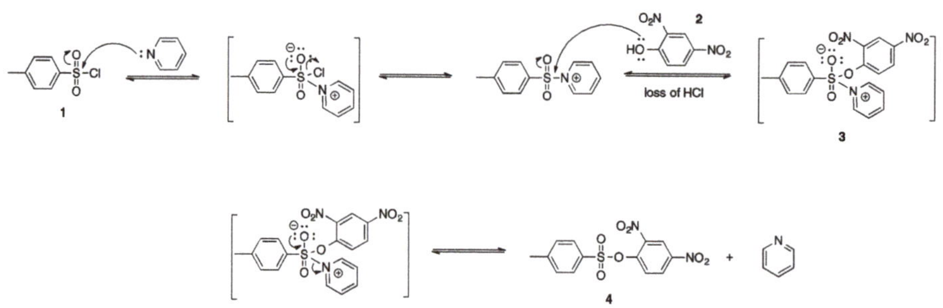

Scheme 3. The proposed mechanism for the treatment of 4-methylbenzenesulfonyl chloride (**1**) with 2,4-dinitrophenol (**2**) in the presence of pyridine and dichloromethane to form the pyridinium adduct of 2,4-dinitrophenyl 4-methylbenzenesulfonate (**3**). Shown below this is the desired rearrangement of compound **3** to give the sulfonate product (**4**).

A new synthetic method was proposed and applied to the synthesis of three x-substituted 2,4-dinitrophenyl-4′-phenylbenzenesulfonates. The resulting products were characterized by crystallographic and spectroscopic means. Crystallographic characterization offers insight into the structural features and inter- and intramolecular interactions of molecules. Herein, we report the facile synthesis and characterization of x-substituted 2,4-dinitrophenyl-4′-phenylbenzenesulfonates.

2. Experimental

The reagents used in the synthesis of x-substituted 2,4-dinitrophenyl-4′-phenylbenzenesulfonates were obtained from commercial sources and used without further purification. Thin-layer chromatography (TLC) was used to track reaction progress and obtain R_f values for the reactions.

Preparation of 2,4-Dinitrophenyl 4-methylbenzenesulfonate in the presence of N,N-diisopropylethylamine (**4**). 2,4-dinitrophenol (2.025 g, 11.0 mmol) was dissolved in a flask containing 11 mL of chilled dichloromethane. 4-Methylbenzenesulfonyl chloride (2.090 g, 11.0 mmol) was added dropwise to the stirred solution. This was followed by the dropwise addition of N,N-diisopropylethylamine (3.8 mL, 21.8 mmol). The stirred solution was left at room temperate for 24 h under N_2 atmosphere. After 24 h, the mixture was diluted with 15 mL of dichloromethane and transferred to a separatory funnel. The organic layer was washed with water (3 × 10 mL). The aqueous layers were combined and back-extracted with 10 mL of dichloromethane. All organic layers were combined, washed with brine (10 mL), and dried over anhydrous sodium sulfate. Evaporation of solvent yielded a solid, yellow residue as a crude product. Purification via column chromatography with dichloromethane as the solvent yielded a yellow powder (1.712 g, 46%). M.p. 196–202 °C. R_f = 0.74 (CH_2Cl_2). ^{1}H-NMR (400 MHz, DMSO-d_6) δ 8.79 (d, *J* = 2.8 Hz, 1H), 8.54 (dd, *J* = 9.1, 2.8 Hz, 1H), 7.78–7.71 (m, 2H), 7.54 (d, *J* = 9.1 Hz, 1H), 7.49 (m, 2H), 2.42 (s, 3H). HRMS (ESI): cald. For $C_{13}H_{10}N_2NaO_7S$ [M + Na]$^+$ 361.0100; found 361.0110.

Preparation of pyridinium adduct of 2,4-Dinitrophenyl 4-methylbenzenesulfonate in the presence of pyridine (**3**). 2,4-dinitrophenol (2.016 g, 11.0 mmol) was dissolved in a flask containing 11 mL of chilled dichloromethane. 4-Methylbenzenesulfonyl chloride (2.088 g, 10.9 mmol) was added dropwise to the stirred solution. This was followed by the dropwise addition of pyridine (1.80 mL, 21.8 mmol). The stirred solution was left at room temperate for 24 h under N_2 atmosphere. After reaction completion, an insoluble precipitate was isolated via vacuum filtration giving a fine, white powder. Recrystallization in 2:1 acetone/water, followed by trituration with hexanes afforded the product as small, translucent crystalline sheets (1.875 g, 41% yield). M.p. 240–248 °C. R_f = 0.70 (80:20 ACN/H_2O w/1 mL NH_4OH). ^{1}H-NMR (400 MHz, DMSO-d_6) δ 9.40–9.33 (m, 2H), 9.05 (d, *J* = 2.5 Hz, 1H), 8.96–8.82 (m, 2H), 8.50–8.35 (m, 3H), 7.43–7.36 (m, 2H), 7.06 (d, *J* = 7.8 Hz, 2H), 2.24 (s, 3H).

Preparation of 2,4-Dinitrophenyl 4-methylbenzenesulfonate in the presence of triethylamine (**4**). 2,4-dinitrophenol (1.999 g, 10.9 mmol) was dissolved in a flask containing 11 mL of chilled dichloromethane. 4-Methylbenzenesulfonyl chloride (2.076 g, 10.9 mmol) was added dropwise to the stirred solution. This was followed by the dropwise addition of triethylamine (3.0 mL, 21.9 mmol). The stirred solution was left at room temperate for 24 h under N_2 atmosphere. After 24 h, the mixture was diluted with 15 mL of dichloromethane and transferred to a separatory funnel. The organic layer was washed with water (3 × 10 mL). The aqueous layers were combined and back-extracted with 10 mL of dichloromethane. All organic layers were combined, washed with brine (10 mL), and dried over anhydrous sodium sulfate. The resulting solution was evaporated to afford a solid, yellow residue. The crude product was recrystallized in 2:1 DCM/ethyl acetate to afford large, pale-yellow, translucent crystals (1.768, 50%).

Preparation of 2,4-Dinitrophenyl 4-methylbenzenesulfonate in the presence of aqueous sodium hydroxide (**4**). 2,4-dinitrophenol (1.021 g, 5.25 mmol) was dissolved in a flask containing 10 mL of THF. 4-Methylbenzenesulfonyl chloride (1.052 g, 5.25 mmol) was then added to the flask, followed by the dropwise addition of 1 M NaOH (10 mL, 10.5 mmol). The solution was stirred at room temperate for 6 h. After reaction completion, a yellow precipitate was isolated via vacuum filtration to afford a yellow powder. The crude product was recrystallized in ethanol to afford large, pale-yellow, translucent crystals (1.204 g, 66%).

Preparation of 2,4-Dinitrophenyl 2,4,6-trimethylbenzenesulfonate (Table 1, *entry 1*). 2,4-dinitrophenol (0.842 g, 4.57 mmol) and 2,4,6-trimethylbenzenesulfonyl chloride (1.002 g, 4.58 mmol) were added to a flask containing 10 mL of tetrahydrofuran. Of aqueous potassium carbonate 0.915 M (10 mL, 9.15 mmol) was added dropwise to the stirred solution. The solution was then stirred at room temperate for 6 h. After reaction completion, a yellow precipitate was isolated via vacuum filtration to afford a yellow powder. The crude product was recrystallized in ethanol to afford large, pale-yellow, translucent crystals (1.470 g, 88%). M.p. 128–132 °C. R_f = 0.86 (CH_2Cl_2). ^{1}H-NMR (400 MHz, Chloroform-*d*) δ 8.75 (d, *J* = 2.8 Hz, 1H), 8.44 (dd, *J* = 9.0, 2.8 Hz, 1H), 7.58 (d, *J* = 9.0 Hz, 1H), 7.03 (s, 2H), 2.57 (s, 6H), 2.35 (s, 3H). HRMS (ESI): cald. For $C_{15}H_{14}N_2NaO_7S$ [M + Na]$^+$ 389.0400; found 389.0410.

Table 1. Solvent and counter-ion effects on the synthesis of 2,4-dinitrophenyl sulfonate derivatives [a].

Entry	R	Base	Solvent	r.t. (h)	Yield (%)
1	2,4,6-trimethyl	0.915 M aq. K_2CO_3	THF	6	88
2	4-phenyl	1.6 M aq. K_2CO_3	THF	8	72
3	4-(4′-methylphenyl)	1.6 M aq. K_2CO_3	THF	8	69
4	4-methyl	1.0 M aq. NaOH	THF	6	66
5	4-(4′-flourophenyl)	1.6 M aq. K_2CO_3	THF	8	59
6 [b]	4-methyl	triethylamine	DCM	24	50
7 [b]	4-methyl	*N,N*-diisopropylethylamine	DCM	24	46
8 [b,c]	4-methyl	pyridine	DCM	24	-

[a] Reaction conditions: 2,4-dinitrophenol (1.0 eq.) was dissolved in 10 mL of solvent. Sulfonyl chloride (1.0 eq) was added dropwise to the solution, followed by the dropwise addition of base (2.0 eq). All reactions were run at room temperature. [b] Reaction performed under N_2 atmosphere. [c] Only pyridinium salt isolated.

Preparation of 2,4-Dinitrophenyl-4′-phenylbenzenesulfonate (**1b**). 2,4-dinitrophenol (0.729 g, 3.96 mmol) and biphenyl-4-sulfonyl chloride (1.005 g, 3.96 mmol) were added to a flask containing 10 mL of tetrahydrofuran. Of aqueous potassium carbonate 1.6 M (5.0 mL, 7.92 mmol) was added dropwise to the stirred solution. The stirred solution was left at room temperate for 8 h. After reaction completion, a yellow precipitate was isolated via vacuum filtration to afford a yellow powder. The crude product was recrystallized in a small amount of dichloromethane to afford pale-yellow translucent crystals (1.136 g, 72% yield). M.p. 137–140 °C, R_f = 0.69 (CH_2Cl_2). ^{1}H-NMR (400 MHz, Chloroform-*d*) δ 8.78 (d, *J* = 2.7 Hz, 1H), 8.50 (dd, *J* = 9.0, 2.7 Hz, 1H), 8.01–7.96 (m, 2H), 7.82–7.75 (m, 3H), 7.64–7.60 (m, 2H), 7.53–7.43 (m, 3H). HRMS (ESI): cald. For $C_{18}H_{12}N_2NaO_7S$ [M + Na]$^+$ 423.0300; found 423.3700.

Preparation of 2,4-Dinitrophenyl-4′-(4-methylphenyl)-benzenesulfonate (**2b**). 2,4-dinitrophenol (0.692 g, 3.75 mmol) and 4′-methylbiphenyl-4-sulfonyl chloride (1.001 g, 3.75 mmol) were added to a flask containing 10 mL of tetrahydrofuran. Of aqueous potassium carbonate 1.6 M (5.0 mL, 7.92 mmol) was added dropwise to the stirred solution. The stirred solution was left at room temperate for 8 h. After reaction completion, a yellow precipitate was isolated via vacuum filtration to afford a yellow powder. The crude product was recrystallized in a small amount of dichloromethane to afford pale-yellow translucent crystals (1.075 g, 69% yield). M.p. 152–155 °C, R_f = 0.66 (CH_2Cl_2). ^{1}H NMR (400 MHz, Chloroform-*d*) δ 8.78 (d, *J* = 2.7 Hz, 1H), 8.50 (dd, *J* = 9.0, 2.7 Hz, 1H), 7.99–7.93 (m, 2H), 7.81–7.74 (m, 3H), 7.56–7.49 (m, 2H), 7.30 (d, *J* = 7.9 Hz, 2H), 2.42 (s, 3H). HRMS (ESI): cald. For $C_{19}H_{14}N_2NaO_7S$ [M + Na]$^+$ 437.0400; found 437.3800.

Preparation of 2,4-Dinitrophenyl-4′-(4-flourophenyl)-benzenesulfonate (**3b**). 2,4-dinitrophenol (0.680 g, 3.69 mmol) and 4′-flourobiphenyl-4-sulfonyl chloride (1.006 g, 3.69 mmol) were added to a flask containing 10 mL of tetrahydrofuran. Of aqueous potassium carbonate 1.6 M (5.0 mL, 7.92 mmol) was added dropwise to the stirred solution. The stirred solution was left at room temperate for 8 h. After reaction completion, a yellow precipitate was isolated via vacuum filtration to afford a yellow powder. The crude product was recrystallized in a small amount of dichloromethane to afford pale-yellow translucent crystals (0.917 g, 59% yield). M.p. 140–143 °C, R_f = 0.64 (CH_2Cl_2). ^{1}H-NMR

(400 MHz, Chloroform-*d*) δ 8.78 (d, *J* = 2.7 Hz, 1H), 8.51 (dd, *J* = 9.0, 2.7 Hz, 1H), 8.02–7.94 (m, 2H), 7.82–7.71 (m, 3H), 7.65–7.55 (m, 2H), 7.24–7.14 (m, 2H). HRMS (ESI): cald. For $C_{18}H_{11}FN_2NaO_7S$ [M + Na]$^+$ 441.0200; found 441.3500.

Spectroscopic and Crystallographic Characterization

^{1}H-NMR spectra (400 MHz) were recorded on a JEOL ECZ400 spectrometer using a DMSO-d_6 or Chloroform-*d* solvent. Chemical shifts are reported in parts per million (ppm, δ) relative to the residual solvent peak, and coupling constants (J) are reported in Hertz (Hz). Results were analyzed and figures were created with the use of MestReNov [8]. The spectra of all compounds synthesized can be found in Figures S1–S8. Spectra were obtained from 16 scans.

The data yielded from the crystallographic characterization of x-substituted 2,4-dinitrophenyl -4′-phenylbenzenesulfonates can be found in Tables S1–S15. Molecular structures of 2,4-dinitrophenyl -4′-phenylbenzenesulfonate (**1b**), 2,4 -dinitrophenyl-4′-(4-methylphenyl)-benzenesulfonate (**2b**), and 2,4 -dinitrophenyl-4′-(4-flourophenyl)-benzenesulfonate (**3b**) can be found in Figures S9–S11. X-ray diffraction was carried out on a Bruker APEXII CCD diffractometer with Mo *K*α radiation. The software used for data collection is as follows: Data collection: *APEX2* [9]; cell refinement: *SAINT* [10]; data reduction: *SAINT* [10]; program used to solve structure: ShelXT [11]; program used to refine structure: *OLEX2* [12,13]; program used to generate figures: Mercury [14–18]; and absorbance correction: *SADABS* [19]. Interactive links for the structures of x-substituted 2,4-dinitrophenyl-4′-phenylbenzenesulfonates are found in the following figures: 2,4-dinitrophenyl -4′-phenylbenzenesulfonate (**1b**), Figure S9; 2,4-dinitrophenyl -4′-(4-methylphenyl)-benzenesulfonate (**2b**), Figure S10; 2,4-dinitrophenyl-4′-(4-flourophenyl)-benzenesulfonate (**3b**), and Figure S11.

3. Results and Discussion

In our work towards developing a facile synthesis of sulfonates, various reaction conditions were investigated. A variety of solvent and base combinations were used to ascertain their effects on the yield and reaction time of sulfonate derivatives, the results of which can be found in Table 1. Entries are presented in order of decreasing yield. Successful formation of the desired product occurred in all cases, except entry 8 where dichloromethane and pyridine were used. This led to the development of a new synthetic method using an aqueous base and the water-miscible solvent, tetrahydrofuran. This single-phase solvent system defeats the need for a phase transfer catalyst and can support a wider range of starting materials. The use of an aqueous base and tetrahydrofuran resulted in higher yields, less environmental impact, and shorter reaction times. Additionally, the desired sulfonate product was isolated directly from the reaction mixture with good purity. An initial concern with this new synthetic method was the hydrolysis of sulfonyl chloride due to the presence of water. However, results show the rate of hydrolysis has little effect on yield.

A comparison of entries 2, 3, and 5 revealed the electronic and steric effects from biphenyl sulfonyl chloride derivatives and their implications in the synthesis of x-substituted 2,4-dinitrophenyl-4′-phenylbenzenesulfonates. The structure and yield of these sulfonates are shown in Figure 1. The highest yielding x-substituted 2,4-dinitrophenyl-4′-phenylbenzenesulfonate was compound **1b**, which has no substituent. Compounds **2b** and **3b**, which contain electron-donating and electron-withdrawing groups, respectively, were slightly lower in yield. In general, the reaction yield may be affected by electronic factors, steric factors or a combination of both. Although it is not clear to us which factor is responsible for the difference in yields, one can imagine a combination of electronic factors and steric factors to be responsible. However, the difference in yields was not statistically significant enough for us to draw clear conclusions on what exactly accounts for the difference in reaction yields.

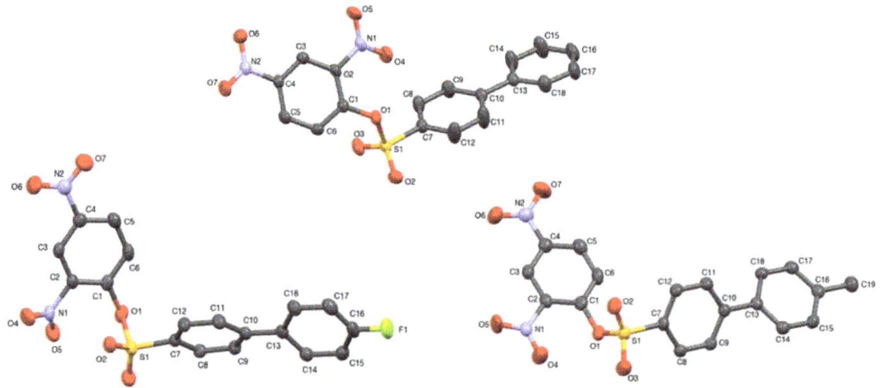

Figure 1. The structure and yield of the synthesized x-substituted 2,4-dinitrophenyl-4′-phenylbenzenesulfonates. Reaction conditions: 2,4-dinitrophenol (1.0 eq) was dissolved in 10 mL of THF. Sulfonyl chloride (1.0 eq) was added dropwise to the solution, followed by the dropwise addition of 1.6 M aq. K_2CO_3 (2.0 eq). All reactions were run at room temperature.

Crystal structures of the x-substituted 2,4-dinitrophenyl-4′-phenylbenzenesulfonates were obtained by single crystal X-ray diffraction. Pertinent data such as bond angles, bond lengths, torsion angles, and other crystallographic parameters can be found in the supplementary material. The asymmetric units of 2,4-dinitrophenyl-4′-phenylbenzenesulfonate (**1b**), 2,4-dinitrophenyl-4′-(4-methylphenyl)-benzenesulfonate (**2b**), and 2,4-dinitrophenyl-4′-(4-flourophenyl)-benzenesulfonate (**3b**) can be found in Figure 2.

Figure 2. The molecular structure of 2,4-dinitrophenyl-4′-phenylbenzenesulfonate (**1b**; **top, middle**), 2,4-dinitrophenyl-4′-(4-methylphenyl)-benzenesulfonate (**2b**; **bottom, right**), and 2,4-dinitrophenyl-4′-(4-flourophenyl)-benzenesulfonate (**3b**; **bottom, left**) with atom labeling scheme. Displacement of ellipsoids is shown at the 50% probability level. Hydrogen atoms have been omitted for clarity.

The x-substituted 2,4-dinitrophenyl-4′-phenylbenzenesulfonates exhibited a two-fold screw axis (−x, 1/2 + y, 1/2 − z), and glide plane geometry (x, 1/2 − y, 1/2 + z) with an inversion center (−x, −y, −z). Screw axis and glide plane geometries are indicative of efficient packing. All three sulfonate structures exhibited a monoclinic system ($P2_1/c$ space group). The central sulfur atom, S1, is a slightly distorted tetrahedron according to the τ_4 descriptor for four-fold coordination [20]. The aryl groups of the sulfonates were oriented gauche about the S1–O1 bond with the following C7–S1–O1–C1 torsion angles: 131.6 (1)°, **1b**; −94.0 (1)°, **2b**; −92.7 (1)°, and **3b** (Table 2). The S1 = O2 and S1 = O3 bond lengths were in good agreement with known values. The S1–C7 and S1–O1 bond lengths were 1.751 (2) Å and 1.626 (1) Å for compound **1b**; 1.745 (2) Å and 1.619 (1) Å for compound **2b**; and 1.743 (2) Å and 1.623 (1) Å for compound **3b**, respectively. The S1–O1–C1 bond angles for compounds **1b**, **2b**, and **3b** were 120.5 (1)°, 120.4 (1)°, and 119.2 (1)°, respectively. Molecules were liked by π–π interactions, C–H⋯O hydrogen bonds, C–H⋯π interactions, and, in the case of compound **3b**, C–H⋯F hydrogen bonds (Figure 3). Hydrogen atoms bonded to carbon atoms were placed in calculated positions and refined as riding: C–H = 0.95–1.00 Å with a fixed $U_{iso}(H) = 1.2U_{eq}(C)$ for all C–H groups.

Table 2. Bond lengths of S1–C7, S1–O1, S1 = O2, and S1 = O3 (Å), bond angles of S1–O1–C1 (°) and torsion angles of C7–S1–O1–C1 (°) for 2,4-dinitrophenyl-4′-phenylbenzenesulfonate (**1b**), 2,4-dinitrophenyl-4′-(4-methylphenyl)-benzenesulfonate (**2b**), and 2,4-dinitrophenyl-4′-(4-flourophenyl)-benzenesulfonate (**3b**).

	Geometric Parameters (Å, °)		
	1b	**2b**	**3b**
S1–C7	1.751 (2)	1.745 (2)	1.743 (2)
S1–O1	1.626 (1)	1.619 (1)	1.623 (1)
S1=O2	1.420 (1)	1.421 (1)	1.424 (1)
S1=O3	1.423 (1)	1.417 (2)	1.413 (1)
S1–O1–C1	120.5 (1)	120.4 (1)	119.2 (1)
C7–S1–O1–C1	131.6 (1)	−94.0 (1)	−92.7 (1)

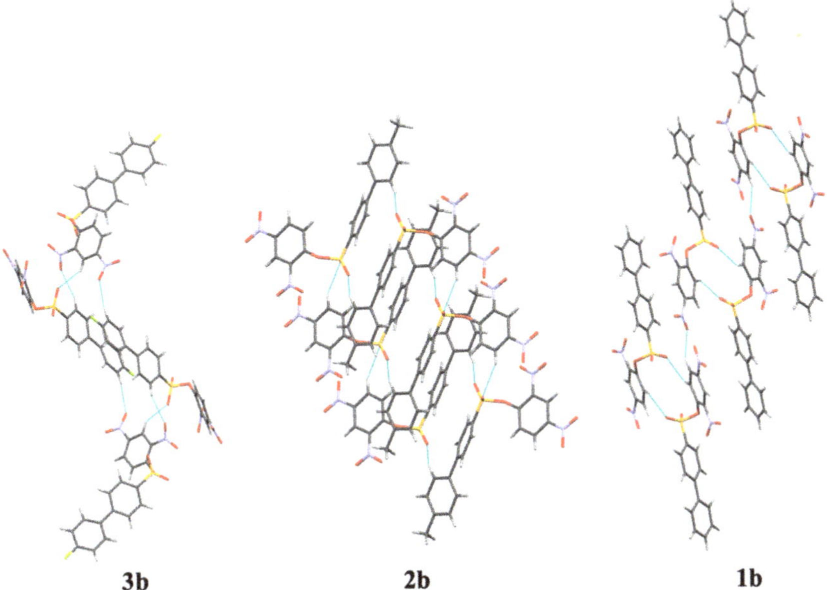

Figure 3. A depiction of the inter- and intramolecular hydrogen bonds present in the crystal structures of 2,4-dinitrophenyl-4′-phenylbenzenesulfonate (**1b**; **right**), 2,4-dinitrophenyl-4′-(4-methylphenyl)-benzenesulfonate (**2b**; **middle**), and 2,4-dinitrophenyl-4′-(4-flourophenyl)-benzenesulfonate (**3b**; **left**) shown as capped sticks with standard CPK colors. Hydrogen bonds and contacts are depicted with cyan dashed lines.

The extent of hydrogen bonding varied significantly in the sulfonate derivatives. In the case of compound 2b, only two C–H⋯O hydrogen bonds were observed (Table 3). A greater number of hydrogen bonds in compounds **1b** and **3b** lend credence to a lattice dependent on hydrogen bond contacts for efficient packing. The F1⋯H17 bond in compound **3b** had a bond length of 2.450 Å, the shortest and therefore strongest hydrogen bond of the sulfonate compounds.

Table 3. Length of hydrogen-bond contacts (Å) and corresponding symmetry codes for 2,4-dinitrophenyl-4′-phenylbenzenesulfonate (**1b**), 2,4-dinitrophenyl-4′-(4-methylphenyl)-benzenesulfonate (**2b**), and 2,4-dinitrophenyl-4′-(4-flourophenyl)-benzenesulfonate (**3b**). Atoms labels follow the atom numbering scheme in Figure 2.

	Bond Length (Å)				
	(1b)	**(2b)**	**(3b)**	**Symmetry Codes**	
H12··· O4	2.534			x, y, z	x, −1 + y, z
H16··· O5	2.650			x, y, z	−x, −1/2 + y, 1.5 − z
H5··· O7	2.480			x, y, z	1 − x, −1/2 + y, 1/2 − z
O3··· H6	2.719			x, y, z	1 − x, 1 − y, 1 − z
O5··· H15	2.602			x, y, z	x, 2.5 − y, −1/2 + z
O6··· H9	2.580			x, y, z	x, 2.5 − y, −1/2 + z
O2··· H5		2.596		x, y, z	−1 + x, y, z
O3··· H14		2.681		x, y, z	−x, 1 − y, 1 − z
O2··· H5			2.478	x, −1 + y, z	−1 + x, −1 + y, z
O4··· H12			2.652	x, −1 + y, z	1 − x, −1.5 + y, 1/2 − z
H3··· O2			2.494	x, −1 + y, z	1 − x, −1.5 + y, 1/2 − z
O3··· H9			2.653	x, −1 + y, z	1 − x, 1 − y, 1 − z
O3··· H14			2.567	x, −1 + y, z	1 − x, 1 − y, 1 − z
F1··· H17			2.450	x, −1 + y, z	2 − x, 3 − y, 1 − z
O6··· H15			2.606	x, −1 + y, z	x, 1.5 − y, −1/2 + z

4. Conclusions

The proposed synthetic method is useful in producing a broad range of sulfonates in good yields with the elimination of chlorinated solvents and organic bases, compared to previous methods. Additionally, the new synthetic method significantly decreased reaction time and work up. It was determined that the reaction favored a single-phase solvent system involving an aqueous base and water-miscible solvent. This facile method produced sulfonate products in high purity. The resulting products were characterized by crystallographic and spectroscopic methods. Crystallographic characterization revealed sulfonate structures exhibiting a monoclinic system ($P2_1/c$ space group) with a two-fold screw axis (−x, 1/2 + y, 1/2 − z) and glide plane geometry (x, 1/2 − y, 1/2 + z) and inversion center (−x, −y, −z). Values obtained for the S = O bond lengths closely resembled known values. Four-fold coordination about the S1 atom reviled a slightly distorted tetrahedron geometry. The crystallographic results supported the successful formation of pure sulfonate products. The resulting products will be used to investigate the reaction conditions for the regioselective cleavage of C–O and S–O bonds in the amination of arylsulfonates.

Supplementary Materials: The following are available online at http://www.mdpi.com/2624-8549/2/2/36/s1, Figures S1–S8 showing ^{1}H-NMR spectra, Figures S9–S11 depicting asymmetric units of x-substituted 2,4-dinitrophenyl-4′-phenylbenzenesulfonates with interactive links, and tables S1–S15 containing the crystallographic data for these compounds.

Author Contributions: Conceptualization, B.A.S. and F.N.N.; Methodology and Practical Chemical and Spectroscopic Studies, B.A.S. and F.N.N.; Crystallography, B.A.S., R.J.S., S.M.B., and F.N.N.; Writing Draft Report, B.A.S. and F.N.N.; Writing of Manuscript and Editing, B.A.S. and F.N.N.; Project Administration and Funding Acquisition, F.N.N. All authors have read and agreed to the published version of the manuscript.

Funding: Funding for this research was provided by the following: National Science Foundation (grant No. MRI CHE-1725699); Grand Valley State University Chemistry Department's Weldon Fund.

Acknowledgments: The authors thank Pfizer, Inc. for the donation of a Varian INOVA 400 FT NMR spectrometer. The CCD-based X-ray diffractometers at Michigan State University were upgraded and/or replaced by departmental funds.

Conflicts of Interest: No potential conflict of interest was declared by the authors.

References

1. Vedovato, V.; Talbot, E.P.A.; Willis, M.C. Copper-Catalyzed Synthesis of Activated Sulfonate Esters from Boronic Acids, DABSO, and Pentafluorophenol. *Org. Lett.* **2018**, *20*, 5493–5496. [CrossRef] [PubMed]
2. Sun, J.; Wang, Q.; Yang, J.; Zhang, J.; Li, Z.; Li, H.; Yang, X. 2,4-Dinitrobenzenesulfonate-functionalized carbon dots as a turn-on fluorescent probe for imaging of biothiols in living cells. *Microchim. Acta* **2019**, *186*, 402. [CrossRef] [PubMed]
3. Zhang, P.; Ding, Y.; Liu, W.; Niu, G.; Zhang, H.; Ge, J.; Wu, J.; Li, Y.; Wang, P. Red emissive fluorescent probe for the rapid detection of selenocysteine. *Sens. Actuators* **2018**, *264*, 234–239. [CrossRef]
4. Hatfield, D.L.; Tsuji, P.A.; Carlson, B.A.; Gladyshev, V.N. Selenium and selenocysteine: Roles in cancer, health, and development. *Trends Biochem. Sci.* **2014**, *39*, 112–120. [CrossRef] [PubMed]
5. Caddick, S.; Wilden, J.D.; Judd, D.B. Direct Synthesis of Sulfonamides and Activated Sulfonate Esters from Sulfonic Acids. *J. Am. Chem. Soc.* **2004**, *126*, 1024–1025. [CrossRef] [PubMed]
6. Jiang, Y.; Alharbi, N.S.; Sun, B.; Qin, H. Facile one-pot synthesis of sulfonyl fluorides from sulfonates or sulfonic acids. *RSC Adv.* **2019**, *9*, 13863–13867. [CrossRef]
7. Tipson, R. On esters of *p*-toluenesulfonic acid. *J. Org. Chem.* **1944**, *3*, 235–241. [CrossRef]
8. Willtott, M.R. MestRe Nova. *J. Am. Chem. Soc.* **2009**, *131*, 13180.
9. *APEX2*, Bruker AXS Inc.: Madison, WI, USA, 2013.
10. *SAINT*, Bruker AXS Inc.: Madison, WI, USA, 2013.
11. Sheldrick, G.M. Crystal Structure Refinement with SHELXL. *Acta Cryst.* **2015**, *71*, 3–8.
12. Dolomanov, O.V.; Bourhis, L.J.; Gildea, R.J.; Howard, J.A.K.; Puschmann, H. OLEX2: A Complete Structure Solution, Refinement and Analysis Program. *J. Appl. Cryst.* **2009**, *42*, 339–341. [CrossRef]
13. Bourhis, L.J.; Dolomanov, O.V.; Gildea, R.J.; Howard, J.A.K.; Puschmann, H. The anatomy of a comprehensive constrained, restrained refinement program for the modern computing environment-Olex2 dissected. *Acta Cryst.* **2015**, *A71*, 59–75.
14. Macrae, C.F.; Sovago, I.; Cottrell, S.J.; Galek, P.T.A.; McCabe, P.; Pidcock, E.; Platings, M.; Shields, G.P.; Stevens, J.S.; Towler, M.; et al. Mercury 4.0: From visualization to analysis, design and prediction. *J. Appl. Cryst.* **2020**, *53*. [CrossRef] [PubMed]
15. Macrae, C.F.; Bruno, I.J.; Chisholm, J.A.; Edgington, P.R.; McCabe, P.; Pidcock, E.; Rodriguez-Monge, L.; Taylor, R.; van de Streek, J.; Wood, P.A. Mercury CSD 2.0-New Features for the Visualization and Investigation of Crystal Structures. *J. Appl. Cryst.* **2008**, *41*, 466–470. [CrossRef]
16. Macrae, C.F.; Edgington, P.R.; McCabe, P.; Pidcock, E.; Shields, G.P.; Taylor, R.; Towler, M.; van de Streek, J. Mercury: Visualization and analysis of crystal structures. *J. Appl. Cryst.* **2006**, *39*, 453–457. [CrossRef]
17. Bruno, I.J.; Cole, J.C.; Edgington, P.R.; Kessler, M.K.; Macrae, C.F.; McCabe, P.; Pearson, J.; Taylor, R. New software for searching the Cambridge Structural Database and visualising crystal structures. *Acta Cryst.* **2002**, *B58*, 389–397. [CrossRef] [PubMed]
18. Taylor, R.; Macrae, C.F. Rules governing the crystal packing of mono- and di-alcohols. *Acta Cryst.* **2001**, *B57*, 815–827. [CrossRef] [PubMed]
19. Krause, L.; Herbst-Irmer, R.; Sheldrick, G.M.; Stalke, D. Comparison of silver and molybdenum microfocus X-ray sources for single-crystal structure determination. *J. Appl. Cryst.* **2015**, *48*, 3–10. [CrossRef] [PubMed]
20. Yang, L.; Powell, D.R.; Houser, R.P. Structural variation in copper(i) complexes with pyridylmethylamide ligands: Structural analysis with a new four-coordinate geometry index, τ_4. *Dalton Trans.* **2007**, *9*, 955–956. [CrossRef] [PubMed]

Article

Structural Elucidation of Enantiopure and Racemic 2-Bromo-3-Methylbutyric Acid [†]

Rüdiger W. Seidel [1,*], Nils Nöthling [2], Richard Goddard [2] and Christian W. Lehmann [2]

[1] Martin-Luther-Universität Halle-Wittenberg, Institut für Pharmazie, Wolfgang-Langenbeck-Str. 4, 06120 Halle (Saale), Germany

[2] Max-Planck-Institut für Kohlenforschung, Kaiser-Wilhelm-Platz 1, 45470 Mülheim an der Ruhr, Germany; noethling@mpi-muelheim.mpg.de (N.N.); goddard@mpi-muelheim.mpg.de (R.G.); lehmann@mpi-muelheim.mpg.de (C.W.L.)

* Correspondence: ruediger.seidel@pharmazie.uni-halle.de

† Dedicated to Dr. Howard Flack (1943–2017).

Received: 27 May 2020; Accepted: 17 July 2020; Published: 30 July 2020

Abstract: Halogenated carboxylic acids have been important compounds in chemical synthesis and indispensable research tools in biochemical studies for decades. Nevertheless, the number of structurally characterized simple α-brominated monocarboxylic acids is still limited. We herein report the crystallization and structural elucidation of (*R*)- and *rac*-2-bromo-3-methylbutyric acid (2-bromo-3-methylbutanoic acid, **1**) to shed light on intermolecular interactions, in particular hydrogen bonding motifs, packing modes and preferred conformations in the solid-state. The crystal structures of (*R*)- and *rac*-**1** are revealed by X-ray crystallography. Both compounds crystallize in the triclinic crystal system with $Z = 2$; (*R*)-**1** exhibits two crystallographically distinct molecules. In the crystal, (*R*)-**1** forms homochiral O–H⋯O hydrogen-bonded carboxylic acid dimers with approximate non-crystallographic C_2 symmetry. In contrast, *rac*-**1** features centrosymmetric heterochiral dimers with the same carboxy *syn*⋯*syn* homosynthon. The crystal packing of centrosymmetric *rac*-**1** is denser than that of its enantiopure counterpart (*R*)-**1**. The molecules in both crystal structures adopt a virtually identical staggered conformation, despite different crystal environments, which indicates a preferred molecular structure of **1**. Intermolecular interactions apart from classical O–H⋯O hydrogen bonds do not appear to have a crucial bearing on the solid-state structures of (*R*)- and *rac*-**1**.

Keywords: 2-bromo-3-methylbutyric acid; 2-bromo-3-methylbutanoic acid; 2-bromoisovaleric acid; halogenated carboxylic acid; hydrogen bonding; chirality; absolute configuration; racemate; crystal structure; X-ray crystallography

---

1. Introduction

Halogenated organic compounds have received considerable research interest for decades, not only in the field of chemical synthesis [1–3] but also because of their biological properties [4,5]. In particular, a vast number of halogenated carboxylic acids have been synthesized and biochemically studied. Since mono-, di- and tricarboxylic acids are important intermediates in many biochemical pathways, their halogenated analogues have become an important research tool for the study of a wide range of biological processes owing to their ability to imitate the properties of the respective carboxylic acids or to inhibit crucial enzymes [6,7]. Despite tremendous research interest in halogenated carboxylic acids, the number of crystal structures of simple α-brominated monocarboxylic acids in the Cambridge Structural Database (CSD) is limited (13 as of June 2020) [8]. An example is bromoacetic acid, which forms a common *syn*⋯*syn* hydrogen-bonded carboxy dimer (Scheme 1) in the crystal (CSD refcode: BRMACA) [9]. Others are (−)-2-bromosuccinamic acid (BRSCAM) [10] and two crystal forms of 2,3-dibromo-3-phenylpropionic acid (CSD refcodes: ROFNOQ and ROFNOQ01) [11,12].

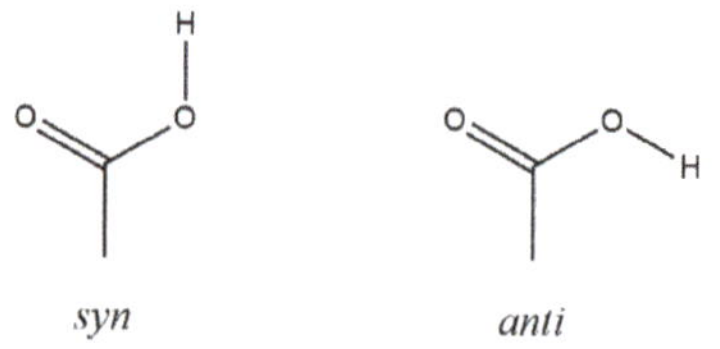

Scheme 1. Carboxy group *syn* and *anti* conformations [13].

2-Bromo-3-methylbutyric acid (2-bromo-3-methylbutanoic acid, **1**), commonly known as 2-bromoisovaleric or α-bromoisovaleric acid is a chiral α-halogenated monocarboxylic acid. Scheme 2 depicts the two enantiomers, (*S*)-**1** and (*R*)-**1**. Their resolution by fractional crystallization was reported almost 100 years ago [14]. Auterhoff and Lang, for example, used this approach to prepare both enantiomers of the hypnotic and sedative agent bromisoval (2-bromo-3-methylbutyrylurea or commonly bromovalerylurea) from (*S*)-**1** and (*R*)-**1** by reaction of the respective acid chlorides with urea [15]. Despite the fact that **1** has long been known and is commercially available, to the best of our knowledge and based on a WebCSD search in May 2020 [16], a crystal structure of **1** has not been reported so far.

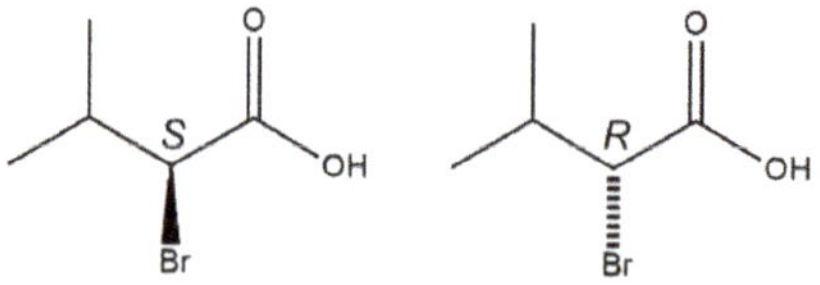

Scheme 2. Chemical diagrams of the enantiomers of the title compound, (*S*)-**1** (**left**) and (*R*)-**1** (**right**) with stereodescriptors.

Chiral carboxylic acids have also attracted research interest in the fields of structural chemistry and crystal engineering, owing to phenomena such as frustration between molecular chirality and centrosymmetric hydrogen bond homosynthon formation in their crystal packing [17]. Enantiomeric mixtures can essentially crystallize as racemic crystals, racemic conglomerates (physical mixture of resolved crystals), inversion twins, disordered solid solutions [18] or, rarely, as kryptoracemates [19]. In this context, **1** attracted our attention. We have crystallized and investigated solvent-free **1** by X-ray crystallography in order to reveal preferred molecular conformations, crystal packing, intermolecular interactions and the outcome of crystallization of an enantiomeric mixture. Herein we report the crystal and molecular structures of (*R*)-**1** and *rac*-**1**.

2. Materials and Methods

(*S*)- and (*R*)-**1** were purchased from Sigma-Aldrich (purity 96%) and used as received. Solvents were of analytical grade and used without further purification. Crystals of (*R*)-**1** suitable for single-crystal X-ray diffraction were obtained from an ethanolic solution by slow evaporation of the solvent at ambient conditions. To obtain *rac*-**1**, equimolar amounts of (*S*)- and (*R*)-**1** were melted together on a Reichert hot-stage (Mikroheiztisch) mounted on a Nikon SMZ 1500 binocular microscope and cooled to room temperature [20]. The material so obtained was dissolved in ethyl acetate. Single-crystals of *rac*-**1** suitable for X-ray analysis appeared when the solvent was allowed to evaporate slowly at ambient conditions.

The X-ray intensity data were collected at 100(2) K on an Enraf–Nonius Kappa CCD for (*R*)-**1** and on a Bruker AXS Apex II for *rac*-**1**, using Mo K_α radiation in both cases. The data were scaled and corrected for absorption effects with SADABS [21]. The crystal structures were solved with SHELXT [22] and refined with SHELXL-2018/3 [23]. The highest residual difference electron density peak each for (*R*)-**1** and *rac*-**1** is ca. 0.7 Å from a bromine atom and can be ascribed to absorption effects. Carbon-bound hydrogen atoms were placed at geometrically calculated positions with $C_{methine}$–H = 1.00 Å, C_{methyl}–H = 0.98 Å and refined using a riding model with $U_{iso}(H) = 1.2\ U_{eq}(C)$ (1.5 for methyl groups). Torsion angles of the methyl groups were initially determined via difference Fourier syntheses and subsequently refined while maintaining tetrahedral angles at the carbon atoms. The carboxy hydrogen atoms in (*R*)-**1** were located in difference electron density maps. In subsequent refinements, the O–H distances were restrained to a target value of 0.84(2) Å. In *rac*-**1**, the carboxy hydrogen atom was placed in an idealized hydrogen bonding position with O–H = 0.84 Å and refined using a riding model. $U_{iso}(H) = 1.2\ U_{eq}(O)$ was used for all carboxy hydrogen atoms. Refined and post-refinement values of the Flack *x* parameter [24] were obtained with SHELXL using TWIN/BASF instructions and Parsons's method [25], respectively. The Hooft parameter [26–28] was calculated with PLATON [29]. Crystal data and refinement details for (*R*)-**1** and *rac*-**1** are listed in Table 1. Representations of the crystal and molecular structures were drawn with DIAMOND [30]. The structure overlay diagram and r.m.s. deviations of molecular structures from one another were obtained with MERCURY [31]. Packing indices were calculated with PLATON.

Table 1. Crystal data and refinement details for (*R*)-**1** and *rac*-**1**.

	(*R*)-**1**	*Rac*-**1**
empirical formula	$C_5H_9BrO_2$	$C_5H_9BrO_2$
M_r	181.03	181.03
T (K)	100(2)	100(2)
λ (Å)	0.71073	0.71073
crystal system	triclinic	triclinic
space group	$P1$	$P\bar{1}$
a (Å)	6.0261(11)	6.5849(14)
b (Å)	6.7000(16)	7.5490(16)
c (Å)	9.900(2)	7.7328(17)
α (°)	102.144(17)	112.283(4)
β (°)	102.477(15)	92.655(4)
γ (°)	107.20(3)	101.085(3)
V (Å^3)	356.34(14)	346.03(13)
Z, Z'	2, 2	2, 1
ρ_{calc} (mg m^{-3})	1.687	1.737
μ (mm^{-1})	5.685	5.854
$F(000)$	180	180
crystal size (mm)	$0.350 \times 0.180 \times 0.100$	$0.279 \times 0.226 \times 0.128$
θ range for data collection (°)	3.411–38.060	2.872–37.221
reflections collected/unique	18,005/7664	12,972/3401
R_{int}	0.0237	0.0371
observed reflections [$I > 2\sigma(I)$]	6888	2743
T_{max}/T_{min}	0.58973/0.2342	0.61929/0.32069
data/restraints/parameters	7664/5/155	3401/0/75
Goodness-of-fit on F^2	1.089	1.041
$R1$ [$I > 2\sigma(I)$]	0.0287	0.0421
$wR2$ (all data)	0.0694	0.1116
Flack *x* parameter (refined)	0.000(8)	-
Flack *x* parameter (from quotients)	−0.006(5) [3068 quotients]	-
Hooft parameter	−0.011(4)	-
$\Delta\rho_{max}/\Delta\rho_{min}$ (e Å^{-3})	1.30/−0.72	2.74/−0.56

3. Results

Both (*R*)-**1** and *rac*-**1** were found to crystallize in the triclinic crystal system with two molecules in the unit cell. As shown in Figure 1, the molecules form O–H···O hydrogen bonded dimers through the carboxy groups in the *syn* conformation with a $R_2^2(8)$ motif [32] in both crystal structures. Geometric parameters of the hydrogen bonds in both (*R*)-**1** and *rac*-**1** are given in Table 2, and selected bond lengths, bond angles and torsion angles are listed in Table 3. The encountered homochiral hydrogen-bonded dimer in (*R*)-**1** comprises two crystallographically unique molecules (*Z′* = 2) and features approximate non-crystallographic C_2 symmetry with the twofold rotation axis passing through the center of the $R_2^2(8)$ hydrogen-bonded set and perpendicular to the mean plane of the two carboxy groups. In contrast, the hydrogen-bonded dimer in *rac*-**1** is heterochiral and lies across a crystallographic inversion center and, thus, is centrosymmetric. The molecule in the chosen asymmetric unit of *rac*-**1** (*Z′* = 1) exhibits *R* configuration (see Figure 1, bottom). The two distinct molecules in (*R*)-**1** and the *R* enantiomer in *rac*-**1** adopt the same staggered conformation, as illustrated by a Newman projection in Scheme 3. A structure overlay diagram for the three molecular structures is depicted in Figure 2. The r.m.s. deviation of the respective non-hydrogen atoms in the two distinct molecules in (*R*)-**1** is 0.0322 Å. Between the respective non-hydrogen atoms and those of the *R* enantiomer in *rac*-**1**, the r.m.s. deviations are 0.0234 and 0.0243 Å.

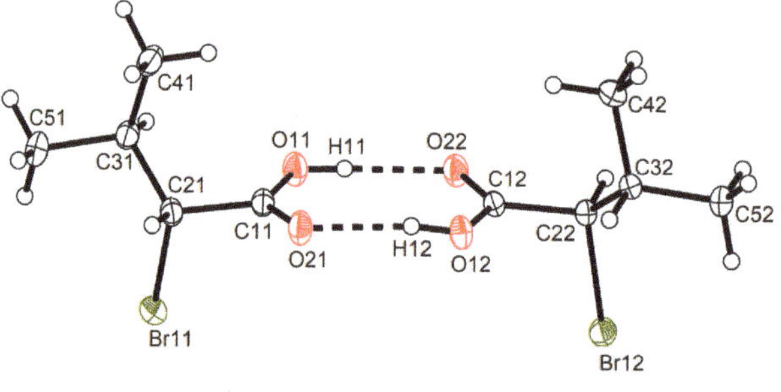

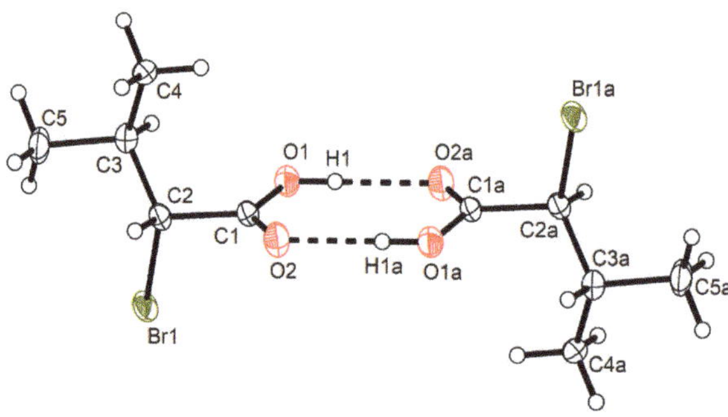

Figure 1. Homochiral and heterochiral hydrogen-bonded dimers respectively in (*R*)-**1** (**top**) and *rac*-**1** (**bottom**) in their crystal structures. Displacement ellipsoids are drawn at the 50% probability level. Hydrogen atoms are represented by small spheres of arbitrary radii. Dashed lines represent hydrogen bonds. Symmetry code: (a) −x + 2, −y + 1, −z + 1.

Table 2. Hydrogen bond geometry for (*R*)-**1** and *rac*-**1** (Å, °) [1].

D–H···*A*	*d*(D–H)	*d*(H···A)	*d*(D···A)	<(DHA)
		(*R*)-**1**		
O11–H11···O22	0.82(2)	1.82(2)	2.636(3)	170(4)
O12–H12···O21	0.82(2)	1.82(2)	2.635(3)	171(4)
		rac-**1**		
O1–H1···O2a	0.84	1.82	2.658(2)	175

[1] Symmetry code: (a) −x + 2, −y + 1, −z + 1.

Table 3. Selected bond lengths, bond angles and torsion angles (°) for (*R*)-**1** and *rac*-**1** (Å, °) [1].

	(*R*)-**1**		*Rac*-**1**
	Molecule 1	Molecule 2	
C2–Br1	1.969(3)	1.966(2)	1.972(2)
C1–O1	1.291(3)	1.282(3)	1.302(3)
C1–O2	1.242(3)	1.248(3)	1.217(3)
O2–C1–O1	124.5(2)	124.4(2)	123.84(19)
O1–C1–C2	116.06(19)	116.07(18)	115.35(19)
O2–C1–C2	119.4(2)	119.51(19)	120.8(2)
C1–C2–C3–C4	−59.2(3)	−61.1(2)	−59.3(2)
C1–C2–C3–C5	179.7(2)	176.70(19)	179.18(19)
O1–C1–C2–C3	−44.8(3)	−46.7(3)	−46.3(3)
Br1–C2–C3–C4	−177.52(15)	−179.30(15)	−177.72(13)
Br1–C2–C3–C5	61.3(2)	58.5(2)	60.8(2)
Br1–C2–C1–O1	78.1(2)	75.9(2)	76.86(19)

[1] Molecule 1 and molecule 2 in (*R*)-**1** are depicted respectively on the left- and right-hand side in Figure 1 (top).

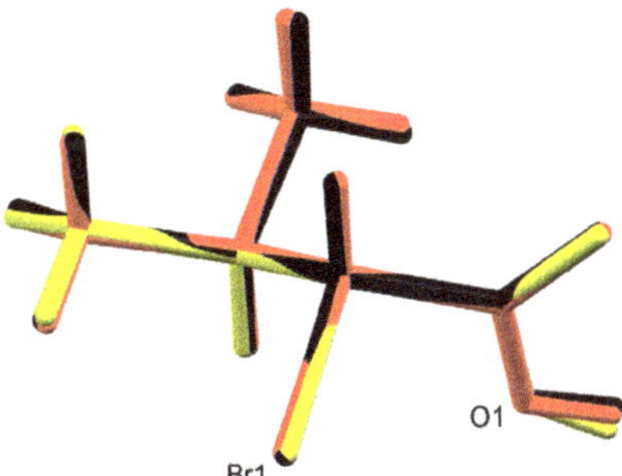

Scheme 3. Newman-projection illustrating the staggered conformation of the *R* enantiomer encountered in the crystal structures of (*R*)-**1** and *rac*-**1**. For the corresponding torsion angles, see Figure 1 and Table 3.

Figure 2. Structure overlay of the two crystallographic distinct molecules in (*R*)-**1** (red and yellow) and the *R* enantiomer in *rac*-**1** (black).

The supramolecular structure of (*R*)-**1** in the crystal features short C–H···O contacts between the α-carbon atom of the carboxylic acid and the (formal) carboxy C=O moiety of an adjacent molecule (Figure 3). The hydrogen bonding motif descriptor is likewise $R_2^2(8)$. In *rac*-**1**, the methine group of the α-carbon atom does not form a similar short C–H···O contact. The (formal) carboxy C=O moiety, however, is approached by a methyl hydrogen atom of the isopropyl group of an adjacent molecule (C···O = 3.58 Å, see Figure S1). The crystal packing in *rac*-**1** is denser than in (*R*)-**1**, which is evident from the volumes of the triclinic unit cells of both $Z = 2$ structures (Table 1). Thus, each molecule in *rac*-**1** occupies 5.2 Å^3 less space in the crystal than in (*R*)-**1**. The calculated densities (Table 1) and Kitaigorodskij packing indices [33] of 68.2% for *rac*-**1** and 66.2% for (*R*)-**1** further indicate a denser crystal packing in *rac*-**1** than in (*R*)-**1**. Short contacts (with respect to the sum of the corresponding van der Waals radii) of the Br···Br or C–H···Br type are neither observed in (*R*)-**1** nor in *rac*-**1**.

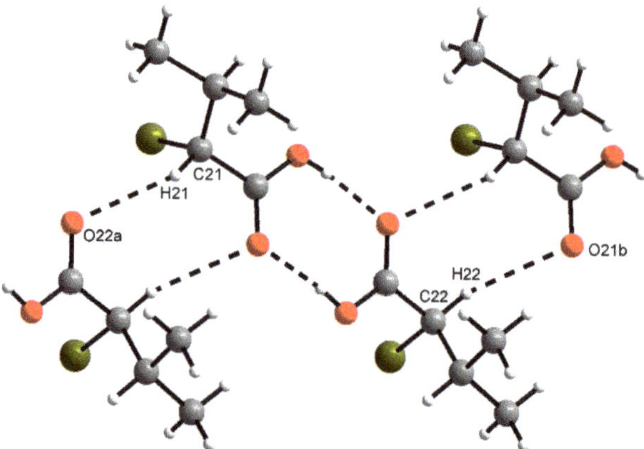

Figure 3. Section of the crystal structure of (*R*)-**1**, showing $R_2^2(8)$ motifs of intermolecular C–H···O and O–H···O contacts (*cf.* Figure 1). C21–H21···O22a: $d(D···A) = 3.384(3)$ Å, $<(DHA) = 145°$; C22–H22···O21b: $d(D···A) = 3.519(3)$ Å, $<(DHA) = 162°$. Symmetry codes: (a) x − 1, y − 1, z; (b) x + 1, y + 1, z.

4. Discussion

The *absolute structure* of a single-crystal of (*R*)-**1** grown from solution was established by anomalous-dispersion effects in the diffraction intensity measurements [34], thereby confirming the *absolute configuration* reported for the purchased bulk material [35,36]. Absolute structure parameters are listed in Table 1. The Flack *x* parameter estimated post-refinement based on quotients [25] and the Hooft parameter based on Bayesian statistics [26–28] are close to zero with adequately small standard uncertainties [18]. By way of comparison, the standard uncertainty of the refined Flack *x* parameter is larger than that of the former two parameters by a factor of ca. two [37].

The X-ray analysis of *rac*-**1** clearly revealed that an equimolar mixture of both enantiomers (Scheme 2) forms a racemic crystal upon crystallization from ethyl acetate and not a racemic conglomerate, which is observed in only ca. 10% of cases [38]. The higher crystallographic density of *rac*-**1** is in accord with Wallach's rule from 1895, which states that racemic crystals tend to be denser than the chiral counterparts [39]. This phenomenon can essentially be explained by the fact that enantiopure compounds can only crystallize in a Sohncke space group, devoid of inversion symmetry. To enable densest packing with $Z' = 1$ in *rac*-**1**, the molecular dimer must be placed across a crystallographic inversion center, which would be impossible for homochiral *R*···*R* and *S*···*S* dimers. (*R*)-**1** crystallizes with $Z' = 2$ in the space group *P*1, which is not among those available for densest packing of molecules of arbitrary shape [33]. Crystallization with $Z' > 1$ is a common phenomenon for chiral carboxylic acids and has been described as frustration between chirality (referring to the whole molecule) and centrosymmetric dimer formation (referring to the hydrogen bond synthon) [17].

According to theoretical studies, the *syn* conformation of a carboxy group, as observed in (*R*)-**1** and *rac*-**1**, is energetically more stable than the *anti* conformation by ca. 21.4–28.9 kJ mol^{-1} [13]. A *syn···syn* dimer (homosynthon) with a $R_2^2(8)$ motif is a hydrogen bonding pattern commonly observed for carboxylic acids [40,41]. Its occurrence in the crystal structures of (*R*)-**1** and *rac*-**1** is thus as expected and also in accord with Etter's rules for hydrogen bonding, whereby all acidic hydrogen atoms and all good hydrogen bond acceptors are involved in hydrogen bonds, and the best donor and the best acceptor are hydrogen-bonded to one another [42]. The short C–H···O contacts observed in (*R*)-**1** (Figure 3) could be interpreted geometrically as weak hydrogen bonds [43]. It reasonable to assume that the hydrogen atom at the α-carbon atom here is prone to weak hydrogen bonds, since the carboxy group as well as the bromine atom should exert an electron-withdrawing effect. Since such contacts are not present in *rac*-**1**, their impact on the overall supramolecular structure in the crystal is probably minor.

The molecular conformations found in the crystal structures of (*R*)-**1** and *rac*-**1** are virtually identical, as evidenced by the calculated r.m.s. deviations of the corresponding heavy atom skeletons and visualized by a structure overlay diagram (Figure 2). It is reasonable to assume that a staggered conformation corresponds to a minimum energy structure, since not only the carbon chains but also the carboxy groups adopt the same orientation in the three molecular structures (Table 3) despite different crystal environments. This suggests that the observed conformation represents a preferred molecular structure of **1**.

5. Conclusions

We have revealed the crystal and molecular structures of (*R*)-**1** and *rac*-**1** by single-crystal X-ray analysis. The absolute configuration of (*R*)-**1** was confirmed by means of anomalous dispersion effects in the diffraction intensity measurements. Not unexpectedly, the ubiquitous carboxy *syn···syn* homosynthon was encountered in both structures. Clearly, O–H···O hydrogen bonds are the dominant intermolecular interaction in both structures. As compared with *rac*-**1**, the more open structure of (*R*)-**1** and the existence of two molecules in its asymmetric unit can be ascribed to frustration between chirality and centrosymmetric homosynthon formation. The observed denser crystal packing of centrosymmetric *rac*-**1** than of its enantiopure counterpart (*R*)-**1** is in accord with Wallach's rule. Short C–H···O contacts, as formed by the α-methine group in (*R*)-**1**, are not encountered in *rac*-**1**. This suggests that these weak intermolecular interactions may not have a crucial bearing on the packing of the hydrogen-bonded carboxylic acid dimers in the solid-state here, which appears to be essentially governed by close packing. A virtually identical molecular conformation in all in total three crystallographically distinct molecules in (*R*)-**1** and *rac*-**1**, despite different crystal environments, suggests that the observed geometry represents the preferred low energy structure.

Supplementary Materials: The following are available online at http://www.mdpi.com/2624-8549/2/3/44/s1, Figure S1: Section of the crystal structure of *rac*-**1**, showing short contacts between methyl hydrogen atoms of the isopropyl groups and the (formal) carboxy C=O moieties of adjacent molecules (C···O = 3.58 Å). CCDC 200603 [(*R*)-**1**] and 2006031 (*rac*-**1**) contain the supplementary crystallographic data for this paper. These data can be obtained free of charge from The Cambridge Crystallographic Data Centre via www.ccdc.cam.ac.uk/structures.

Author Contributions: Conceptualization, N.N., R.G. and R.W.S.; methodology, N.N. and R.G.; validation, R.W.S. and R.G.; formal analysis, R.W.S. and R.G.; investigation, N.N. and R.G.; resources, C.W.L.; data curation, R.W.S. and R.G.; writing—original draft preparation, R.W.S.; writing—review and editing, R.G.; visualization, R.W.S. and R.G.; supervision, C.W.L.; project administration, R.W.S.; All authors have read and agreed to the published version of the manuscript.

Funding: This research received no external funding.

Acknowledgments: The authors would like to thank Alois Fürstner for providing laboratory resources for this project. R.W.S. is grateful to Peter Imming for his support.

Conflicts of Interest: The authors declare no conflict of interest.

References and Note

1. Czekelius, C.; Tzschucke, C.C. Synthesis of Halogenated Carboxylic Acids and Amino Acids. *Synth. Stuttg.* **2010**, *543*–566. [CrossRef]
2. Kubitschke, J.; Lange, H.; Strutz, H. Carboxylic Acids, Aliphatic. *Ullmann's Encycl. Ind. Chem.* **2014**, 1–18. [CrossRef]
3. *Ullmann's Fine Chemicals*; Wiley: New York, NY, USA, 2014.
4. Kirk, K.L. Biochemistry of halogenated organic compounds. In *PATAI'S Chemistry of Functional Groups*; Rappoport, Z., Ed.; Wiley: New York, NY, USA, 2009. [CrossRef]
5. Yue, Y.; Chen, J.; Bao, L.; Wang, J.; Li, Y.; Zhang, Q. Fluoroacetate dehalogenase catalyzed dehalogenation of halogenated carboxylic acids: A QM/MM approach. *Chemosphere* **2020**, *254*, 126803. [CrossRef] [PubMed]
6. Whitehouse, S.; Cooper, R.H.; Randle, P.J. Mechanism of activation of pyruvate dehydrogenase by dichloroacetate and other halogenated carboxylic acids. *Biochem. J.* **1974**, *141*, 761–774. [CrossRef] [PubMed]
7. Kirk, K.L. Biochemistry of Halogenated Carboxylic Acids. In *Biochemistry of Halogenated Organic Compounds*; Kirk, K.L., Ed.; Springer: Boston, MA, USA, 1991. [CrossRef]
8. Groom, C.R.; Bruno, I.J.; Lightfoot, M.P.; Ward, S.C. The Cambridge Structural Database. *Acta Crystallogr. B Struct. Sci. Cryst. Eng. Mater.* **2016**, *72*, 171–179. [CrossRef] [PubMed]
9. Vor der Bruck, O.; Leiserowitz, L. Molecular packing modes. Two crystalline modifications of bromoacetic acid, $C_2H_3BrO_2$. *Cryst. Struct. Commun.* **1975**, 647–651.
10. Murakami, Y.; Iitaka, Y. Determination of the Absolute Configuration of (-)-2-Bromosuccinamic Acid by X-Ray Diffraction Method. *Chem. Pharm. Bull.* **1969**, *17*, 2397–2404. [CrossRef]
11. Thong, P.Y.; Lo, K.M.; Ng, S.W. 2,3-Dibromo-3-phenylpropionic acid. *Acta Crystallogr. Sect. E* **2008**, *64*, o1946. [CrossRef]
12. Howard, T.R.; Mendez-deMello, K.A.; Cardenas, A.J.P. 2,3-Dibromo-3-phenylpropanoic acid: A monoclinic polymorph. *IUCrData* **2016**, *1*, x161885. [CrossRef]
13. D'Ascenzo, L.; Auffinger, P. A comprehensive classification and nomenclature of carboxyl-carboxyl(ate) supramolecular motifs and related catemers: Implications for biomolecular systems. *Acta Crystallogr. Sect. B* **2015**, *71*, 164–175. [CrossRef]
14. Levene, P.A.; Mori, T.; Mikeska, L.A. On Walden inversion: X. On the oxidation of 2-thiolcarboxylic acids to the corresponding sulfonic acids and on the Walden inversion in the series of 2-hydroxycarboxylic acids. *J. Biol. Chem.* **1927**, *75*, 337–365.
15. Auterhoff, H.; Lang, W. Darstellung und Eigenschaften der optisch aktiven Bromisovale. *Arch. Pharm.* **1970**, *303*, 49–52. [CrossRef] [PubMed]
16. Thomas, I.R.; Bruno, I.J.; Cole, J.C.; Macrae, C.F.; Pidcock, E.; Wood, P.A. WebCSD: The online portal to the Cambridge Structural Database. *J. Appl. Crystallogr.* **2010**, *43*, 362–366. [CrossRef] [PubMed]
17. Steed, K.M.; Steed, J.W. Packing problems: High Z' crystal structures and their relationship to cocrystals, inclusion compounds, and polymorphism. *Chem. Rev.* **2015**, *115*, 2895–2933. [CrossRef] [PubMed]
18. Thompson, A.L.; Watkin, D.J. X-ray crystallography and chirality: Understanding the limitations. *Tetrahedron Asymmetry* **2009**, *20*, 712–717. [CrossRef]
19. Rekis, T. Crystallization of chiral molecular compounds: What can be learned from the Cambridge Structural Database? *Acta Crystallogr. Sect. B* **2020**, *76*, 307–315. [CrossRef]
20. A preliminary X-ray analysis of the crystals obtained from the melt revealed a severely disordered structure with unit cell parameters similar to those of *rac*-1 crystallized from ethyl acetate.
21. *SADABS*; Bruker AXS Inc.: Madison, WI, USA, 2012.
22. Sheldrick, G.M. SHELXT—Integrated space-group and crystal-structure determination. *Acta Crystallogr. A Found. Adv.* **2015**, *71*, 3–8. [CrossRef]
23. Sheldrick, G.M. Crystal structure refinement with SHELXL. *Acta Crystallogr. C Struct. Chem.* **2015**, *71*, 3–8. [CrossRef]
24. Flack, H. On enantiomorph-polarity estimation. *Acta Crystallogr. Sect. A* **1983**, *39*, 876–881. [CrossRef]
25. Parsons, S.; Flack, H.D.; Wagner, T. Use of intensity quotients and differences in absolute structure refinement. *Acta Crystallogr. B Struct. Sci. Cryst. Eng. Mater.* **2013**, *69*, 249–259. [CrossRef]
26. Hooft, R.W.W.; Straver, L.H.; Spek, A.L. Determination of absolute structure using Bayesian statistics on Bijvoet differences. *J. Appl. Crystallogr.* **2008**, *41*, 96–103. [CrossRef] [PubMed]

27. Hooft, R.W.W.; Straver, L.H.; Spek, A.L. Probability plots based on Student's t-distribution. *Acta Crystallogr. Sect. A* **2009**, *65*, 319–321. [CrossRef] [PubMed]
28. Hooft, R.W.W.; Straver, L.H.; Spek, A.L. Using the t-distribution to improve the absolute structure assignment with likelihood calculations. *J. Appl. Crystallogr.* **2010**, *43*, 665–668. [CrossRef]
29. Spek, A.L. checkCIF validation ALERTS: What they mean and how to respond. *Acta Crystallogr. E Crystallogr. Commun.* **2020**, *76*, 1–11. [CrossRef]
30. Brandenburg, K. *DIAMOND, 3.2k3*; Crystal Impact GbR: Bonn, Germany, 2018.
31. Macrae, C.F.; Sovago, I.; Cottrell, S.J.; Galek, P.T.A.; McCabe, P.; Pidcock, E.; Platings, M.; Shields, G.P.; Stevens, J.S.; Towler, M.; et al. Mercury 4.0: From visualization to analysis, design and prediction. *J. Appl. Crystallogr.* **2020**, *53*, 226–235. [CrossRef]
32. Bernstein, J.; Davis, R.E.; Shimoni, L.; Chang, N.L. Patterns in hydrogen bonding—Functionality and graph set analysis in crystals. *Angew. Chem. Int. Ed.* **1995**, *34*, 1555–1573. [CrossRef]
33. Kitajgorodskij, A.I. *Molecular Crystals and Molecules*; Academic Press: New York, NY, USA, 1973.
34. Parsons, S. Determination of absolute configuration using X-ray diffraction. *Tetrahedron Asymmetry* **2017**, *28*, 1304–1313. [CrossRef]
35. Flack, H.D.; Bernardinelli, G. Absolute structure and absolute configuration. *Acta Crystallogr. Sect. A* **1999**, *55*, 908–915. [CrossRef]
36. Flack, H.D.; Bernardinelli, G. Reporting and evaluating absolute-structure and absolute-configuration determinations. *J. Appl. Crystallogr.* **2000**, *33*, 1143–1148. [CrossRef]
37. Spek, A. Absolute structure determination: Pushing the limits. *Acta Crystallogr. Sect. B* **2016**, *72*, 659–660. [CrossRef]
38. Collet, A.; Ziminski, L.; Garcia, C.; Vigné-Maeder, F. Chiral Discrimination in Crystalline Enantiomer Systems: Facts, Interpretations, and Speculations. In *Supramolecular Stereochemistry*; Siegel, J.S., Ed.; Springer: Dordrecht, The Netherlands, 1995; pp. 91–110. [CrossRef]
39. Brock, C.P.; Schweizer, W.B.; Dunitz, J.D. On the validity of Wallach's rule: On the density and stability of racemic crystals compared with their chiral counterparts. *J. Am. Chem. Soc.* **1991**, *113*, 9811–9820. [CrossRef]
40. Leiserowitz, L. Molecular packing modes. Carboxylic acids. *Acta Crystallogr. Sect. B* **1976**, *32*, 775–802. [CrossRef]
41. Corpinot, M.K.; Bučar, D.-K. A Practical Guide to the Design of Molecular Crystals. *Cryst. Growth Des.* **2018**, *19*, 1426–1453. [CrossRef]
42. Etter, M.C. Encoding and decoding hydrogen-bond patterns of organic-compounds. *Acc. Chem. Res.* **1990**, *23*, 120–126. [CrossRef]
43. Thakuria, R.; Sarma, B.; Nangia, A. 7.03—Hydrogen Bonding in Molecular Crystals. In *Comprehensive Supramolecular Chemistry II*; Atwood, J.L., Ed.; Elsevier: Oxford, UK, 2017; pp. 25–48. [CrossRef]

chemistry

Article

Absolute Configuration of In Situ Crystallized (+)-γ-Decalactone[†]

Michael Patzer, Nils Nöthling, Richard Goddard and Christian W. Lehmann *

Max-Planck-Institut für Kohlenforschung, Kaiser Wilhelm Platz 1, 45470 Mülheim an der Ruhr, Germany; patzer@mpi-muelheim.mpg.de (M.P.); noethling@mpi-muelheim.mpg.de (N.N.); goddard@mpi-muelheim.mpg.de (R.G.)

* Correspondence: lehmann@mpi-muelheim.mpg.de

† Dedicated to Dr. Howard Flack (1943–2017).

Abstract: Knowledge about the absolute configuration of small bioactive organic molecules is essential in pharmaceutical research because enantiomers can exhibit considerably different effects on living organisms. X-ray crystallography enables chemists to determine the absolute configuration of an enantiopure compound due to anomalous dispersion. Here, we present the determination of the absolute configuration of the flavoring agent (+)-γ-decalactone, which is liquid under ambient conditions. Single crystals were grown from the liquid in a glass capillary by in situ cryo-crystallization. Diffraction data collection was performed using Cu-Kα radiation. The absolute configuration was confirmed. The molecule consists of a linear aliphatic non-polar backbone and a polar lactone head. In the solid state, layers of polar and non-polar sections of the molecule alternating along the c-axis of the unit cell are observed. In favorable cases, this method of absolute configuration determination of pure liquid (bioactive) agents or liquid products from asymmetric catalysis is a convenient alternative to conventional methods of absolute structure determination, such as optical rotatory dispersion, vibrational circular dichroism, ultraviolet-visible spectroscopy, use of chiral shift reagents in proton NMR and Coulomb explosion imaging.

Keywords: γ-(+)-decalactone; absolute configuration; in situ cryo-crystallization; flavoring agent; lactone; hydrogen bonding; crystal structure; X-ray crystallography

check for updates

Citation: Patzer, M.; Nöthling, N.; Goddard, R.; Lehmann, C.W. Absolute Configuration of In Situ Crystallized (+)-γ-Decalactone. *Chemistry* **2021**, *3*, 578–584. https://doi.org/10.3390/chemistry3020040

Academic Editors: Catherine Housecroft and Katharina M. Fromm

Received: 30 March 2021
Accepted: 18 April 2021
Published: 21 April 2021

Publisher's Note: MDPI stays neutral with regard to jurisdictional claims in published maps and institutional affiliations.

1. Introduction

Knowledge of the absolute configuration of small bioactive organic molecules is essential for understanding the significant different pharmacological effects of enantiomers on organisms [1,2]. The synthesis and characterization of chiral drug compounds is of considerable interest, especially in the pharmaceutical industry [3]. Additionally, enantiomers can differ in taste and smell perception [4,5]. This particularly applies to the natural compound discussed in this paper. γ-(+)-Decalactone [γ-**lac**], a compound that is liquid under ambient conditions, can be found in fruits, for example, strawberries or peaches and is suitable as a fruit flavoring agent [6,7]. In nature, the R-configuration of γ-**lac** is predominant. The absolute configuration of γ-**lac** has a significant influence on olfactory and taste perception. While the aroma of the *R*-enantiomer is reminiscent of peaches, the *S*-enantiomer smells of coconuts [8]. X-ray crystallography is able to determine the absolute configuration of enantiopure crystalline compounds by measuring intensity differences of *Bijvoet pairs* that are caused by anomalous dispersion [9]. One challenge is to determine the absolute configuration of organic drugs without atoms heavier than oxygen owing to the small anomalous scattering contribution of these elements [7]. For γ-**lac**, this is the case. The molecule contains solely carbon, oxygen, and hydrogen. The anomalous scattering contribution of an atom is not only dependent on the atom type but also on the wavelength of the radiation used. To get more accurate results, Cu-Kα radiation has an advantage in comparison with Ag-Kα or Mo-Kα radiation because of the enhanced anomalous scattering factor at this wavelength [10]. Currently, there exist several approaches to circumvent these difficulties of absolute configuration determination. For example, co-crystallization with compounds that contain heavy atoms or co-crystallization with a chiral compound with known configuration are possible solutions to obtain the absolute configuration of the target molecule [11]. Similarly, in case of an amine, the formation of a hydrochloride can serve this purpose [9]. The object of this study was to determine the absolute configuration of an enantiopure compound that is liquid under ambient conditions. We chose to look at γ-**lac** because of its industrial relevance. In situ crystallization is a powerful tool to grow crystals direct on the diffractometer from pure liquid compounds in a glass capillary [12,13]. In this paper, the determination of the absolute configuration of a liquid natural organic compound is presented, together with a further technique for in situ crystal growth. This procedure can easily be transferred to other liquid or gaseous organic compounds.

2. Materials and Methods

(+)-γ-Decalactone was purchased from Sigma-Aldrich (purity > 97%) and used as received $[\alpha]_D^{24} = +41.6$ (c 0.087, CHCl$_3$). The liquid compound was crystallized in a capillary with a diameter of 0.3 mm directly (in situ) on the diffractometer (volume of approximately 0.5 μL). The low-temperature phase behavior was established beforehand by differential scanning calorimetry (DSC). Experiments were performed on a METTLER TOLEDO DSC 820 (Mettler-Toledo GmbH, Gießen, Germany). Two cycles with different cooling and heating rates were carried out in the range of −150−+20 °C. The first and second scan employed temperature gradients of 10 and 5 K per minute, respectively.

For single crystal growth, the compound was cooled below its liquid–solid phase transition temperature of 258 K (determined by LT-DSC) on the diffractometer using a stream of cold nitrogen gas delivered by an Oxford Cryosystems Cryostream 700 (Oxford Cryosystems, Long Hanborough, Oxford, UK). The crystalline powder thus obtained was used as starting material for crystal growth. By translation of the capillary through the cold nitrogen gas stream, a suitable single crystal was grown at the liquid–solid phase boundary (inverse zone melting) following a newly developed procedure [12].

For this purpose, a small attachment to a standard Huber goniometer head (model 1004) was constructed, in order to move the capillary along its axis at a controlled speed over a distance of several millimeters during a time period of several hours (Figure 1). Since there is a temperature gradient across the cold gas stream of the Cryostream 700, it is possible to slowly grow a single crystal at the solid–liquid interface and to continually transfer the crystallized portion of the compound into the colder part of the gas stream, until such time that a nearly perfect single crystal is located in the X-ray beam. The attachment comprises a bracket, which holds a miniaturized stepper motor equipped with a gear box and connected at one end to an indexer board and power supply. The motor axis is coupled to a square nut, which fits snugly over the height adjustment drive of the goniometer head. This attachment can be slid off after crystal growth has been accomplished, thus allowing subsequent free rotation of all goniometer axes.

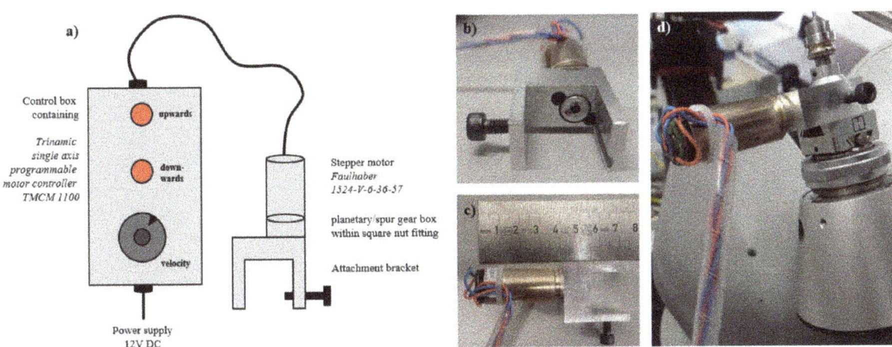

Figure 1. (a) Sketch of the attachment showing its main components, (b) front view of attachment bracket with square nut fitting, (c) side view and size relations of attachment, (d) device attached to a Huber goniometer head (model 1004) mounted on the φ-axis of a Mach-III four circle goniometer.

Once crystal growth is completed, there are in some cases several crystals next to each other in the capillary. The procedure can be repeated and if the problem subsists, the major single crystal component can be selected, or the twinning identified in a three-dimensional representation of the reciprocal space.

X-ray intensity data were collected at 100(2) K on a Bruker-AXS Kappa Mach3 goniometer equipped with an APEX-II detector, using Cu-K_α radiation produced by a FR591 rotating anode X-ray source (BrukerNonius B. V., Delft, The Netherlands). Scaling and absorption correction were performed with SADABS (Bruker AXS Inc., Madison, WI, USA). The crystal structure was solved by SHELXT and refined using SHELXL-2018/3. No further constraints or restraints were applied. A summary of the crystallographic details is given in Table 1, while further structural details including bond lengths and angles can be found in the Supplementary Material. CCDC 2072278 contains the supplementary crystallographic data for this crystal structure. These data can be obtained free of charge from the Cambridge Crystallographic Data Centre via www.ccdc.cam.ac.uk/structures.

Table 1. Crystal data and refinement details for (+)-γ-Decalactone.

Empirical formula	$C_{10}H_{18}O_2$
M_r	170.24
T (K)	100(2)
λ (Å)	1.54178
crystal system	orthorhombic
Space group	$P2_12_12_1$
a (Å)	5.0543(6)
b (Å)	5.3683(6)
c (Å)	37.143(4)
V (Å)	1007.8(2)
Z, Z'	4, 1
ρ_{calc} (mg m^{-3})	1.122
μ (mm^{-1})	0.603
F(000)	376
Crystal size (mm)	1.078 × 0.441 × 0.411
θ range for data collection (°)	2.379 to 71.877°
Reflections collected/unique	33819/1902
R_{int}	0.0389
observed reflections [$I > 2\sigma(I)$]	1880
T_{max} / T_{min}	0.84565/0.69343
data/restraints/parameters	1902/0/128

Table 1. *Cont.*

Goodness-of-fit on F^2	1.089
R1 $[I > 2\sigma(I)]$	0.0302
ωR2 (all data)	0.0688
Flack x parameter (refined)	$-0.025(285)$
Flack x parameter (from quotients)	$-0.019(60)$
Hooft parameter	$-0.03(5)$
$\Delta\rho_{max}/\Delta\rho_{min}\ \left(e\ \text{Å}^{-3}\right)$	$0.106/-0.144$

3. Results

The absolute configuration of γ-**lac** was determined using anomalous dispersion effects. The derived *Flack parameter* calculated from 717 quotients of *Bijvoet pairs* according to Parsons' method (SHELXL) is $-0.019(60)$ and thus the R-configuration is verified (Figure 2). γ-**lac** consists of a non-polar alkyl chain and a more polar cyclic ester. The polar parts appear to interact via non-classical C-H···O hydrogen bonds involving the carbonyl group of the lactone. The supramolecular structure comprising these intermolecular hydrogen bonds between the polar groups is shown in Figure 3 (numerical values are given in Table 2). In the crystal layers consisting of the non-polar and the polar parts alternate along the c-axis of the unit cell. The structure in solid state appears to be governed by the polar ends of the molecule, since there are no short contacts between atoms in the non-polar region. The five-membered lactone ring adopts an envelope conformation with the apex at C3, which is 0.5 Å above the mean plane (r. m. s. 0.01 Å) formed by the other four atoms of the ring. The ring puckering parameter φ equals 291.8° (see Supplementary Material) [14]. The aliphatic alkyl chain adopts an all-trans conformation in the crystal structure (Figure 2) [15].

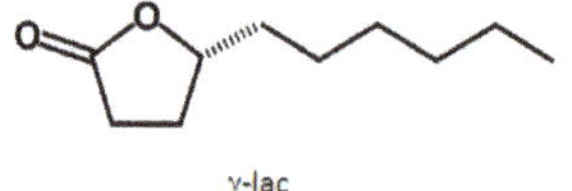

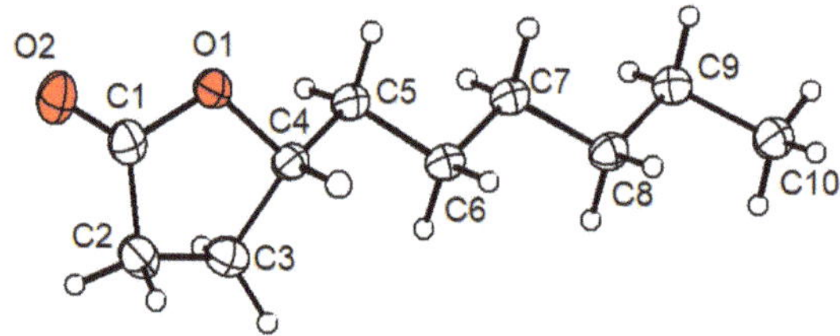

Figure 2. Molecular structure of the chiral compound (+)-γ-decalactone ((R)-5-hexyloxolan-2-one) determined by single crystal X-ray crystallography. The probability level of the displacement ellipsoids is 50%. Hydrogen atoms are shown with arbitrary sizes.

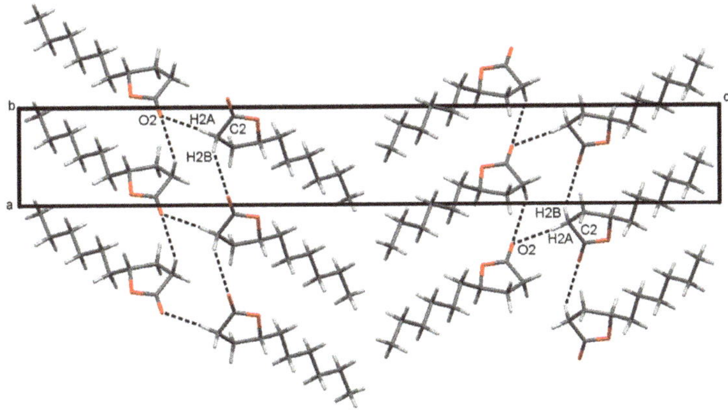

Figure 3. Crystal structure of γ-**lac**, view along the b-axis, C-H···O hydrogen bonds marked with black dashed lines.

Table 2. Distance/Å and angles/° of the hydrogen bonds in γ-**lac**.

D-H···A	d(D-H)	d(H···A)	D(D···A)	<(DHA)
C2-H2A···O2 [i]	0.99	2.44	3.424(2)	172.9
C2-H2B···O2 [ii]	0.99	2.59	3.423(2)	142.0
rule for H-bonds [16,17]	-	<2.5	<3.9	>90
symmetry code [i] −x,y+1/2,−z+1/2; [ii] 1+x,y,z				

4. Discussion

For molecules without atoms heavier than oxygen, the determination of the *absolute structure* becomes difficult due to the low anomalous scattering signal. Two strategies were pursued. First, Cu-Kα radiation was chosen because of the larger resonant scattering factors. This resulted in larger measured intensity differences of the Bijvoet pairs as compared with shorter wavelength radiation. Second, the diffracted intensities were measured with a high redundancy. This enabled outliers to be more easily identified and errors in the measured intensities to be reduced. Both techniques allowed the absolute structure parameters to be determined more accurately. The Flack parameter obtained from 718 quotients with its resultant standard deviation verifies the absolute configuration of the molecule and hence that of the structure (Table 1). The Flack parameter in absolute structure determination has to be evaluated carefully. There can be significant differences between the classical Flack parameter derived from fitting scale factors to the structure factors of both antipodes and the Flack parameter according to Parsons' method, which uses quotients [18]. The Flack parameter determined by Parsons' method usually has a smaller standard uncertainty than the classical Flack parameter and is implemented in standard crystallography programs like SHELXL [19,20]. A standard uncertainty for the Flack parameter smaller than 0.1 is necessary for accurate absolute structure determination in order to ensure that its value is zero within error and hence can be differentiated reliably from the configuration with the opposite hand, for which a value of one would be expected. Here, we report an example in which the Flack parameter is very sensitive to the measured intensity of just one outlier. By chance, we discovered that the reflection with Miller indices -5 2 14 and a d-spacing of 0.89 Å has a significant influence on the determination of the absolute configuration. The measured intensity was far too large (19 standard uncertainties from the expected value) and was subsequently attributed to a diffracted intensity of a second crystal in the capillary. If this reflection is included in the intensity data, the Flack parameter is −1.62(22) (determined from 733 Bijvoet pairs according to the procedure described in [18]). On omitting this reflection, the Flack parameter assumes a meaningful value of 0.02(6) (see Supplementary Material). In the capillary, multiple crystals cannot always be avoided when a crystal is grown by translation perpendicular to the nitrogen gas stream and extreme caution is advised, as this example illustrates.

To the best of our knowledge, there is no report in the Cambridge Structural Database of the crystal structure determination of a pure γ-lactone that has only an aliphatic substituent with no functional groups in the chain. The bipolar structure of γ-**lac** can be compared with non-ionic surfactants, which have a polar head and a non-polar backbone. The packing of the molecules in the crystal is likely to be a consequence of the polarity of the molecular structure of γ-**lac**. A layer structure or in more general a structure in which some parts of a molecule agglomerate may be expected when the molecule has a hydrophobic chain and polar head.

One of the advantages of the goniometer head attachment described above is the simplicity of operation. The set-up is initialized by retracting the height adjustment until the stall guard functionality of the stepper motor controller signals that the end of travel has been reached. A 10-turn potentiometer connected to the analogue input of the controller allows setting the speed of the translational movement over a time range between minutes and hours, while two illuminated push buttons allow the direction of travel to be input.

5. Conclusions

We have shown that in situ cryo-crystallization allows one to determine the absolute configuration of the enantiopure liquid compound γ-**lac** by using single crystal x-ray crystallography. Especially for medicinal and pharmaceutical research, in situ cryo-crystallization combined with X-ray crystallography promises to be a powerful and easy tool to determine the absolute configuration of drugs and small-molecule compounds that are liquid under ambient conditions. In addition, the technique allows intermolecular interactions to be studied in more detail. In this example, the polarity and molecular geometry of γ-**lac** leads to a layered structure. Intermolecular C-H⋯O hydrogen bonds between the lactone-part of γ-**lac** can be observed. This procedure can easily be transferred to other liquid organic drug compounds that crystallize at low temperature.

Supplementary Materials: The following are available online at https://www.mdpi.com/article/10.3390/chemistry3020040/s1, Figure S1: ORTEP-Plot of the molecular structure of (+)-γ-Decalactone. Figure S2: Screen shots of the capillary, face indexing and unit cell indexing. Figure S3: Recorded ambient to low temperature DSC curve of γ-**lac**. Figure S4: ATR-FT-IR spectra of γ-**lac**. Figure S5: Determination of the Flack-Parameter by Parsons' method; (a) The full data-set of 733 Bijvoet-Pairs was used for determination of the Flack-Parameter (variable p in linear regression), the quotient with highest difference from theory is marked with a red circle; (b) Linear regression of 732 Bijvoet pairs, the red marked quotient in (a) was omitted. Figure S6: Excerpt from the HKLF4 file for γ-**lac**. Symmetry-equivalent reflections are marked with colored boxes (point group 2 2 2: h k l = -h-k l = -h k -l = h -k -l). The reflection (-5 -2 -14) in the blue box strongly deviates from its Bijvoet-pairs (red box). Table S1: Crystal data and structure refinement. Table S2: Bond lengths [Å] and angles [°]. Table S3: Anisotropic displacement parameters (Å2). Table S4: Hydrogen coordinates and isotropic displacement parameters (Å2). Table S5: Torsion angles [°]. Table S6: Hydrogen bonds [Å and °]. Table S7: Asymmetry Parameters of the Five Membering Ring (Puckering Coordinates Analysis).

Author Contributions: Conceptualization, N.N.; methodology, M.P., N.N., and C.W.L.; writing—original draft preparation, M.P.; writing—review and editing, all; visualization, M.P.; supervision, C.W.L. All authors have read and agreed to the published version of the manuscript.

Funding: This research received no external funding.

Acknowledgments: We are grateful to M. Felderhoff for providing access to low temperature DSC and A. Fürstner for access to ATR-IR spectroscopy and optical rotation measurements.

Conflicts of Interest: The authors declare no conflict of interest.

References

1. Maher, T.J.; Johnson, D.A. Review of chirality and its importance in pharmacology. *Drug Dev. Res.* **1991**, *24*, 149–156. [CrossRef]
2. Jayakumar, R.; Vadivel, R.; Ananthi, N. Role of Chirality in Drugs. *Org. Med. Chem. Int. J.* **2018**, *5*, 555661.
3. Sheldon, R.A. Chirotechnology: Designing economic chiral syntheses. *J. Chem. Technol. Biotechnol. Int. Res. Process Environ. AND Clean Technol.* **1996**, *67*, 1–14.
4. Bassoli, A.; Borgonovo, G.; Drew, M.G.B.; Merlini, L. Enantiodifferentiation in taste perception of isovanillic derivatives. *Tetrahedron Asymmetry* **2000**, *11*, 3177–3186. [CrossRef]
5. Zawirska-Wojtasiak, R. Chirality and the Nature of Food Authenticity of Aroma. *Acta Sci. Pol. Technol. Aliment.* **2006**, *5*, 21–36.
6. Ulrich, D.; Hoberg, E.; Rapp, A.; Kecke, S. Analysis of strawberry flavour—Discrimination of aroma types by quantification of volatile compounds. *Z. Lebensm. Forsch. A* **1997**, *205*, 218–223. [CrossRef]
7. Guillot, S.; Peytavi, L.; Bureau, S.; Boulanger, R.; Lepoutre, J.-P.; Crouzet, J.; Schorr-Galindo, S. Aroma characterization of various apricot varieties using headspace–solid phase microextraction combined with gas chromatography–mass spectrometry and gas chromatography–olfactometry. *Food Chem.* **2006**, *96*, 147–155. [CrossRef]
8. Ternes, W. Naturwissenschaftliche Grundlagen der Lebensmittelzubereitung. Behr: Santa Ana, CA, USA, 2008.
9. Parsons, S. Determination of absolute configuration using X-ray diffraction. *Tetrahedron Asymmetry* **2017**, *28*, 1304–1313. [CrossRef]
10. Linden, A. Best practice and pitfalls in absolute structure determination. *Tetrahedron Asymmetry* **2017**, *28*, 1314–1320. [CrossRef]
11. Bhatt, P.M.; Desiraju, G.R. Co-crystal formation and the determination of absolute configuration. *CrystEngComm* **2008**, *10*, 1747–1749. [CrossRef]
12. Boese, R. Special issue on In Situ Crystallization. *Z. Krist. Cryst. Mater.* **2014**, *229*, 595–601. [CrossRef]
13. Veith, M.; Frank, W. Low-temperature x-ray structure techniques for the characterization of thermolabile molecules. *Chem. Rev.* **1988**, *88*, 81–92. [CrossRef]
14. Cremer, D.; Pople, J.A. General definition of ring puckering coordinates. *J. Am. Chem. Soc.* **1975**, *97*, 1354–1358. [CrossRef]
15. Thomas, S.A.; Agbaji, E.B. Molecular conformations of γ-lactone rings from crystal structure data. *J. Crystallogr. Spectrosc. Res.* **1989**, *19*, 3–23. [CrossRef]
16. Torshin, I.Y.; Weber, I.T.; Harrison, R.W. Geometric criteria of hydrogen bonds in proteins and identification of 'bifurcated' hydrogen bonds. *Protein Eng. Design Sel.* **2002**, *15*, 359–363. [CrossRef] [PubMed]
17. Wood, P.A.; Allen, F.H.; Pidcock, E. Hydrogen-bond directionality at the donor H atom—Analysis of interaction energies and database statistics. *CrystEngComm* **2009**, *11*, 1563–1571. [CrossRef]
18. Parsons, S.; Flack, H.D.; Wagner, T. Use of intensity quotients and differences in absolute structure refinement. *Acta Crystallogr. Sect. B* **2013**, *69*, 249–259. [CrossRef] [PubMed]
19. Escudero-Adan, E.C.; Benet-Buchholz, J.; Ballester, P. The use of Mo K[alpha] radiation in the assignment of the absolute configuration of light-atom molecules; the importance of high-resolution data. *Acta Crystallogr. Sect. B* **2014**, *70*, 660–668. [CrossRef] [PubMed]
20. Sheldrick, G. SHELXT—Integrated space-group and crystal-structure determination. *Acta Crystallogr. Sect. A* **2015**, *C71*, 3–8. [CrossRef] [PubMed]

Article

Novel Ansa-Chain Conformation of a Semi-Synthetic Rifamycin Prepared Employing the Alder-Ene Reaction: Crystal Structure and Absolute Stereochemistry [†]

Christopher S. Frampton [1,*], James H. Gall [2] and David D. MacNicol [2,*]

[1] Experimental Techniques Centre, Brunel University London, Kingston Road, Uxbridge, Middlesex UB8 3PH, UK
[2] School of Chemistry, University of Glasgow, Glasgow G12 8QQ, UK; james.gall@glasgow.ac.uk
* Correspondence: chris.frampton@brunel.ac.uk (C.S.F.); david.macnicol@glasgow.ac.uk (D.D.M.); Tel.: +44-7841-373969 (C.S.F.)
† Dedicated to Dr. Howard Flack (1943–2017).

Abstract: Rifamycins are an extremely important class of antibacterial agents whose action results from the inhibition of DNA-dependent RNA synthesis. A special arrangement of unsubstituted hydroxy groups at C21 and C23, with oxygen atoms at C1 and C8 is essential for activity. Moreover, it is known that the antibacterial action of rifamycin is lost if either of the two former hydroxy groups undergo substitution and are no longer free to act in enzyme inhibition. In the present work, we describe the successful use of an Alder-Ene reaction between Rifamycin O, **1** and diethyl azodicarboxylate, yielding **2**, which was a targeted introduction of a relatively bulky group close to C21 to protect its hydroxy group. Many related azo diesters were found to react analogously, giving one predominant product in each case. To determine unambiguously the stereochemistry of the Alder-Ene addition process, a crystalline zwitterionic derivative **3** of the diethyl azodicarboxylate adduct **2** was prepared by reductive amination at its spirocyclic centre C4. The adduct, as a mono chloroform solvate, crystallized in the non-centrosymmetric Sohnke orthorhombic space group, $P2_12_12_1$. The unique conformation and absolute stereochemistry of **3** revealed through X-ray crystal structure analysis is described.

Keywords: Rifamycin O; ansamysin; antibacterial; semi-synthesis; Alder-Ene; conformation; zwitterionic; hydrogen bonding; absolute configuration; chirality; crystal structure; X-ray crystallography

Citation: Frampton, C.S.; Gall, J.H.; MacNicol, D.D. Novel Ansa-Chain Conformation of a Semi-Synthetic Rifamycin Prepared Employing the Alder-Ene Reaction: Crystal Structure and Absolute Stereochemistry . *Chemistry* **2021**, *3*, 734–743. https://doi.org/10.3390/chemistry3030052

Academic Editors: Catherine Housecroft and Katharina M. Fromm

Received: 26 May 2021
Accepted: 9 July 2021
Published: 11 July 2021

Publisher's Note: MDPI stays neutral with regard to jurisdictional claims in published maps and institutional affiliations.

1. Introduction

The rifamycins constitute an important class of ansamycin antibiotic active against mycobacteria and other bacterial pathogens, also exhibiting antiviral properties. These molecules are comprised of a substituted naphthalene or naphthoquinone core spanned by a seventeen-membered aliphatic ansa bridge. A vast number of semi-synthetic rifamycins have been produced by structural modification of the aromatic region of naturally occurring rifamycins [1]. The important bridging ansa moiety has not been so intensively studied, though recent highlights are the excellent antibacterial activity found for 24-desmethylrifampicin [2]; and the synthesis of C25 carbamate derivatives which are resistant to ADP-ribosyl transferases [3]. The present work was directed at the introduction of a bulky group close to the hydroxy group on C21 of Rifamycin O, **1**, Scheme 1, to inhibit transferase deactivation [4,5]. Attempts to carry out a Diels–Alder reaction with dimethyl acetylenedicarboxylate, hoping to exploit the *cisoid* diene arrangement of **1**, torsion angle 36°, in the crystal [6] proved unsuccessful; however, gratifyingly, we found that diethyl azodicarboxylate and related diesters reacted readily and quantitatively giving a major product, along with a by-product in each case. The disappearance of a methyl doublet from the proton NMR spectrum with introduction of an allylic methyl singlet at lower field (see

189

methyl assignments in Experimental) clearly pointed to an Alder-Ene reaction [7,8] rather than a Diels–Alder reaction for which four methyl doublets would have been expected. MS confirmed a 1:1 adduct had been formed and inspection of a tactile Dreiding model of 1 revealed that azo nitrogen attack at the C18 alkene face giving an *S* configuration at C18 would, to effect hydrogen abstraction from C20, lead to a *trans* (*E*) double bond between C19 and C20. Corresponding alternative attack on the opposite alkene face would lead to an *R* configuration at the C18 stereogenic centre with a predicted *cis* (*Z*) double bond formed between C19 and C20. Although product **2** was obtained stereochemically pure at C4 (single AB quartet for the diastereotopic methylene protons of the spirolactone at C4), suitable single crystals for X-ray analysis could not be obtained. However, crystals were obtained from chloroform for **3**, which was derived from **2** by reductive amination as described in the Experimental section below and this resolved the stereochemical question and also revealed an unprecedented ansa-chain conformation.

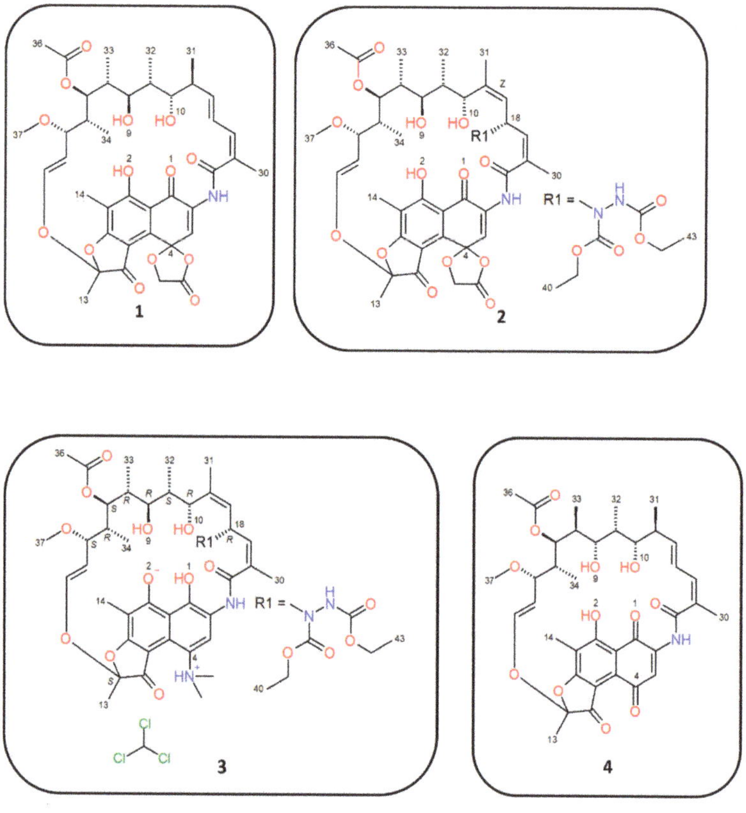

Scheme 1. Rifamycin structures referred to in the text.

2. Materials and Methods

Typical reaction conditions for **2**: compound **1** (1g, 0.00132 mmol) and diethyl azodicarboxylate (0.69 g, 0.00396 mmol) were refluxed in toluene (40 mL) under argon for 5 h and then left at 50 °C for one week. The toluene was removed, and the reaction product boiled in iso-propanol and then cooled and filtered, yielding **2** as a yellow powder. Assignments for the methyl resonances of **2**: ^{1}HNMR (400 MHz, CDCl$_3$), δ_H, C34, 0.16, 3H, *d*, *J* = 7 Hz; C33, 0.60, 3H, *d*, *J* = 7 Hz; C32, 1.02, 3H, *d*, *J* =7 Hz; C40 or C43, 1.22, 3H, *t*, *J* = 7 Hz; C40 or C43, 1.28, 3H, *t*, *J* = 7 Hz; C13, 1.68, 3H, *s*; C31, 1.77, 3H, *s*; C30,1.96, 3H, *s*; C36, 2.06,

3H, s; C14, 2.20, 3H, *s*; C37, 3.06, 3H, *s*. It may be noted that the spectrum, relevant to a future analysis of the conformational situation in solution, not considered here, shows retention of all functional groups including the unaltered (*E*) vinyl ether bridge component. Dimethyl azodicarboxylate and related, di*iso*propyl and dibenzyl esters, for example, all exhibited similar reactivity with respect to the Alder-Ene reaction with Rifamycin O **1**. The Alder-Ene reactions were quantitative, a single minor by-product being formed in each case; typical ratios being approximately 5:1. ^{1}HNMR data were collected on a Bruker AV III 400 MHz spectrometer.

Compound **2** was converted with modest yield into **3** by employing the general reductive amination procedure of Cricchio and Tamborini as described in [9], in which, interestingly, the amine acts a reducing agent. An excess of dimethylamine methanol solution was added by syringe to compound **2** in dry degassed THF and this was then left in the dark at 50 °C for a week. The THF was removed and the reaction product was dissolved in ethyl acetate and shaken with 7.4 pH phosphate buffer. The acetate layer was washed with water, dried, and the solvent evaporated to give **3**. Crystallisation of **3** proved challenging. However, small colourless single crystals of a plate morphology suitable for X-ray analysis were obtained from a $CHCl_3$ solution.

X-ray intensity data for **3** were collected at 100(1)K on a Rigaku Oxford Diffraction SuperNova Dual-flex AtlasS2 diffractometer equipped with an Oxford Cryosystems Cobra cooler using Cu $K\alpha$ radiation ($\lambda = 1.54178$ Å). The crystal structure was solved with SHELXT-2018/2 [10] and refined with SHELXL-2018/3 [11]. Hydrogen atoms bound to carbon were placed at geometrically calculated positions with $C_{methine}$-H = 1.00 Å, $C_{methylene}$-H = 0.99 Å, C_{methyl}-H = 0.98 Å, $C_{aromatic}$-H = 0.95 Å. These hydrogen atom positions were refined using a riding model with U_{iso}(H) = 1.2 U_{eq} (C) (1.5 U_{eq} (C) for methyl groups). Methyl group torsion angles were allowed to refine whilst maintaining an idealized tetrahedral geometry. Heteroatom (N-H, O-H) hydrogen atoms were located via a difference Fourier synthesis and their positions and isotropic temperature factors were allowed to refine freely. Values of the Flack *x* parameter [12] were obtained from the final refinement cycle of SHELXL. Two values were calculated, the first using the TWIN and BASF instructions and the second using the Parsons method of Intensity Quotients [13]. The Hooft *y* parameter [14–16] was calculated through the implementation in the program PLATON [17]. Details of the sample, data collection and structure refinement are given in Table 1. Crystal packing and structural overlay figures were produced using the CCDC program Mercury [18].

Table 1. Sample, data collection and structure refinement for compound **3**.

	Compound 3
Empirical formula	$C_{45}H_{62}N_4O_{15}$, $CHCl_3$
M_r	1018.35
T (K)	100(1)
Wavelength	Cu$K\alpha$ (1.54178 Å)
Crystal system	Orthorhombic
Space group	$P2_12_12_1$
a (Å)	14.4057(4)
b (Å)	14.9409(3)
c (Å)	22.8735(7)
α (°)	90
β (°)	90
γ (°)	90
V, (Å^3)	4923.2(2)
Z', Z	1, 4
ρ_{calc} (Mg m^{-3})	1.374
μ (mm^{-1})	2.287

Table 1. *Conts.*

	Compound 3
$F(000)$	2152
Crystal colour, shape	Colourless, plate
Size (mm)	$0.251 \times 0.250 \times 0.071$
Diffraction limit	0.80 Å
Coverage, %	99.9
Friedel coverage, %	81.0
Friedel fraction max %	99.8
Reflections collected/unique	26,683/10,050
R_{int}	0.0406
Observed reflections, $I > 2\sigma(I)$	9228
T_{min}, T_{max}	0.748, 1.000
Data/restraints/parameters	10,050/0/650
GOF, (S) on F^2	1.035
$R_1 [I > 2\sigma(I)]$	0.0422
wR^2 (all data)	0.1140
Flack x parameter (refined)	$-0.003(18)$
Flack x parameter (from 3814 quotients)	$-0.009(9)$
Hooft y parameter	$-0.008(6)$
Min/max residual density (e Å^{-3})	$0.658/-0.481$
CCDC deposition number	2,045,594

CCDC 2045594 contains the supplementary crystallographic data for compound **3**, which can be obtained free of charge from The Cambridge Crystallographic Data Centre, see www.ccdc.cam.ac.uk/structures.

3. Results

Small colourless crystals of **3** exhibiting a plate morphology were obtained from slow evaporation of a chloroform solution. The asymmetric unit of the structure consists of a single fully ordered molecule of compound **3** and a single fully ordered molecule of chloroform as a solvate. The structure refined very well in the non-centrosymmetric Sohnke orthorhombic space group, $P2_12_12_1$ and gave a final residual R-factor based on the observed data of $R_1 [I > 2\sigma(I)] = 4.22$ %. Figure 1 shows the asymmetric unit viewed obliquely from below the plane of the basal naphthenic moiety. Figure 2 shows a view of molecule of compound **3** with -CH hydrogen atoms removed for clarity and intramolecular contacts as green dashed lines; this view is obliquely down onto the plane of the basal naphthenic moiety. Selected torsion angle and intermolecular contact distances are listed in Table 2 along with comparative data for Rifamycin O, **1** and Rifamycin S, **4** (CSD codes PUTDUD [1] and PAFRAP [19]). Geometric hydrogen bond data are given in Table 3. The structure is zwitterionic, reflecting the high acidity of the OH group on C8, see for example [20–22]. The substituted 1,2-Dihydro-naphtho[2,1-*b*]furan moiety defined by atoms C1 to C12, O3 is planar with an r.m.s. deviation of the fitted atoms of 0.0586 Å, with atom C2 showing the greatest deviation from planarity, $-0.124(3)$ Å. The single chloroform solvate molecule in the symmetric unit forms two short C-H···O interactions of [H46···O1, 2.395 Å] and [H46···O14, 2.255 Å]. There is possibly a small rotational disorder component to the solvent molecule, as evidenced by the small difference density maxima located near the chlorine atoms.

Table 2. Selected torsion angles (°) and intramolecular contact distances (Å) for compound **3** and related structures *.

Torsion Angle	Compound 3	Rifamycin O PUTDUD, 1	Rifamycin S PAFRAP, 4
C2-N1-C15-C16	−173.29(3)	−176.4	−177.3
N1-C15-C16-C17	59.8(4)	63.6	92.8
O3-C12-O5-C29	−85.2(3)	−81.7	−61.6
C12-O5-C29-C28	66.0(4)	65.9	−117.2
C21-C22-C23-C24	55.0(4)	63.2	60.4
C20-C21-C22-C23	176.0(3)	−172.1	−174.8
C25-C26-C27-C28	−171.6(3)	56.0	−172.3
C16-C17-C18-C19	146.4(3)	36.5	178.8
C17-C18-C19-C20	−136.4(3)	−178.6	−175.3
C18-C19-C20-C21	−1.2(5)	117.4	−45.6
C19-C20-C21-C22	92.8(4)	172.5	−175.6
Intramolecular Contact Distance	**Compound 3**	**Rifamycin O PUTDUD, 1**	**Rifamycin S PAFRAP, 4**
O1⋯⋯O2	2.436(4)	2.538	2.566
O1⋯⋯O9	7.622(4)	4.300	7.245
O1⋯⋯O10	6.906(4)	2.912	6.205
O2⋯⋯O9	8.276(4)	3.613	6.166
O2⋯⋯O10	7.811(4)	3.980	7.836
O9⋯⋯O10	2.711(4)	2.702	2.689
C2⋯⋯C33	3.474(5)	6.601	6.314
C3⋯⋯C33	3.383(5)	6.442	5.858

* Structural data for Rifamycin O and Rifamycin S, CSD codes PUTDUD and PAFRAP are taken from [1] and [19], respectively.

Table 3. Intra and intermolecular hydrogen bond data (Å,°) *.

D-H⋯⋯A	d(D⋯⋯H)	d(H⋯⋯A)	d(D⋯⋯A)	<(DHA)
O1-H1A⋯⋯O2	0.96(8)	1.54(8)	2.436(3)	154(7)
O9-H9⋯⋯O4′	0.73(6)	2.10(6)	2.750(4)	149(6)
O10-H10⋯⋯O9	0.75(6)	2.04(6)	2.711(4)	148(6)
N1-H1B⋯⋯O1	0.89(5)	2.22(5)	2.666(4)	111(4)
N1-H1B⋯⋯O14	0.89(5)	2.54(5)	3.390(4)	159(4)
N1-H1A⋯⋯N2	0.89(5)	2.68(5)	3.279(4)	126(4)
N3-H3⋯⋯O11″	0.86(5)	1.94(5)	2.771(4)	163(4)
N4-H4⋯⋯O4	1.00(5)	1.60(5)	2.595(4)	171(4)

* Symmetry operations; $'-x, y-1/2, -z+1/2$, $''-x+1, y-1/2, -z+1/2$.

The packing of molecules in the crystal is governed by the formation of two intermolecular hydrogen bond interactions. The first interaction is a hydroxy hydrogen, -OH, acting as a donor to a furanone carbonyl oxygen atom acting as an acceptor [O9-H9⋯O4, 2.750(4) Å]. The second interaction is an amide hydrogen, -NH, acting as a donor to a carbonyl oxygen atom acting as an acceptor [N3-H3⋯O11, 2.771(4) Å]. Both interactions use the same 2_1 screw axis symmetry operation along the *b*-axis, the second interaction is translated by one-unit cell along the *a*-axis, thus linking the molecules into a crosslinked infinite chain parallel to the *b*-axis of the unit cell, as shown in Figure 3. Details of the hydrogen bond interactions are given in Table 3.

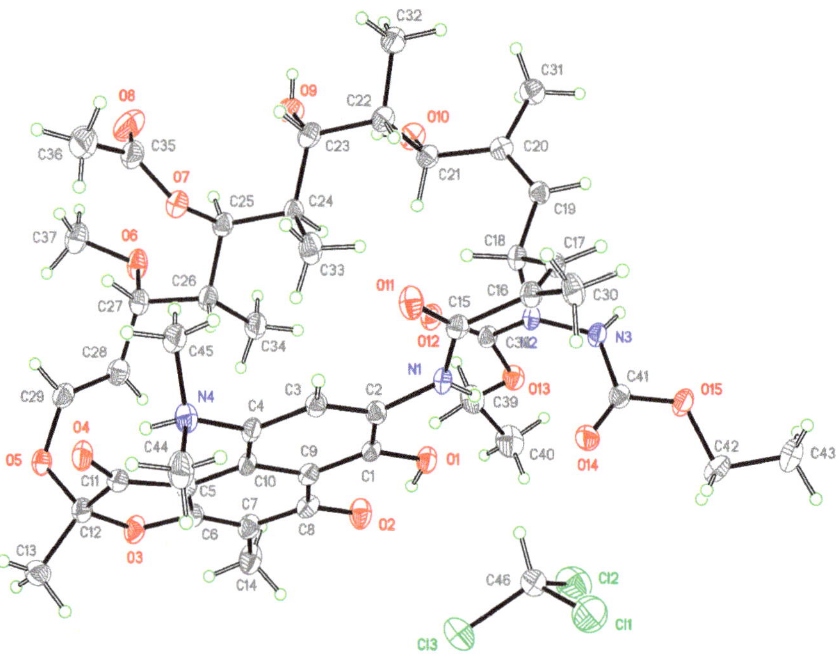

Figure 1. A view of the asymmetric unit of the crystal structure showing the atom numbering scheme employed. In this figure, the structure is viewed obliquely from below the naphthalene ring. Anisotropic atomic displacement ellipsoids for the non-hydrogen atoms are shown at the 50% probability level. Hydrogen atoms are displayed with an arbitrary small radius.

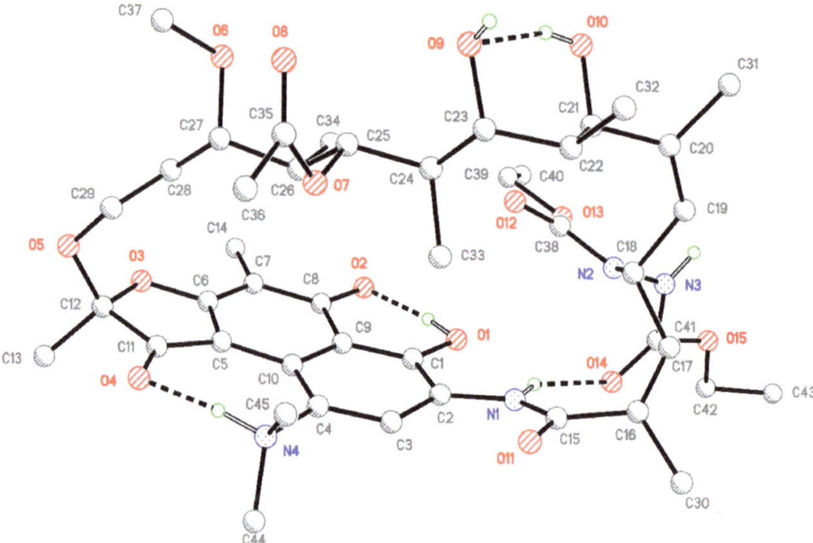

Figure 2. A view of a molecule of compound **3** from the crystal structure showing the atom numbering scheme employed and the intramolecular hydrogen bonds as dashed lines. In this figure, the structure is viewed obliquely from above the naphthalene ring. C-H hydrogen atoms have been removed for clarity.

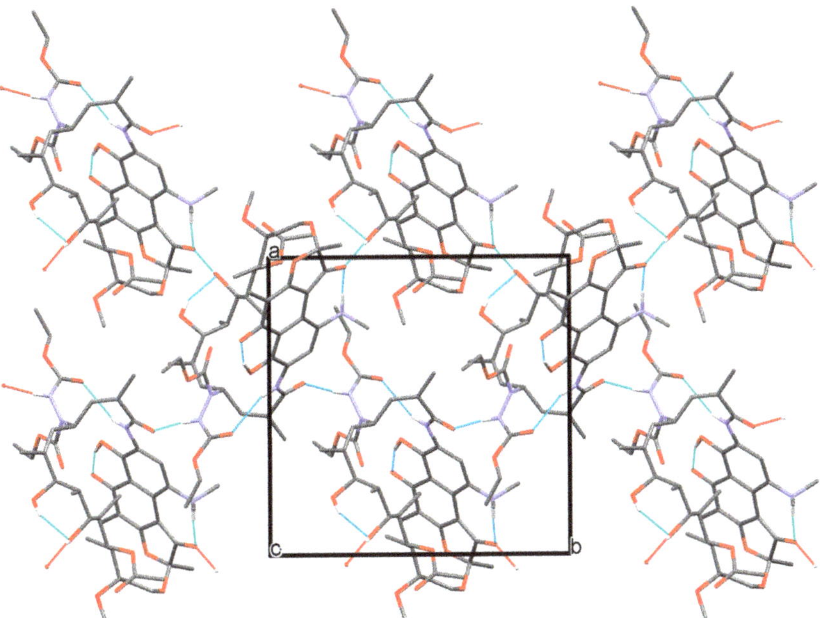

Figure 3. A view of part of the crystal packing of compound **3**, showing the formation of a layer of two crosslinked infinite chains parallel to the *b*-axis of the unit cell. Inter- and intramolecular hydrogen bond interactions are shown as dark and light blue dashed lines, respectively. Incomplete hydrogen bonds are shown as red dashed lines.

4. Discussion

The absolute stereochemistry of a single crystal of **3** has been determined through the anomalous dispersion effect on the diffracted beam intensities. This result was greatly enhanced by the fact that the crystal was a mono chloroform solvate since the anomalous scatting coefficients for the chlorine atoms are much larger than those for C, N and O for Cu *K*α radiation. For the structure as presented with the chiral centres C12, C18, C21, C22, C23, C24, C25, C26, C27 in the *S, R, R, S, R, R, S, R, S* configuration, respectively, the Flack parameter = −0.003(18). Determination of the absolute structure using Bayesian statistics on Bijvoet differences (Hooft method), reveals that the probability of the absolute structure as presented being correct is 1.000, while the probabilities of the structure being a racemic twin or false are both 0.000. The Flack equivalent and its uncertainty calculated through this program was *y* = −0.008(6). This calculation was based on the values of 4497 Bijvoet differences. The post refinement method based on 3814 intensity quotients (Parsons method) gave a value of *x*= −0.009(9). It can be seen that all three methods are in good agreement (with the exception of the standard uncertainty value which is approximately a factor of two greater for the refined parameter) and that the absolute stereochemistry of compound **3** is well defined. As can be seen, the molecule has an *R* configuration at C18 and the introduced double bond between C19 and C20 has a *cis* (Z) configuration. Since **3** came directly from the pure major product of the Alder-Ene reaction, this establishes that the major product has the structure **2** as formulated. A salient feature is the wide separation of O1-O10 and of O2-O9, these distances having increased by 3.994 and 4.663 Å from those of Rifamycin O, **1**, in the crystal. Also striking is the location of the methyl group, (C33), attached to C24; the shortest contacts to naphthalene ring atoms are 3.474(5) Å to C2 and 3.383(5) Å to C3. A comparison of the seventeen ring torsion angles for **3** [23], with those of its ultimate precursor **1** shows that the ring torsion angles close to the aromatic ring are only modestly changed; and values (those for **1** given first) for C2-N1-C15-C16,

N1-C15-C16-C17, O3-C12-O5-C29, C12-O5-C29 C28 are $-176.4°$, $-173.3(3)°$; $63.6°$, $59.8(4)°$; $-81.7°$, $-85.2(3)°$; $65.9°$, $66.0(4)°$, respectively. The key 1,3-diol component of the ansa chain maintains its stereochemical integrity with values for C21-C22-C23-C24 of $63.2°$ and $55.0(4)°$, along with accompanying values for C20-C21-C22-C23 of $-172.1°$ and $-176.0(3)°$. For the structurally unaltered part of the ansa chain running from C21 to O5, the most dramatic change is found for torsion C25-C26-C27-C28, $56.0°$ and $-171.6(3)°$, respectively. Close to C18, massive changes are consequent upon double-bond migration, and for C16-C17-C18-C19, C17-C18-C19-C20, C18-C19-C20-C21, C19-C20-C21-C22 the corresponding values are: $36.5°$, $146.4(3)°$; $-178.6°$, $-136.4(3)°$; $117.4°$, $-1.2(5)°$; $-172.5°$, $92.8(4)°$.

A calculated overlay of compound **3**, Rifamycin O, **1** and Rifamycin S, **4** (CSD codes PUTDUD and PAFRAP, respectively) is shown in Figure 4. The overlay was computed based on all ten of the naphthalene core carbon atoms and yielded a r.m.s deviation of 0.0642Å for compound **3** and Rifamycin O, **1** and 0.0630Å for compound **3** and Rifamycin S, **4** [18]. Overlay figures for compound **3** and Rifamycin O, **1** and compound **3** and Rifamycin S, **4** [active conformation] can be found in the supplementary data. See below for details.

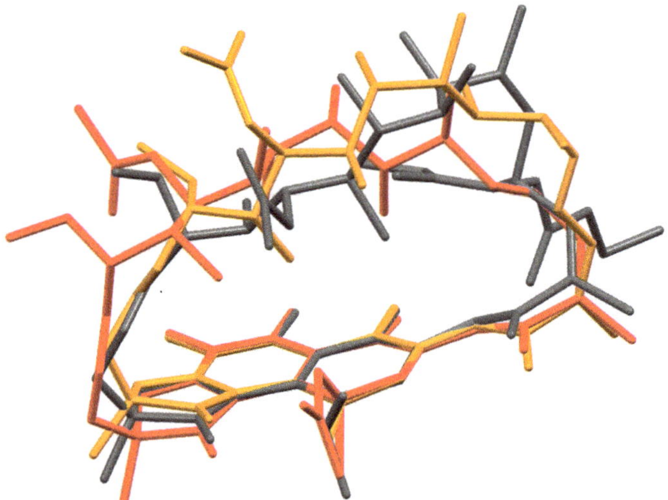

Figure 4. Overlay of compound **3** (grey), Rifamycin O (red) and Rifamycin S (orange) [active conformation]. See text for details.

5. Conclusions

New synthetic access to modified ansamycins is important for combatting mutant strains of bacterial pathogens. We have shown that whilst pure Rifamycin O is totally resistant to Diels–Alder addition, it reacts smoothly in an Alder-Ene process with diethyl azodicarboxylate and related azo compounds to give a new series of semi-synthetic rifamycins. The stereochemistry of the predominant product, **2**, of the addition reaction has been defined by conversion into a zwitterionic derivative **3** whose structure has been defined by single-crystal X-ray analysis, which established the absolute stereochemistry at C18 as having an *R* configuration, and a *Z* configuration at the introduced double bond between C19 and C20. A secondary product has been observed though not yet isolated and we have provisionally assigned to it a structure, isomeric with **2**, having an *S* configuration at C18 and a *trans* (*E*) double bond between C19 and C20. The potential anti-bacterial properties of **2** (and its isomer) and related compounds as well as those of zwitterionic **3** still remain to be determined. It is interesting to note that Rifamycin O has itself recently been found to show promise as an alternative anti-*Mycobacterium abscessus* agent [24].

Supplementary Materials: The following are available online at https://www.mdpi.com/article/10.3390/chemistry3030052/s1, **Figure S1:** Structure overlay for compound **3** and Rifamycin O, **1**. **Figure S2:** Structure overlay for compound **3** and Rifamycin S, **4**.

Author Contributions: Conceptualization, D.D.M. and J.H.G.; methodology, C.S.F., D.D.M. and J.H.G.; validation, C.S.F.; formal analysis, C.S.F. and D.D.M.; investigation, C.S.F., D.D.M. and J.H.G.; resources, C.S.F. and D.D.M.; data curation, C.S.F.; writing—original draft preparation, C.S.F. and D.D.M.; writing—review and editing, C.S.F. and D.D.M.; visualization, C.S.F. and D.D.M.; funding acquisition, D.D.M. All authors have read and agreed to the published version of the manuscript.

Funding: This research was funded Financial support from the Malaysia HIR MOHE, Grant No.F000009-21001 is gratefully acknowledged.

Data Availability Statement: The data is available from the CCDC program. CCDC 2045594 contains the supplementary crystallographic data for compound **3**, which can be obtained free of charge from The Cambridge Crystallographic Data Centre, see www.ccdc.cam.ac.uk/structures.

Acknowledgments: We thank F. Johnson (Stony Brook) for kindly providing a pure sample of Rifamycin O.

Conflicts of Interest: The authors declare that there are no conflicts of interest.

Sample Availability: Samples of the compounds are unavailable from the authors.

References

1. Floss, H.G.; Yu, T.-W. RifamycinMode of Action, Resistance, and Biosynthesis. *Chem. Rev.* **2005**, *105*, 621–632. [CrossRef]
2. Nigam, A.; Almabruk, K.H.; Saxena, A.; Yang, J.; Mukherjee, U.; Kaur, H.; Kohli, P.; Kumari, R.; Singh, P.; Zakharov, L.N.; et al. Modification of Rifamycin Polyketide Backbone Leads to Improved Drug Activity against Rifampicin-resistant Mycobacterium tuberculosis. *J. Biol. Chem.* **2014**, *289*, 21142–21152. [CrossRef]
3. Combrink, K.; Denton, D.A.; Harran, S.; Ma, Z.; Chapo, K.; Yan, D.; Bonventre, E.; Roche, E.D.; Doyle, T.B.; Robertson, G.T.; et al. New C25 carbamate rifamycin derivatives are resistant to inactivation by ADP-ribosyl transferases. *Bioorganic Med. Chem. Lett.* **2007**, *17*, 522–526. [CrossRef]
4. Dabbs, E.R.; Yazawa, K.; Mikami, Y.; Miyaji, M.; Morisaki, N.; Iwasaki, S.; Furihata, K. Ribosylation by mycobacterial strains as a new mechanism of rifampin inactivation. *Antimicrob. Agents Chemother.* **1995**, *39*, 1007–1009. [CrossRef]
5. Spanogiannopoulos, P.; Waglechner, N.; Koteva, K.; Wright, G.D. A rifamycin inactivating phosphotransferase family shared by environmental and pathogenic bacteria. *Proc. Natl. Acad. Sci. USA* **2014**, *111*, 7102–7107. [CrossRef]
6. Bacchi, A.; Pelizzi, G.; Nebuloni, M.; Ferrari, P. Comprehensive Study on Structure−Activity Relationships of Rifamycins: Discussion of Molecular and Crystal Structure and Spectroscopic and Thermochemical Properties of Rifamycin O. *J. Med. Chem.* **1998**, *41*, 2319–2332. [CrossRef] [PubMed]
7. Hoffmann, H.M.R. The Ene Reaction. *Angew. Chem. Int. Ed.* **1969**, *8*, 556–577. [CrossRef]
8. Singleton, D.A.; Hang, C. Isotope Effects and the Mechanism of Triazolinedione Ene Reactions. Aziridinium Imides Are Innocent Bystanders. *J. Am. Chem. Soc.* **1999**, *121*, 11885–11893. [CrossRef]
9. Cricchio, R.; Tamborini, G. New series of semisynthetic rifamycins. N-substituted derivatives of 4-amino-4-deoxyrifamycin SV. *J. Med. Chem.* **1971**, *14*, 721–723. [CrossRef] [PubMed]
10. Sheldrick, G.M. SHELXT—Integrated space-group and crystal-structure determination. *Acta Crystallogr. Sect. A Found. Adv.* **2015**, *71*, 3–8. [CrossRef] [PubMed]
11. Sheldrick, G.M. Crystal structure refinement with SHELXL. *Acta Crystallogr. Sect. C Struct. Chem.* **2015**, *71*, 3–8. [CrossRef]
12. Flack, H.D. On enantiomorph-polarity estimation. *Acta Crystallogr. Sect. A Found. Crystallogr.* **1983**, *39*, 876–881. [CrossRef]
13. Parsons, S.; Flack, H.D.; Wagner, T. Use of intensity quotients and differences in absolute structure refinement. *Acta Crystallogr. Sect. B Struct. Sci. Cryst. Eng. Mater.* **2013**, *69*, 249–259. [CrossRef] [PubMed]
14. Hooft, R.W.W.; Straver, L.H.; Spek, A.L. Determination of absolute structure using Bayesian statistics on Bijvoet differences. *J. Appl. Crystallogr.* **2008**, *41*, 96–103. [CrossRef] [PubMed]
15. Hooft, R.W.W.; Straver, L.H.; Spek, A.L. Probability plots based on Student's t-distribution. *Acta Crystallogr. Sect. A Found. Crystallogr.* **2009**, *65*, 319–321. [CrossRef]
16. Hooft, R.W.W.; Straver, L.H.; Spek, A.L. Using the t-distribution to improve the absolute structure assignment with likelihood calculations. *J. Appl. Crystallogr.* **2010**, *43*, 665–668. [CrossRef]
17. Spek, A.L.; Program PLATON. *A Multipurpose Crystallographic Tool.* © 2021–2020; Utrecht University: Utrecht, The Netherlands, 1980–2020.
18. Macrae, C.F.; Sovago, I.; Cottrell, S.J.; Galek, P.T.A.; McCabe, P.; Pidcock, E.; Platings, M.; Shields, G.P.; Stevens, J.S.; Towler, M.; et al. Mercury 4.0: From visualization to analysis, design and prediction. *J. Appl. Crystallogr.* **2020**, *53*, 226–235. [CrossRef]
19. Arora, S.K.; Arjunan, P. Molecular structure and conformation of rifamycin S, a potent inhibitor of DNA-dependent RNA polymerase. *J. Antibiot.* **1992**, *45*, 428–431. [CrossRef] [PubMed]

20. Bujnowski, K.; Synoradzki, L.; Darłak, R.C.; Zevaco, T.A.; Dinjus, E. Semi-synthetic zwitterionic rifamycins: A promising class of antibiotics; survey of their chemistry and biological activities. *RSC Adv.* **2016**, *6*, 114758–114772. [CrossRef]
21. Pyta, K.; Wicher, B.; Stefanska, J.; Przybylski, P.; Gdaniec, M. Intramolecular proton transfer impact on antibacterial properties of ansamycin antibiotic rifampicin and its new amino analogues. *Org. Biomol. Chem.* **2012**, *10*, 2385–2388. [CrossRef]
22. Wicher, B.; Pyta, K.; Przybylski, P.; Tykarska, E.; Gdaniec, M. Redetermination of rifampicin pentahydrate revealing a zwitterionic form of the antibiotic. *Acta Crystallogr. Sect. C Cryst. Struct. Commun.* **2012**, *68*, o209–o212. [CrossRef] [PubMed]
23. Bacchi, A.; Carcelli, M.; Pelizzi, G. Sampling rifamycin conformational variety by cruising through crystal forms: Implications for polymorph screening and for biological models. *New J. Chem.* **2008**, *32*, 1725–1735. [CrossRef]
24. Hanh, B.T.B.; Park, J.-W.; Kim, T.H.; Kim, J.-S.; Yang, C.-S.; Jang, K.; Cui, J.; Oh, D.-C.; Jang, J.; Rifamycin, O. An Alternative Anti-Mycobacterium abscessus Agent. *Molecules* **2020**, *25*, 1597. [CrossRef] [PubMed]

chemistry

MDPI

Article

Erdmann's Anion—An Inexpensive and Useful Species for the Crystallization of Illicit Drugs after Street Confiscations [†]

Matthew R. Wood [1], Sandra Mikhael [1], Ivan Bernal [1,2] and Roger A. Lalancette [1,*

[1] Carl A. Olson Memorial Laboratories, Department of Chemistry, Rutgers University, 73 Warren St., Newark, NJ 07102, USA; mwood@co.ocean.nj.us (M.R.W.); tug90124@temple.edu (S.M.); bernalibg@gmail.com (I.B.)

[2] Molecular Sciences Institute, School of Chemistry, University of the Witwatersrand, Private Bag 3, Johannesburg 2050, South Africa

[*] Correspondence: roger.lalancette@gmail.com; Tel.: +1-973-353-5646

[†] In Memorian: Howard D. Flack (1943–2017)—he, and his work, will be remembered.

Abstract: Erdmann's anion [1,6-diammino tetranitrocobaltate(III)] is useful in the isolation and crystallization of recently confiscated street drugs needing to be identified and catalogued. The protonated form of such drugs forms excellent crystals with that anion; moreover, Erdmann's salts are considerably less expensive than the classically used $AuCl_4^-$ anion to isolate them, while preparation of high-quality crystals is equally easy in both cases. We describe the preparation and structures of the $K^+CoH_6N_6O_8^-$ and $NH_4^+CoH_6N_7O_8^-$, salts of Erdmann's. In addition, herein are described the preparations of this anion's salts with cocaine ($C_{17}H_{28}CoN_7O_{12}$), with methamphetamine ($C_{10}H_{22}CoN_7O_8$), and with methylone ($C_{22}H_{34}CoN_8O_{14}$), whose preparation and stereochemistry had been characterized by the old $AuCl_4^-$ salts methodology. For all species in this report, the space groups and cell constants were determined at 296 and 100 K, looking for possible thermally induced polymorphism—none was found. Since the structures were essentially identical at the two temperatures studied, we discuss only the 100 K results. Complete spheres of data accessible to a Bruker ApexII diffractometer with Cu–Kα radiation, λ = 1.54178 Å, were recorded and used in the refinements. Using the refined single crystal structural data for the street drugs, we computed their X-ray powder diffraction patterns, which are beneficial as quick identification standards in law enforcement work.

Keywords: Flack test; Erdmann's anion; bath salts; street drugs; cocaine; methamphetamine; methylone; π–π interactions; racemic mimics; kryptoracemic crystallization

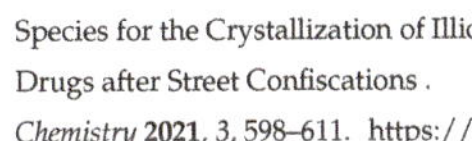

check for
updates

Citation: Wood, M.R.; Mikhael, S.; Bernal, I.; Lalancette, R.A. Erdmann's Anion—An Inexpensive and Useful Species for the Crystallization of Illicit Drugs after Street Confiscations . *Chemistry* **2021**, *3*, 598–611. https://doi.org/10.3390/chemistry3020042

Academic Editors: Katharina M. Fromm and Catherine Housecroft

Received: 9 March 2021
Accepted: 26 April 2021
Published: 30 April 2021

Publisher's Note: MDPI stays neutral with regard to jurisdictional claims in published maps and institutional affiliations.

1. Introduction

Notes: (a) Erdmann's salt should not be confused with Erdmann's reagent (sulfuric acid containing dilute nitric acid), which has been used as an alkaloid color test [1]. (b) It also should not be confused with the cis–diamino (1,2-diamino) derivative that was described by Shintani, et al. [2]. (c) For the reader's convenience, the six letter acronyms used in the references provide easy access to the Cambridge Crystallographic Database [3] information and CIF documents.

In collaboration with the Ocean County Sheriff's Office Forensic Science Laboratory (NJ, USA), we have been engaged in studies of the nature of the street drugs commonly known as bath salts [4,5], the addictive principle of which are positively charged amino species, per se, or have been converted into hydrohalides (Cl^- or Br^-, or mixtures thereof) in order to make them water soluble. Some of the samples used were from police seizures, which in most cases are of unknown provenance. Because an effective method of isolating and identifying them has, traditionally, been to crystallize them as salts using the expensive $AuCl_4^-$ anion, we decided to find alternative, inexpensive anions, which would be simple to make even by our first-year chemistry major or nonmajor, students. Those salts should provide equally good, hopefully better, microscopic and X-ray diffraction quality crystals

Chemistry **2021**, *3*, 598–611. https://doi.org/10.3390/chemistry3020042

https://www.mdpi.com/journal/chemistry

with those of the traditional gold anion samples. Given that all of the street drugs are amines, it is not difficult to assume that, in cationic form, they will readily interact, via hydrogen bonding, with moieties that can act as proton donors–acceptors.

Since a number of these drugs contain oxygen moieties that can act as bases to proton donors, an ideal crystallization partner would be one that can function equally well as either an acid or a base. Such a reagent is Erdmann's anion, which is simple to prepare in multigram quantities at a very low cost and can act both as a proton acceptor and as a proton donor to various cationic drugs. Representative samples (cocaine, methamphetamine, methylone) were selected in order to demonstrate the practical use of the reagent. They were crystallized as Erdmann's salts by the addition of a 5% aqueous solution of either ammonium or potassium Erdmann's anion and a few milligrams of the target drug compound. As the potassium salt, Erdmann's salt was first described in 1866 [6]; later, Jørgensen improved the synthesis of the ammonium salt [7]. The crystal structure of $K[Co(NH_3)_2(NO_2)_4)]$ was initially determined at room temperature by X-ray diffraction using FeK_α radiation ($\lambda = 1.937 \text{Å}$) in 1956 [8]. Here, we describe the crystal structures of both the potassium and ammonium salts at 100K using complete spheres of data and give a detailed description of the structures of complexes of three cationic drugs with Erdmann's anion.

2. Materials and Methods

Note: Origin of the drugs used in this study: Cocaine (**3**) and methylone (**5**) were obtained from drug seizures. Methamphetamine (**4**) was of pharmaceutical grade (enantiopure), purchased to set up a standard for the forensics laboratory. All other chemicals were of analytical reagent grade and were obtained from Sigma Aldrich (St. Louis, MO, USA), Fisher Scientific (Waltham, MA, USA), or VWR (Radnor, PA, USA) and used without purification. Any law enforcement seizures were of unknown provenance but were characterized by GC/MS analysis.

2.1. Syntheses and Crystallization

2.1.1. Syntheses of (**1**) and (**2**)

In order to have a common source of this reagent (Erdmann's anion), it was prepared in a large scale as follows: The potassium salt of Erdmann's anion $K[Co(NH_3)_2(NO_2)_4]$, complex (**1**) (MW = 316.12 g/mol) was prepared by weighing 40.0 g of $CoCl_2 \cdot 6H_2O$ (MW = 237.93 g/mol) (0.168 moles) dissolved in 100 mL of distilled water with stirring. In a separate beaker, 60.0 g $NaNO_2$ (MW = 69.01 g/mol) (0.869 moles) and 35.0 g NH_4Cl (MW = 53.492 g/mol) (0.654 moles) were dissolved in 288 mL of distilled water with stirring and slight heating. This second solution was filtered through a glass frit filter. To the second solution, 12 mL (0.180 moles) of "fresh" conc. NH_4OH (15 M) was added with stirring. Both solutions were combined in a side-arm flask fitted with a rubber stopper and a glass tube (1 cm in diameter) to allow air to be drawn into the mixture. Air was bubbled vigorously through the mixture for 90 min. To the mixture was added 30g KCl (MW = 74.55 g/mol) (0.402 moles), after which it turned from brown to brownish red. The product was transferred to an evaporating dish, where it was left for 2–3 days. It yielded a yellow–brown precipitate and a red–orange liquid. The solid was filtered using glass frit filter, and the precipitate was dissolved in 300 mL of distilled H_2O at 60 °C. After 2 min, the brown solution was filtered through a glass frit filter and then cooled in an ice bath. The resulting crystals were recovered using a glass frit filter, dissolved in 300 mL of hot water, and allowed to crystallize, yielding 28.4 g (53% yield based on Co). For the ammonium salt, complex (**2**), in a different preparation, 21.5 g NH_4Cl were added at the end of the procedure, instead of KCl.

2.1.2. Preparation of the Drug Crystals

Complex (**3**): A sample of a few milligrams of crystalline cocaine·HCl was dissolved on a glass slide in H_2O, and a single drop of a 5% Erdmann's potassium salt solution in water was added. Crystals of cocaine–trans–diamino–tetranitrocobaltiate(III) began to form as yellow needles through slow evaporative condensation at room temperature. A suitable crystal was chosen for single crystal X-ray analysis.

Complex (**4**): Several milligrams of crystalline methamphetamine·HCl were reacted with a drop of the previously prepared 5% solution of potassium Erdmann's salt on a pre-cleaned microscope slide. Yellow rods precipitated from solution and were allowed to grow at room temperature until they reached a size necessary for X-ray diffraction.

Complex (**5**): The synthetic cathinone (methylone) was also crystallized using the potassium Erdmann's salt reagent. A few crystals of crystalline methylone·HCl in water were mixed on a glass slide with the Erdmann's salt test reagent, and small yellow rods quickly grew out of the solution. A sample suitable for the single crystal X-ray diffraction experiment was chosen for analysis.

The identities of all three illicit drug specimens were previously confirmed by standard gas chromatography–mass spectrometry practices at the Ocean County Sheriff's Office Forensic Science laboratory.

2.2. Crystallographic Studies

Each of the crystals (**1–5**) was mounted on a Cryoloop using Paratone–N and subsequently mounted on a Bruker Smart ApexII diffractometer. Complete spheres of data were recorded at 100 K using Cu–Kα radiation, λ = 1.54178 Å. Data processing, Lorentz polarization, and face-indexed numerical absorption corrections were performed using SAINT, APEX, and SADABS computer programs [9–11]. The structures were all solved by direct methods and refined by full matrix least squares methods on F^2 using the SHELXTL V6.14 program package. All nonhydrogen atoms were refined with anisotropic displacement parameters; all of the H atoms were found in difference electron density maps. The methylene, methine, aromatic, and amine H atoms were placed in geometrically idealized positions and constrained to ride on their parent C atoms, with C–H = 0.99, 1.00, and 0.95 Å, respectively; the H atoms of the nitrogen and oxygen atoms were refined positionally, and their thermal parameters were fixed to be $1.2U_{iso}N$ and $1.5U_{iso}O$, respectively. For (**1**) and (**2**), structural and refinement parameters and the CCDC deposition numbers can be found in Table 1; for (**3–5**), the parameters and CCDC numbers are found in Table 2. The hydrogen bonding results are all found in Tables T1–T5 in the Supplementary Information.

Table 1. X-ray Experimental Details for the K^+ and NH_4^+ Erdmann's Salts.

	Crystal Data	
	(**1**) = Potassium salt	(**2**) = Ammonium salt
Chemical formula	$CoH_6KN_6O_8$	$CoH_{10}N_7O_8$
M_r	316.14	295.08
Crystal system, space group	Orthorhombic, $P2_12_12_1$	Orthorhombic, $P2_12_12_1$
Temperature (K)	100	100
a, b, c (Å)	6.6678(4), 11.1459(7), 12.7205(9)	6.6760(2), 11.3965(4), 12.7830(4)
α, β, γ (°)	90., 90., 90.	90., 90., 90.
V (Å^3)	945.37(11)	972.57(5)
Z, Z'	4, 1	4, 1
Radiation type	Cu $K\alpha$	Cu $K\alpha$
μ (mm^{-1})	18.733	14.415
Crystal size (mm)	0.186 × 0.230 × 0.351	0.100 × 0.131 × 0.152

Table 1. *Cont.*

Data Collection		
Diffractometer	Bruker APEX2	Bruker APEX2
Absorption correction	numerical	numerical
T_{min}, T_{max}	0.053, 0.175	0.210, 0.409
No. of measured, independent, and observed [$I > 2\sigma(I)$] reflections	7865, 1473, 1459	8624, 1632, 1569
R_{int}	0.034	0.027
$(\sin\theta/\lambda)_{max}$ (Å^{-1})	0.610	0.618
Refinement		
$R[F > 2\sigma(F)]$, $wR(F)$, S	0.019, 0.047, 1.06	0.021, 0.048, 1.02
No. of refl., params., restraints	1473, 164, 0	1632, 175, 4
H−atom treatment	refxyz	refxyz
Flack parameter	−0.011(5)	0.030(4)
CCDC number	2047594	2047595

Table 2. X-ray Experimental Details for the Three Drug Complexes with Erdmann's Anion.

	(3) = cocaine salt	**(4)** = methamphetamine salt	**(5)** = methylone salt
Crystal Data			
Chemical formula	$C_{17}H_{28}CoN_7O_{12}$	$C_{10}H_{22}CoN_7O_8$	$C_{22}H_{34}CoN_8O_{14}$
M_r	581.39	427.27	693.50
Crystal system, space group	Triclinic, $P1$	Monoclinic, $P2_1$	Triclinic, $P-1$
Temperature (K)	100	100	100
a, b, c (Å)	6.2403(3), 11.0319(4), 18.9421(7)	6.3873(2), 13.0182(3), 21.6772(5)	7.0437(4), 10.3155(7), 10.7697(7)
α, β, γ (°)	106.450(2), 93.831(2), 92.655(2)	90., 94.6700(17), 90.	90.712(5), 106.330(4), 107.985(4)
V (Å^3)	1244.94(9)	1796.50(8)	710.02(8)
Z, Z'	2, 2	4, 2	1, 1
Radiation type	Cu $K\alpha$	Cu $K\alpha$	Cu $K\alpha$
μ (mm^{-1})	6.07	8.01	5.50
Crystal size (mm)	$0.04 \times 0.09 \times 0.23$	$0.06 \times 0.10 \times 0.50$	$0.11 \times 0.18 \times 0.65$
Data Collection			
Diffractometer	Bruker APEX2	Bruker APEX2	Bruker APEX2
Absorption correction	numerical	numerical	numerical
T_{min}, T_{max}	0.347, 0.772	0.440, 0.619	0.220, 0.682
No. of measured, independent, and observed [$I > 2\sigma(I)$] refl.	9894, 5432, 4513	15772, 5875, 4066	6733, 1302, 1100
R_{int}	0.030	0.073	0.029
$(\sin\theta/\lambda)_{max}$ (Å^{-1})	0.607	0.610	0.497
Refinement			
$R[F > 2\sigma(F)]$, $wR(F)$, S	0.038, 0.087, 0.95	0.057, 0.073, 0.90	0.079, 0.170, 1.06
No. of refl., params., restraints	5432, 675, 3	5875, 471, 1	1302, 209, 24
H−atom treatment	mixed	mixed	constr
Flack parameter	−0.003(4)	0.024(6)	−
$\Delta\rho_{max}$, $\Delta\rho_{min}$ (e Å^{-3})	0.43, −0.39	0.65, −0.48	0.77, −0.43
CCDC number	2042087	2042090	2042091

3. Results

3.1. Structures **(1)** *and* **(2)**

3.1.1. The Potassium Salt of Erdmann's Anion (1)

Since the potassium and ammonium salts are isomorphous and isostructural, only the packing diagram of the potassium salt **(1)** is displayed in Figure 1.

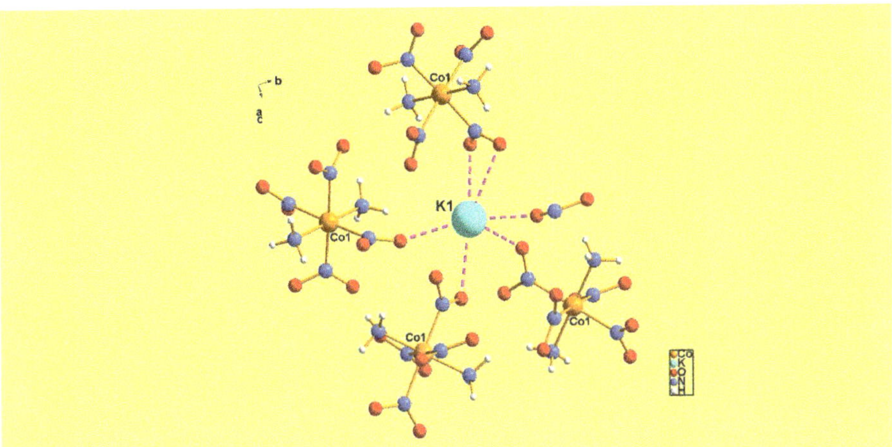

Figure 1. The surroundings around the potassium cation present in (**1**). To avoid cluttering, not all of the bonded interactions are shown here. In the ammonium salt, the nitrogen is located exactly at the site of the potassium shown above, but linkages between cation and anion are via NH_4^+ hydrogens and the $-NO_2^-$ oxygens on the anion.

Figures 2 and 3, below, display a segment of the packing of the cations and anions in the ammonium salt. As mentioned above (Figure 1's caption), stereochemical information for the ammonium and potassium salts are interchangeable. Note that Figures 2 and 3 are, respectively, *c* and *a* projections, chosen on purpose to display the packing from different angles.

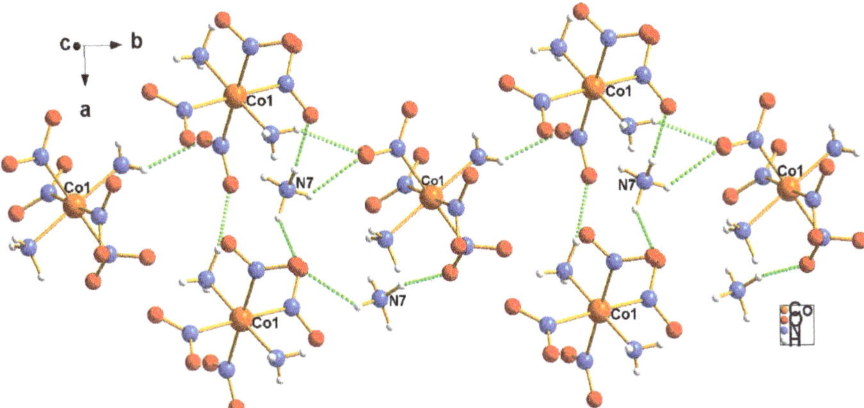

Figure 2. The anions form rows parallel to the *b*-axis in the structure of the ammonium salt (**2**), which are linked by the ammonium cations. Identical rows above and below the one shown here continue ad infinitum. For clarity, the complexity of cationic–anionic interactions present is minimally illustrated here. Figures 2 and 3 also illustrate the amphoteric nature of Erdmann's anion, which is the origin of its usefulness as a co-crystallization agent.

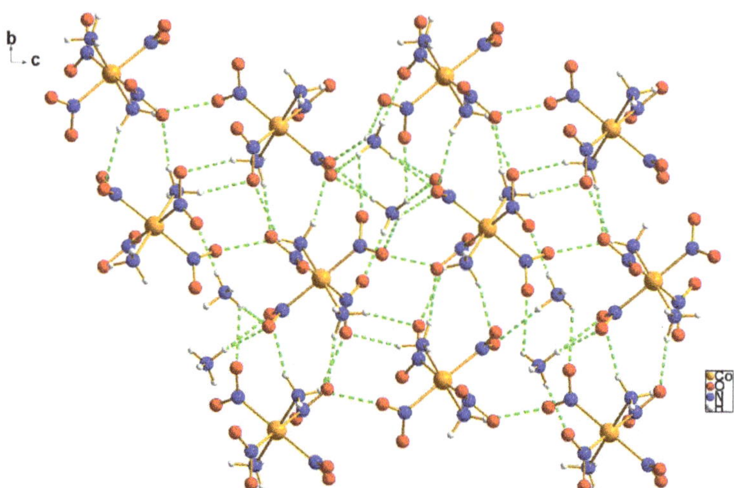

Figure 3. In this complicated diagram, the dotted lines define the hydrogen bonds in the ammonium salt of Erdmann's anion (**2**) in which the anions are linked, not only to the ammonium cations, but also to one another.

3.1.2. Structure of the Ammonium Salt of Erdmann's Anion (**2**)

In order to show what a versatile and powerful hydrogen-bonding moiety Erdmann's anion is, we present in Figure 3 the *a*-projection of the packing diagram of its ammonium salt.

In Figure 3, many hydrogen bonds were omitted, because they either (a) clutter the picture or (b) point up or down or both. Note that the anions engage into hydrogen-bonded interactions while acting both as acids (via their –NH$_3$ ligands) or bases (via –NO$_2$ oxygens). In fact, it is just such a versatility that makes Erdmann's anion so attractive for the purification and crystallization of street drugs, which are often contaminated or adulterated for maximizing street profit. Thus, efficient precipitating counter anions serve the dual role of purifying the adulterated material and of providing high quality crystals for X-ray analysis.

3.2. Structures of the Erdmann's Complexes with Various Street Drugs (**3**), (**4**), (**5**)

The structures of cocaine, methamphetamine, and methylone, forensically important drugs, were determined by precipitation with the anion of Erdmann's salt, followed by single crystal X-ray diffraction analyses. The former two ((**3**) and (**4**)) crystallize in Sohncke space groups, *P*1 and *P*2$_1$, respectively; thus, their absolute configurations were determined via the Flack Parameter test (see Table 2 and the deposited CIF files for details). The Erdmann's derivative of methylone (**5**) crystallized as a racemate in space group *P*−1. In all three cases, the Erdmann's anion from either the potassium or ammonium salt readily replaced the original anion (either Cl$^-$ or Br$^-$) when a few milligrams of the drug compound were reacted with one drop of 5% aqueous solution of Erdmann's anion on a microscope slide. The resulting precipitates are the salts formed by the protonated drug cation and the Erdmann's anion, in which the four NO$_2$ groups and the two *trans*–NH$_3$ groups act as good bases and acids.

A Caveat: It is possible, and sometimes likely, that a crystalline sample, such as methylone, prepared as described above, may give a Flack Parameter value close to 1.0 or 0.0 [12,13], suggesting a pure chiral substance is present despite the fact that this street drug is a manmade racemate. Such result would be due to (a) crystallization in a Sohncke space group as a result of packing, as in the case of NaClO$_3$ or sodium uranyl acetate (both space group *P*2$_1$3), or (b), if Z′ = 2.0, and a pair of *near-racemic* molecules caused by small differences in dissymmetry of flexible fragments caused by packing forces; in that case, the space group may be Sohncke, and the molecules crystallize as kryptoracemates. (For a discussion of the concept of kryptoracemic crystallization, see [14–16]). Additionally, the crystalline material may simply be a case of conglomerate crystallization with Z′ = 2—a widely known phenomenon since Pasteur's day. In all cases, additional measurements, such as CD (circular dichroism) in the solution, etc., would have to be made to correctly interpret the results.

3.2.1. Erdmann's Salt of Cocaine, $C_{17}H_{28}CoN_7O_{12}$ (3)

The Erdmann's salt of cocaine, $C_{17}H_{28}CoN_7O_{12}$ (**3**), crystallizes in the triclinic Sohncke space group *P*1, with two cocaine cations and two Erdmann's anions in the asymmetric unit (Figure 4). That the space group is *P*1, and not *P*−1 is guaranteed by the fact that the sample is a natural product and that the Flack Parameter test (−0.003(4)) verifies such is the case (see Table 2). There is an intramolecular hydrogen bond in each cation from the quaternary N to the carbonyl O of the methoxy carbonyl moiety: N13–H13⋯O19 is 2.838(7) Å and N14–H14⋯O23 is 2.805(7) Å. One of the cations has an H bond to an O atom on a nitro group on Co1 [N14–H14⋯O8] = 3.052(7) Å. There also exists an H bond from the nearest Erdmann's anion to the ketone [O23] to the cation N6–H6⋯O23[x + 1, y, z] = 3.103(7) Å.

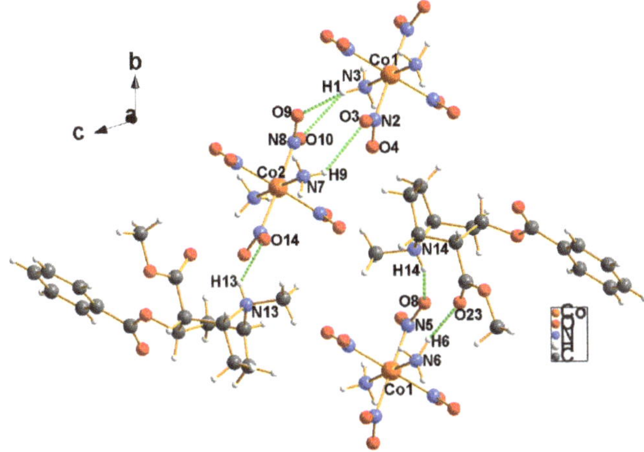

Figure 4. The interactions between the cations and anions and cations among themselves in the cocaine–Erdmann's salt (**3**). Note that the –NH₃ ligand to Co1 [N6] acts as an acid toward the—C=O oxygen base of the drug [O23], while the drug's ammonium hydrogen atoms [N13] and [N14] act as acids to the –NO₂ oxygen atoms [O14 and O8] of the anionic ligands.

3.2.2. The Methamphetamine–Erdmann's Complex (4)

The Erdmann's salt of pharmaceutical grade enantiopure methamphetamine, $C_{10}H_{22}Co N_7O_8$, crystallizes in the monoclinic space group *P*2₁ with Z = 4 and Z′ = 2. Had the sample been a chiraly mixed, manmade sample, these crystals would constitute a case of conglomerate crystallization, because C1 (from cation 1) and C11 (from cation 2) are both (S) (see Figure 5 below). This is a case in which the Flack Parameter test [13,14] would be of great value to the authorities as a warning of the presence of a meth lab—a nontrivially useful datum for enforcing institutions.

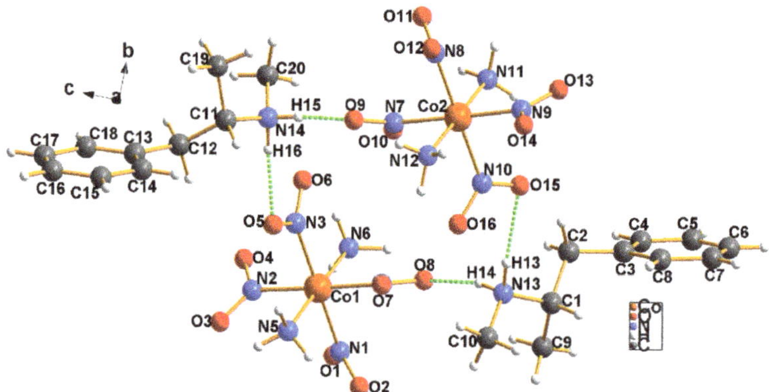

Figure 5. There are two independent cation–anion pairs in the asymmetric unit of the methamphetamine specimen we examined (**4**). Note that the cations and anions are linked largely by the NH$_2$ moieties of both cations [N13 and N14], given that the drug has no oxygen atoms of its own. Thus, the hydrogen bonding network is not as robust as it was in the case of the cocaine (see Figure 4 above and Table T4 in the Supplementary Materials). Nonetheless, the fact that the entire lattice is strongly hydrogen bonded leads to crystals of very fine quality.

In Figure 5, one of the Erdmann's anions, with the central metal Co1, makes a hydrogen bond with both proton atoms on N13 of one of the methamphetamine cations in the asymmetric unit [N13–H14 ⋯ O8 = 2.883(8)] and [N13–H14 ⋯ O7 = 3.207(8)] Å. Moreover, there are two other hydrogen bonds from N13 to O15 and O16: N13–H13 ⋯ O15[x + 1, y, z] = 3.070(9) and N13–H13 ⋯ O16[x + 1, y, z] = 3.077(8) Å, and one from N14–H15 ⋯ O9[x, y, z − 1] = 2.851(8) Å. The second anion makes similar H bonds to a symmetry−related cation N14-H16 ⋯ O6[x − 1, y, z − 1] = 3.182(9) and N14-H16 ⋯ O5[x − 1, y, z − 1] = 2.921(8) Å.

3.2.3. The Methylone–Erdmann's Complex (5)

Methylone, C$_{22}$H$_{34}$CoN$_8$O$_{14}$ (**5**), also forms attractive crystalline lattices with Erdmann's anions, which are useful for its detection and examination; its packing is shown in Figure 6.

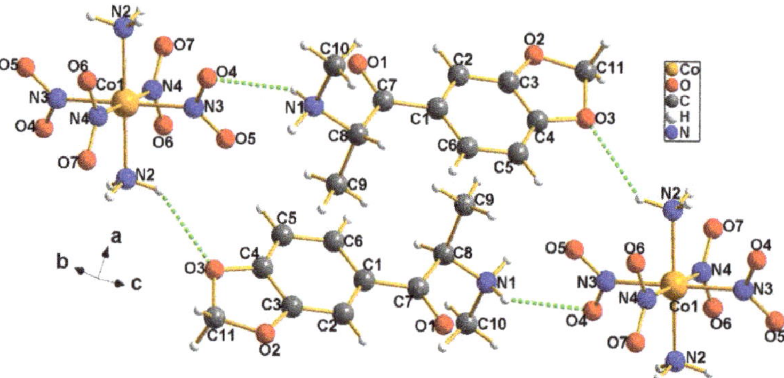

Figure 6. Our crystals contain racemic pairs of the drug methylone (**5**), indicating that it is a manmade product, a synthesized analog of cathinone, a stimulant found in *Catha edulis*. Again, note the robust hydrogen bonding network present, in which the anion is displaying its amphoteric nature by linking the equally amphoteric cations. That feature is absent in the case of the classically used AuCl$_4^-$ anions, which are also considerably more expensive.

Methylone crystallizes with Erdmann's anion in $P-1$ triclinic space group (**5**). A pair of symmetry-related anions are joined across the inversion center of the unit cell by a hydrogen bond from N2–H4 to O3[$1 - x, 2 - y, -z$] = 3.28(1) Å. Additionally, the O4 atoms of the symmetric pair are joined by hydrogen bonds: N1–H1 ⋯ O4[$1 - x, 1 - y, 1 - z$] = 3.16(1) Å. See Table T5 for bond distances and angles. Figure 6, above, shows the asymmetric unit with an additional symmetry–related anion and cation [$-x, -y, z-$] present to show the hydrogen bonds and the close contacts and to demonstrate the infinite propagation of anions these close contacts allow.

3.3. Packing Considerations in the Three Drug Complexes

Methylone (**5**) crystallizes in $P-1$, with $Z' = 1$; thus, no special comments are needed in this case, as is obvious from Figure 6 and comments above. The complexes with cocaine and methamphetamine, however, deserve considerably more careful examination, as illustrated in what follows:

3.3.1. Overlay Diagrams of the Drug Fragments for the Complexes with Cocaine (3) and with Methamphetamine (4)

Given that cocaine and methamphetamine crystallize with $Z' = 2$, it was interesting to inquire in what way the two independent components differ; therefore, we resorted to the MERCURY routine of CSD [3]. The resulting overlay Figures 7 and 8 were created with *DIAMOND* [17]. Cocaine (**3**) is shown below:

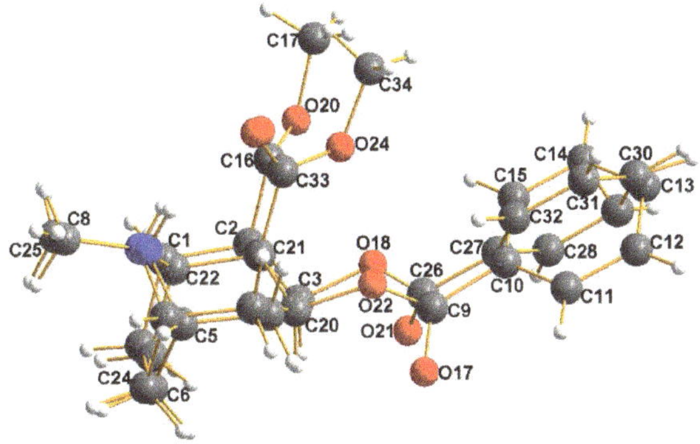

Figure 7. Cocaine–Erdmann complex (**3**). This is an overlay of the cationic cocaine molecule2 onto molecule1. The space group is $P2_12_12_1$, and $Z' = 2$. The program MERCURY [3] was used to overlay cation 2 onto cation 1; then, the fit was optimized. The result displayed above amply justifies the need for $Z' = 2$, given the significant differences in torsional angles observed.

This is a simple case of a pure optically active natural product crystallizing in a Sohncke space group with $Z' = 2$ because the two molecules are stereochemically flexible and, upon crystallizing, they pack more densely this way. The Flack Parameter [12,13] properly recognizes this, given the fact that the value is $-0.003(4)$. However, there is more to this packing mode, which will be elaborated upon in the section on Racemic Mimics.

Next, we consider the case of methamphetamine (**4**):

Again, as in the case of cocaine, the sample of methamphetamine was known to be chiraly pure, (since it was purchased as a standard material). Therefore, the same comments concerning racemic mimics apply in this case, and relevant comments will be made next.

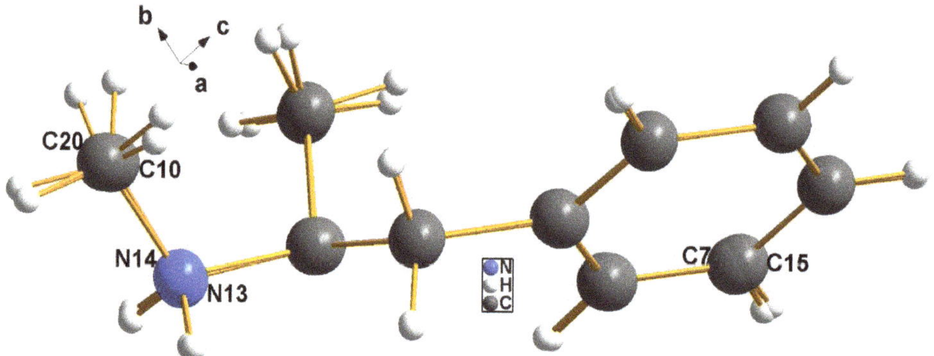

Figure 8. Methamphetamine–Erdmann complex (**4**). The overlay here is nearly perfect; the only nonhydrogen atoms that are barely separated enough to discern are shown above. As in the case of the cocaine complex, MERCURY was used to overlay cation 2 onto cation 1, and the fit was optimized. The few labels shown are for those atoms for which the fit was poor enough to allow the observer to note the presence of both atoms.

3.3.2. Racemic Mimics

Historically, it appears that an awareness of the existence of this type of crystalline material was first published in papers by a) Furberg and Hassel, who studied the crystal structure of phenyl glyceric acid slowly grown from water [18]; b) Schouwstra, who studied crystals of DL–methylsuccinic acid grown by sublimation [19] and from water solution [20]; and c) Mostad, who examined o–tyrosine crystals grown from methanol containing small amounts of ammonia to increase its solubility [21]. In all those cases, crystals of the racemate and of the optically pure material crystallized with identical cell constants; this leads to values of Z' = 1 for the racemic samples and Z' = 2 for the pure enantiomorphs.

[Caveat: because some of those lattices contained racemic pairs and had Z' = 2.0, the authors of those days [16–19] labeled them racemates. In fact, the proper term today would be kryptoracemates, but because we do not want, at this stage, to branch out into that topic, a brief but suitable discussion of this issue is given in Supplementary Materials 2, below. We thank the referee for bringing this issue to our attention.]

Given that the two lattices (kryptoracemates and Sohncke space groups), Furberg and Hassel [18] asked, "why," and, "how?" In a remarkably clear and simple answer, they indicated that the pure chiral material seemed to crystallize *"as if a twin resembling in its packing that of the true racemate"*: in other words, as a "racemic twin"; thus, the name *Racemic Mimics* that later evolved. They also proposed that substances containing flexible (dissymmetric) fragments whose torsional barriers were low would make ideal candidates for the existence of such a phenomenon, and they documented additional cases [18].

(The overlay diagrams shown in this document show the extent to which torsional differences are associated with the observed Z' value of 2.0). That was a remarkably advanced concept for its day and happens to conform to what we describe in our presentation, since we have two cases of racemic mimics in the cases of the cocaine derivative and of the methamphetamine derivative of Erdmann's salts. For readers interested in more extended commentary on this and related topics, we recommend Herbstein's authoritative compendium [22].

a.   The Case of Cocaine (**3**)

Figure 9 shows the asymmetric unit for the structure of the cocaine-Erdmann's complex.

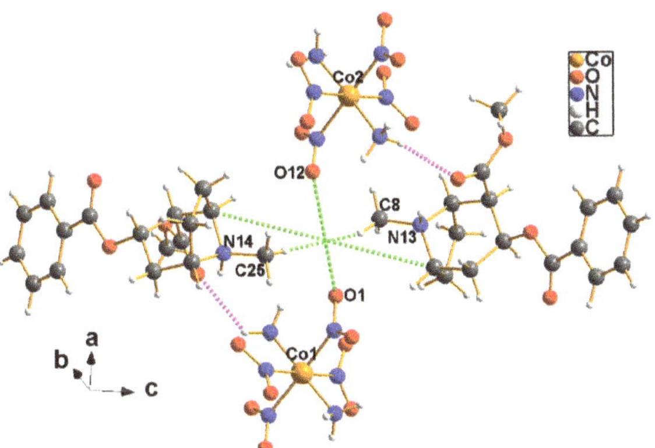

Figure 9. The center of mass (0.4741, 0.4173, 0.4689) of the cocaine–Erdmann lattice is located very near to $\frac{1}{2}, \frac{1}{2}, \frac{1}{2}$, but in $P1$, the origin is totally arbitrary, which renders the issue moot for this case. Note, however, that is not the case for methamphetamine (see Figure 10, next).

b.   The Case of Methamphetamine (**4**)

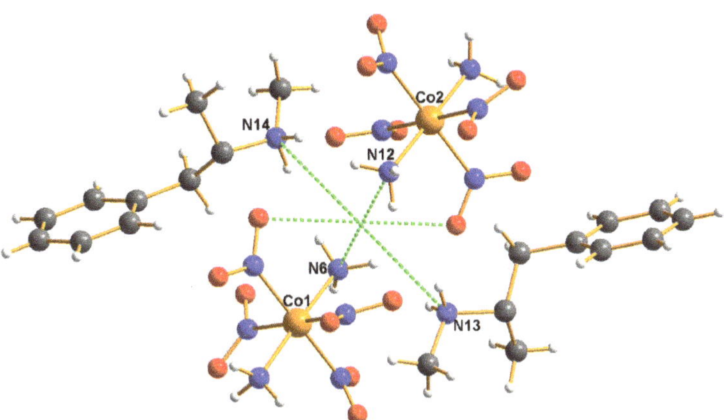

Figure 10. The pair of cations and anions observed in the case of methamphetamine. The intersection of the dotted lines is located at 0.4749, 0.5239, 0.5182, which is also very close to $\frac{1}{2}, \frac{1}{2}, \frac{1}{2}$, as expected for a case of a racemic mimic. Overlays, above, provide a rationale for the reason why both the cocaine and amphetamine cations can function thus.

3.3.3. π–π Contacts

The criterion for meaningful contacts between aromatic fragments labeled "π–π" interactions" in the report by Janiak [23] suggests that, given the experimental data available (see Figure 7 and relevant commentary in that paper), the range of 3.3–4.6 Å is reasonable. Using that as an acceptable gauge, our compounds do not have acceptable "contacts" in that range and should be ignored, because, in both cases considered here, they are closer to 6.0 Å. However, we think it is worth pointing out that that some meaningful "residual contacts" exist and depict them below in Figures 11 and 12. This caveat is in the same spirit as that in past discussions of the existence, or lack thereof, in hydrogen bonding discussions.

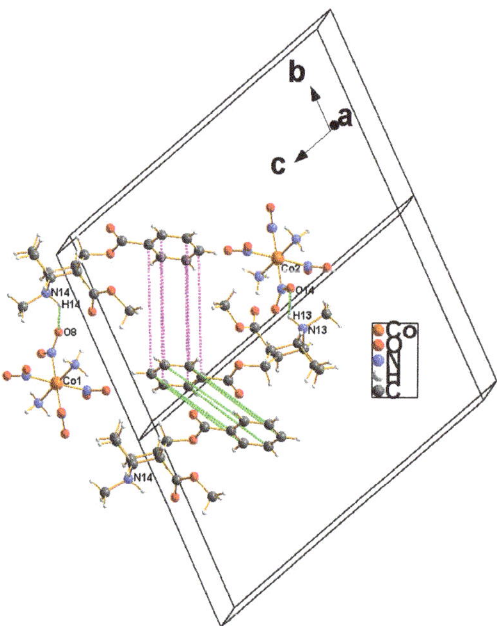

Figure 11. π–π interactions observed in the cocaine–Erdmann's complex crystals that are not separated by simple lattice translations, in which case the aromatic fragments in question are parallel to each other; thus, the latter are ideally suited for such electronic intermolecular interactions are and ignored here.

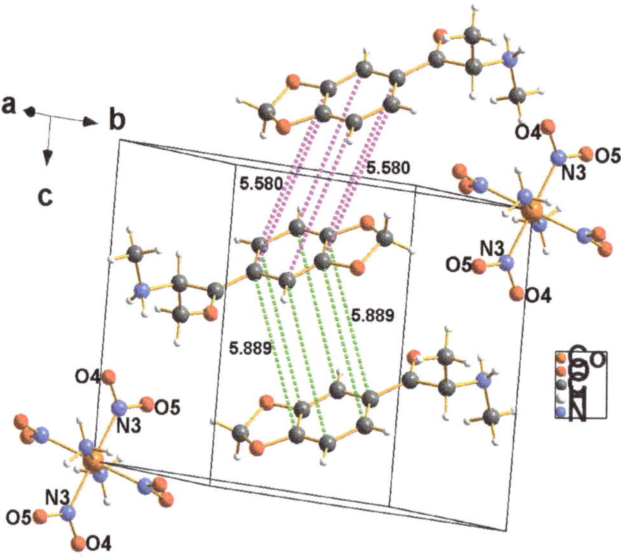

Figure 12. A packing diagram for the methylone complex with Erdmann's anion (**5**) is shown above. The same comments about π–π interactions in the case of cocaine apply here.

a. Cocaine Cation with Erdmann's Anion (**3**)

The distances between the central ring (C10–C15) and the closest ring (C27′–C32′) range between 5.78 and 5.94 Å. The second ring atoms are generated through symmetry [x, y, z + 1]. The centroid-to-centroid distance = 5.868 Å [symm = x, y, z + 1], and the angle between the ring normal and the centroid-to-centroid vector is = 85.1° for this pair.

The other close π–π interaction is the central ring to the upper ring in the diagram, generated through symmetry [x, y, z + 1]; these distances range between 6.79 and 6.97 Å. The centroid-to-centroid distance = 6.868 Å [symm = x, y − 1, z], and the angle between the ring normal and the centroid-to-centroid vector is 51.7° for this one.

b. Methamphetamine Cation with Erdmann's Anion

In crystals of methamphetamine cation with Erdmann's anion (**4**), there are no close π–π interactions other than those dictated by translations; thus, we saw no need to illustrate those in this instance.

c. Methylone Cation with Erdmann's Anion (**5**)

The distances between the central ring (C10–C15) and the closest ring (C10′–C15′) range between 5.86 and 5.92 Å. The second ring atoms are generated through symmetry [1 − x, 1 − y, 1 − z]. The centroid-to-centroid distance = 5.889 Å [symm = 1 − x, 1 − y,1 − z], and the angle between the ring normal and the centroid-to-centroid vector is = 54.7° for this one.

The other close π–π interaction is the central ring to the upper ring in the diagram, generated through symmetry [1 − x, −1 − y, −z]. These distances range between 5.58 and 5.62 Å. The centroid-to-centroid distance is 5.889 Å [symm = 1 − x, −1 − y, −z], and the angle between the ring normal and the centroid-to-centroid vector is 49.5° for this one.

In conclusion, it seems that, whenever aromatic bearing fragments are not sterically hindered, π–π interactions, other than those dictated by lattice translations short enough to be meaningful, are useful in forming sturdier lattices such as observed in the cases of (**3**) and (**5**).

4. Summary of Experimental Results

The adoption of, and continued investigation of, the utility of Erdmann's salt in forensic analytical testing schemes will aid the analyst by reducing sample preparation time, as well as reduced reagent cost, and will allow easy transfer of a sample to a confirmation technique, such as infrared spectroscopy or X-ray powder diffraction. Finally, testing with Erdmann's salt is essentially a nondestructive testing technique, preserving the resulting precipitate for further analysis or courtroom presentation.

The powder diffraction patterns of the precipitation products of the potassium Erdmann's salt with each of the street drugs described here (structures (**3**), (**4**), (**5**)) (Figures S1–S3) will provide the necessary analytical confirmation to any forensic lab that is using powder X-ray diffraction in conjunction with crystal tests. These powder patterns are calculated from the single crystal structures using the powder pattern generating routine in the SHELX package. X-ray diffraction is considered a "Category A" technique by the Scientific Working Group for the Analysis of Seized Drugs (SWGDrug), due to its high discrimination capabilities [24].

5. Conclusions

It seems we have justified the use of Erdmann's anion as a valuable, easily accessible, and inexpensive reagent for the purification and co-crystallization of samples of illicit drugs confiscated in the streets or police raids. Large amounts of the ammonium and/or potassium salts can be prepared by very simple methods described above; the advantage of such a procedure is that a single, purified, large source can then be used for future forensic studies with the confidence of uniformity.

In the cases of crystallographic tests (single crystal or powder), the resulting test specimens we tested have been very satisfactory, especially when the results are subjected to the Flack Parameter test.

Supplementary Materials: The supplementary materials are available online at https://www.mdpi.com/article/10.3390/chemistry3020042/s1.

Author Contributions: S.M. was an undergraduate student who prepared the ammonium and potassium salts of Erdmann's complexes under direct supervision of R.A.L. M.R.W. obtained the drug samples, and he prepared the drug complexes under direct supervision of R.A.L., I.B., R.A.L. and M.R.W. together wrote the manuscript. All authors have read and agreed to the published version of the manuscript.

Funding: This research received no external funding.

Data Availability Statement: All data are deposited in CCDC.

Acknowledgments: We acknowledge the National Science Foundation for NSF–CRIF Grant No. 0443538 for part of the purchase of the X-ray diffractometer.

Conflicts of Interest: The authors have declared that no competing interests exist.

References

1. Dymock, W.; Warden, C.H.J.; Hooper, D. *Pharmacographia Indica: A History of the Principal Drugs of the Vegetable Origin*; Kegan, Paul, Trench, Trubner & Co., Ld.: London, UK, 1893; pp. 261–265.
2. Shintani, H.; Sato, S.; Saito, Y. The crystal structure of $(-)_{589}$ dinitrobis(ethylenediamine)-cobalt(III)$(-)_{589}$-dinitrooxolatodiamminecobaltate(III) monohydrate. *Acta Cryst.* **1976**, *B32*, 1184–1188. [CrossRef]
3. Groom, C.R.; Bruno, J.; Lightfoot, M.P.; Ward, S.C. CSD = CCDC = The Cambridge Structural Database. (v4.0.0) of 2019. *Acta Cryst.* **2016**, *B72m*, 171–179. [CrossRef]
4. Wood, M.R.; Thompson, H.W.; Brettell, T.A.; Lalancette, R.A. The hydrated and anhydrous gold(III) tetrachloride salts of L-ecgonine, and important forensic toxicology marker for cocaine. *Acta Cryst.* **2010**, *C66*, m4–m8. [CrossRef]
5. Wood, M.R.; Lalancette, R.A.; Bernal, I. Crystallographic investigations of select cathinones: Emerging illicit street drugs known as 'bath salts'. *Acta Cryst.* **2015**, *C71*, 32–38. [CrossRef]
6. Erdmann, O.L. Ueber einige salpetrigsaure Nickel- und Kobalt- verbindungen. *J. Prakt. Chem.* **1866**, *97*, 385–413. [CrossRef]
7. Jorgensen, S.M. Preparation of cobalt-ammonia salts. *Zeit. Anorg. Chem.* **1898**, *17*, 455–479.
8. Komiyama, Y. The crystal structure of potassium tetranitro-diammine-cobaltiate(III), K[Co(NH$_3$)$_2$(NO$_2$)$_4$]. *Bull. Chem. Soc. Jpn.* **1956**, *29*, 300–304. [CrossRef]
9. Bruker. *SADABS, SAINT and SMART*; Bruker AXS, Inc.: Madison, WI, USA, 2008.
10. Sheldrick, G.M. SHELXT. *Acta Cryst.* **2015**, *A71*, 3–8. [CrossRef]
11. Sheldrick, G.M. SHELXL. *Acta Cryst.* **2015**, *C71*, 3–8. [CrossRef]
12. Flack, H.D. On enantiomorph-polarity estimation. *Acta Cryst.* **1983**, *A39*, 876–881. [CrossRef]
13. Flack, H.D. Physical and spectrometric analysis: Absolute configuration determination by X-ray crystallography. *Compr. Chirality* **2012**, *8*, 648–656.
14. Bernal, I. In Proceedings of the ACA Annual Meeting, Montreal, QC, Canada, 1995. (Abstract 4a.1.e).
15. Bernal, I.; Watkins, S.F. A list of organoetallic kryptoracemates. *Acta Cryst.* **2015**, *C71*, 216–221.
16. Clevers, S.; Coquerel, G. Kryptoracemic compound hunting and frequency in the Cambridge Structural Database. *Cryst. Eng. Comm.* **2020**, *22*, 7407–7419. [CrossRef]
17. Putz, H.; Brandenburg, K. *DIAMOND, Version 8.5.10.*; GbR, Kreuzherrenstr.: Bonn, Germany, 2019.
18. Furberg, S.; Hassel, O. Remarks on the crystal structure of asymmetric molecules. The phenylglyceric acids. *Acta Chem. Scand.* **1950**, 1020–1023. [CrossRef]
19. Schouwstra, Y. Crystal and molecular structures of DL-Methylsuccinic acid, I. *A modification obtained by sublimation at 70 °C in vacuo. Acta Cryst.* **1973**, *B29*, 1–4.
20. Schouwstra, Y. Crystal and molecular structures of DL-methylsuccinic acid, II. Two modifications obtained by slow evaporation of aqueous solutions. *Acta Cryst.* **1973**, *B29*, 1636–1640. [CrossRef]
21. Mostad, A.; Rømming, C.; Tressum, L. The crystal and molecular structure of (2-Hydroxyphenyl)alanine (oTyrosine). *Acta Chem. Scand.* **1975**, *B29*, 171–176. [CrossRef]
22. Herbstein, F.H. *Crystalline Molecular Complexes and Compounds*; Oxford University Press: New York, NY, USA, 2005. [CrossRef]
23. Janiak, C. A critical account of π-π stacking in metal complexes with aromatic nitrogen- containing ligands. *J. Chem. Soc. Dalton Trans.* **2000**, 3885–3896. [CrossRef]
24. Scientific Working Group for the Analysis of Seized Drugs. SWGDrug Guidelines. Available online: http://www.swgdrug.org (accessed on 30 July 2020).

MDPI

St. Alban-Anlage 66

4052 Basel

Switzerland

Tel. +41 61 683 77 34

Fax +41 61 302 89 18

www.mdpi.com

Chemistry Editorial Office

E-mail: chemistry@mdpi.com

www.mdpi.com/journal/chemistry